Biomedical Visualization

Volume 6

Series Editor

Paul M. Rea, Anatomy Facility, School of Medicine, Dentistry and Nursing, College of Medical, Veterinary and Life Sciences, University of Glasgow, Glasgow, UK

The Biomedical Visualization book series invites contributions related to visualization and imaging in the biomedical sciences and related fields such as medicine, dentistry, veterinary surgery, informatics, and the allied health professions.

This can encompass work in image analysis, workflow methodologies, photogrammetry in science, animations, digital reconstructions and applications, big data and visualizations, educational methodologies, usability and evaluations, augmented and virtual reality, and 3D and 4D technologies, including 3D printing, as well as informatics, e-tutorials, MOOCs, HCI and public engagement. It can be from macroscopic to microscopic but show how we can view data and information related to the biomedical field in a much more accessible, innovative and engaging way using technology.

This series is intended for researchers, clinicians, students, and teaching staff in biomedical and medical schools who use and develop visualization and imaging techniques in medical education, patient care, and biomedical and medical research.

Leonard Shapiro
Editor

How to use 3D Printing Innovations and Digital Storage to Democratize Anatomy Education

Springer

Editor
Leonard Shapiro
Department of Family, Community and Emergency Care
Faculty of Health Sciences, University of Cape Town
Cape Town, South Africa

ISSN 2731-6130 ISSN 2731-6149 (electronic)
Biomedical Visualization
ISBN 978-3-031-68500-2 ISBN 978-3-031-68501-9 (eBook)
https://doi.org/10.1007/978-3-031-68501-9

Cover Illustration: Catherine MacRobbie.

This Springer imprint is published by the registered company Springer Nature Switzerland AG
The registered company address is: Gewerbestrasse 11, 6330 Cham, Switzerland

Preface

This volume in the Springer Biomedical Visualization book series contains chapters that cover bespoke three-dimensional (3D) printing for both human and veterinary anatomical education, a detailed 'how-to' guide with hyperlinks to open-source resources which democratize 3D printing, a digital protocol for the generation of a 3D acellular dermal scaffold for 3D bioprinting, and the 3D printing of a 'femur flute' by an artist-patient working in collaboration with an orthopedic surgeon.

The long history of anatomical models as educational resources is explored in fascinating detail, from historic wax models to the latest, cutting-edge 3D printers. A comprehensive and detailed guide is provided with examples of innovative 3D printing techniques specific to anatomical education, with links to open-source scans for download and 3D printing. A further guide is included to help the reader decide on specific 3D printer models to suit their specific educational needs. A veterinary anatomy educator describes a transformation in teaching and learning methods in veterinary education using augmented reality (AR), virtual reality (VR), and 3D visualization methods like CT or MRI images which can be used to reconstruct complete 3D virtual models, as well as 3D prints from these scans. Following on the 3D printing theme, we have a chapter on the digital protocol for the bioprinting of a three-dimensional acellular dermal scaffold (ADS) for use in wound healing, as an alternative to skin grafting for secondary intention wound healing. The development of bio-engineered ADS constructs is integral to the advancement of wound healing technology. We then follow the personal journey of an artist-patient who worked in collaboration with an orthopedic surgeon to 3D print his femur from digital scans of his femur, which they then transformed into a flute (yes, a musical one). The artist writes about his exploration of his 'disembodied' femur and the fashioning of it into a flute: a reference to bone flutes found at ancient Neanderthal sites, the earliest dating back 53,000 years. A chapter offers a complete and extensive guide to applied anatomy for acupuncture. This is a chapter in four parts that covers the upper limb, lower limb, trunk, head and neck. Each part of the chapter is replete with photographic images of meticulously prepared cadaveric prosections that illustrate and complement the text. An extensive 'bone library'—in the form of a digitized and categorized, online, cloud-based bone repository—is the first digital human skeletal repository in southern Africa. We then look at vertebral artery variations and its role in clinical conditions, current insights into polycystic ovarian syndrome, and visual interpretation using multiplex immunoassay of serum samples.

Cape Town, South Africa Leonard Shapiro

Acknowledgements

My gratitude to each and every author for your valuable contributions to this volume. You have shared your expert knowledge as well as your passion for research and writing with us, your readers.

It is a pleasure to work with our team: Series Editor Paul Rea and Ina Stoeck, Editor at Springer Nature. Thank you for your expertise, enthusiasm, and support throughout the months of work that went into preparing the contents of this book for publication. My thanks to the Springer design and production team for their attention to detail and the efficiency with which they produced this book. You all made this journey a pleasure!

Contents

About the Editors and Contributors

About the Editor

Leonard Shapiro is honorary lecturer in the Department of Family, Community and Emergency Care, Faculty of Health Sciences, University of Cape Town (UCT). He is an artist and educator working in the health sciences and designs tailored, art-based exercises for application within the anatomical education curriculum, specifically aimed at improving observation and three-dimensional (3D) spatial awareness. Leonard has developed a novel, multi-sensory observation method, which specifically employs the sense of touch (haptics) coupled with the simultaneous act of drawing. It is called the haptico-visual observation and drawing (HVOD) method.

In anatomy education, the benefits of using the HVOD method include (1) the enhanced observation of the 3D form of anatomical parts, (2) the cognitive memorization of anatomical parts as a 3D mental picture, (3) improved spatial orientation within the volume of anatomical parts, and (4) an ability to draw. Leonard teaches the HVOD method to medical students and clinicians in South Africa and abroad.

Professor Iain Keenan (Newcastle University) and Leonard Shapiro (UCT) have designed a Massive Open Online Course (MOOC) consisting of art-based exercises for improving student 3D spatial awareness and their observation of human anatomy. This online course is titled *Exploring 3D Anatomy* and is available to medical students globally, as a supplement to their anatomy curriculum.

A second online course titled *Exploring 3D Anatomy Plus* was developed for clinicians and surgeons for improving their 3D spatial awareness, leading to improved 3D spatial skills.

Leonard contributes to the anatomy education discourse via publications and articles and by presenting at anatomy conferences.

About the Series Editor

Paul M. Rea is Professor of Digital and Anatomical Education at the University of Glasgow. He is Director of Innovation, Engagement and Enterprise within the School of Medicine, Dentistry and Nursing. He is also Senate Assessor for Student Conduct, is Council Member on Senate, coordinates the day-to-day running of the Body Donor Program, and is Licensed Teacher of Anatomy, licensed by the Scottish Parliament.

He is qualified with a medical degree (MBChB), an MSc (by research) in craniofacial anatomy/surgery, a PhD in neuroscience, the Diploma in Forensic Medical Science (DipFMS), and an MEd with Merit (Learning and Teaching in Higher Education). He is Senior Fellow of the Higher Education Academy, Fellow of the Institute of Medical Illustrators (FIMI), and a registered medical illustrator with the Academy for Healthcare Science.

Paul has published widely and presented at many national and international meetings, including invited talks. He has been the Lead Editor for the first 17 volumes of *Biomedical Visualisation*, part of Advances in Experimental Medicine and Biology. In 2022, this then became its own standalone volume becoming *Biomedical Visualization*, for which Paul is the Series Editor. It has now published four additional volumes, with several currently in press. These volumes have resulted in over 170,000 online accesses, more than 850 citations, and over 470 social media mentions. It has published 212 chapters, from 434 different authors across 133 different institutes in 23 countries across the globe, and with readership in 75 countries. He is also Associate Editor for the *European Journal of Anatomy* and has reviewed for 25 different journals/publishers.

He is the Public Engagement and Outreach lead for anatomy, coordinating collaborative projects with the Glasgow Science Centre, NHS, and Royal College of Physicians and Surgeons of Glasgow. Paul is also a STEM ambassador and has visited numerous schools to undertake outreach work.

His research involves a long-standing strategic partnership with the School of Innovation and Technology, the Glasgow School of Art. This has led to multi-million pound investment in creating world-leading 3D digital datasets to be used in undergraduate and postgraduate teaching to enhance learning and assessment. This successful collaboration resulted in the creation of the world's first taught *MSc Medical Visualisation and Human Anatomy* combining anatomy and digital technologies of which he was the Founding Director. The Institute of Medical Illustrators also accredits it. It has created college-wide, industry, multi-institutional, and NHS research linked projects for students.

Contributors

Jan H. P. Ackermann Sefako Makgatho Health Sciences University, Pretoria, South Africa

Amanda Alblas Stellenbosch University, Cape Town, South Africa

Lunga Bam The South African Nuclear Energy Corporation, Elias Motsoaledi Street Extension (Church Street West), Brits, South Africa

Amélie Beaudet The Centre National de la Recherche Scientifique (CNRS), Paris, France

Frikkie de Beer The South African Nuclear Energy Corporation, Elias Motsoaledi Street Extension (Church Street West), Brits, South Africa

Ralph Borland HUMA - Institute for Humanities in Africa, University of Cape Town, Cape Town, South Africa

Pearl N. Bothma Forensic Anthropology Research Centre, University of Pretoria, Pretoria, South Africa

Steven Buddle Keele Anatomy and Surgical Training Centre, School of Medicine, Keele University, Keele, UK

Angelique Carmichael Forensic Anthropology Research Centre, University of Pretoria, Pretoria, South Africa

Marine Cazenave The Centre National de la Recherche Scientifique (CNRS), Paris, France

Eugenia Cunha Faculty of Medicine, Universidade de Coimbra, Polo I – Edifício Central Rua Larga, Coimbra, Portugal

Meg-Kyla Erasmus Forensic Anthropology Research Centre, University of Pretoria, Pretoria, South Africa

Thomas C. Flanagan Anatomy, School of Medicine, University College Dublin, Dublin, Ireland

Nalini Govender Department of Basic Medical Sciences Faculty of Health Sciences, Durban University of Technology, Durban, South Africa

Department of Basic Medical Sciences, Durban University of Technology, Durban, South Africa

Anuradha Gupta Guru Angad Dev Veterinary and Animal Sciences University, Ludhiana, India

Alieske C. Hagg Forensic Anthropology Research Centre, University of Pretoria, Pretoria, South Africa

Rohen Harrichandparsad Department of Neurosurgery, School of Clinical Medicine, University of KwaZulu-Natal, Durban, South Africa

Miksha Harripershad Forensic Anthropology Research Centre, University of Pretoria, Pretoria, South Africa

Patria C. Hartman Sefako Makgatho Health Sciences University, Pretoria, South Africa

Yann Heuzé The Centre National de la Recherche Scientifique (CNRS), Paris, France

Univ. Bordeaux, CNRS, Ministère de la Culture, PACEA, UMR5199, Pessac Cedex, France

Jon Hobbs Acupuncture Association of Chartered Physiotherapists, Acupuncture Training Providers, Peterborough, UK

Jakobus Hoffman The South African Nuclear Energy Corporation, Elias Motsoaledi Street Extension (Church Street West), Brits, South Africa

Princess Ijeoma Department of Pharmacology, University of Pretoria, Hatfield, South Africa

Edwin de Jager Forensic Anthropology Research Centre, University of Pretoria, Pretoria, South Africa

Nhlanhla Japhta Sefako Makgatho Health Sciences University, Pretoria, South Africa

Kritima Kapoor Guru Angad Dev Veterinary and Animal Sciences University, Ludhiana, India

Department of Veterinary Anatomy, College of Veterinary Science, Guru Angad Dev Veterinary and Animal Sciences University, Ludhiana, India

Reabetswe Khonyane Forensic Anthropology Research Centre, University of Pretoria, Pretoria, South Africa

Ruth Kobedi Sefako Makgatho Health Sciences University, Pretoria, South Africa

Gabriele C. Krüger Forensic Anthropology Research Centre, University of Pretoria, Pretoria, South Africa

Ericka N. L'Abbé Forensic Anthropology Research Centre, University of Pretoria, Pretoria, South Africa

Lelika Lazarus Discipline of Clinical Anatomy, School of Laboratory Medicine and Medical Science, College of Health Sciences, University of KwaZulu-Natal, Durban, South Africa

Leandi Liebenberg Forensic Anthropology Research Centre, University of Pretoria, Pretoria, South Africa

Maritza Liebenberg Forensic Anthropology Research Centre, University of Pretoria, Pretoria, South Africa

Marius Loots Forensic Anthropology Research Centre, University of Pretoria, Pretoria, South Africa

Chantelle Marais Stellenbosch University, Cape Town, South Africa

Thandolwethu Mbonani Forensic Anthropology Research Centre, University of Pretoria, Pretoria, South Africa

Clarisa van der Merwe Forensic Anthropology Research Centre, University of Pretoria, Pretoria, South Africa

Michaela Middleton Stellenbosch University, Cape Town, South Africa

Indheresan Govindsamy Moodley Jackpersad and Partners Inc.,, Durban, South Africa

Jagidesa Moodley Women's Health and HIV Research Group, College of Health Sciences, University of KwaZulu-Natal, Durban, South Africa

Samantha Muller Forensic Anthropology Research Centre, University of Pretoria, Pretoria, South Africa

Thajasvarie Naicker Optics and Imaging Centre, Doris Duke Medical Research Institute, College of Health Sciences, University of KwaZulu-Natal, Durban, South Africa

Optics and Imaging Centre, Doris Duke Medical Research Institute, College of Health Sciences, University of KwaZulu-Natal, Durban, South Africa

Girija Naidoo Department of Basic Medical Sciences, Durban University of Technology, Durban, South Africa

Sagaren Naidu Durban Fertility Clinic, Chelmsford Medical Centre 3, St Augustine Hospital, Durban, South Africa

Robert Nshimirimana The South African Nuclear Energy Corporation, Elias Motsoaledi Street Extension (Church Street West), Brits, South Africa

Anna C. Oettlé Sefako Makgatho Health Sciences University, Pretoria, South Africa

Bukola Omotoso Discipline of Clinical Anatomy, School of Laboratory Medicine and Medical Science, College of Health Sciences, University of KwaZulu-Natal, Durban, South Africa

Hafiza Parkar Department of Pharmacology, University of Pretoria, Hatfield, South Africa

Mark Pickering Anatomy, School of Medicine, University College Dublin, Dublin, Ireland

Rachel Pieterse Stellenbosch University, Cape Town, South Africa

Sapna Ramdin Department of Basic Medical Sciences, Durban University of Technology, Durban, South Africa

Alison F. Ridel Forensic Anthropology Research Centre, University of Pretoria, Pretoria, South Africa

Alison Fany Ridel Department of Anatomy, University of Pretoria, Hatfield, South Africa

Okuhle Sapo Forensic Anthropology Research Centre, University of Pretoria, Pretoria, South Africa

Nirmal Singh Guru Angad Dev Veterinary and Animal Sciences University, Ludhiana, India

Shoohana Singh Discipline of Human Physiology, College of Health Sciences, University of KwaZulu-Natal, Durban, South Africa

Franci Swanepoel Sefako Makgatho Health Sciences University, Pretoria, South Africa

Charlotte E. G. Theye Forensic Anthropology Research Centre, University of Pretoria, Pretoria, South Africa

Dirk Vandermeulen ESAT, Center for Processing Speech and Images, KU Leuven, Leuven, Belgium

Rudolph Venter Stellenbosch University, Cape Town, South Africa

Luke Welsh Keele Anatomy and Surgical Training Centre, School of Medicine, Keele University, Keele, UK

Part I

3D Printing, Digitization, Anatomical Education

1 Democratizing Anatomy Education with Bespoke 3D-Printed Models as Visualization Tools

Mark Pickering and Thomas C. Flanagan

Abstract

Anatomical models have been used for centuries to support anatomical dissection and anatomy learning. Such models have evolved to levels of exceptional anatomical detail that provide an accurate representation of anatomical structure and relations within the human body. Despite the evolution of these anatomical models in terms of realism, complexity, and their widespread commercial availability, anatomical models have essentially remained as static depictions of the human form. To support the teaching and visualization of more complex, dynamic anatomical processes, and to fill the gaps in the available commercial model catalog, many anatomical educators, including ourselves, have resorted to developing low-tech, bespoke models in the classroom, which can be extremely effective at explaining difficult concepts. However, such models are, more often than not, single-use items, perishable items, or generally have a limited life span without any significant reusability. With the technological advances and increasing accessibility associated with rapid fabrication devices such as 3D printers, the development of more advanced, durable, and reusable bespoke anatomical resources is now within reach for most anatomy educators. In this chapter, we will discuss how we have harnessed the accessibility offered by 3D printing to develop a sustainable and reusable portfolio of bespoke anatomical teaching resources that aid the understanding of both static and dynamic anatomical concepts beyond the scope of currently available commercial anatomical models. We will describe how integrating additional simple fabrication tools and processes, including resin casting, basic electronics, and do-it-yourself (DIY) techniques, can enliven static models and demonstrations without reliance on large budgets for expensive machinery or for multiple traditional commercial anatomical models. We conclude the chapter with a "how to" beginner's guide on 3D printing and model making, in order to assist anyone with an interest in developing their own models to begin the creative process.

Keywords

Anatomy · Education · Anatomical models · 3D printing · Beginner's guide

M. Pickering · T. C. Flanagan (✉)
Anatomy, School of Medicine, University College Dublin, Dublin, Ireland
e-mail: thomas.flanagan@ucd.ie

L. Shapiro (ed.), *How to use 3D Printing Innovations and Digital Storage to Democratize Anatomy Education*, Biomedical Visualization 6, https://doi.org/10.1007/978-3-031-68501-9_1

1.1 Anatomical Models: A Short History

Anatomical models have been used for centuries to support, or even replace, anatomical dissection as a means for students to understand the complexities of body structure, tissue and organ function, and anatomical relations. The development of anatomical models as educational tools has evolved alongside advances in scientific understanding and material technology. From rudimentary wooden figures to modern-day commercial plastic replicas, a constant has remained in this evolution by efforts to maximize the anatomical realism of the product using the available materials. Early anatomical models from the fifteenth–eighteenth centuries, made from wood, ivory, or wax, were crafted by skilled artists who collaborated closely with anatomists to enhance anatomical accuracy and were generally prohibitively expensive to commission. The need for realistic models during this period was driven by poor tissue preservation techniques, as well as the scarcity of cadavers to dissect, owing largely to an absence of formalized body donation processes; by way of example, sourcing of cadavers was not legalized in the UK prior to the enactment of the Anatomy Act of 1832 (Burrows 2019). Prior to the development of modern embalming methods and refrigeration, dissected corpses could only be used for teaching for brief periods (a matter of days) and dissection was a particularly unpleasant experience, which by necessity was confined mainly to the colder winter months (Ballestriero 2010). Accurate anatomical models were therefore essential resources for early anatomy teaching, although the quest for anatomical realism often came at the expense of model functionality and interactivity.

1.1.1 Wax, Wood, and Ivory Models as Anatomy Teaching Resources

1.1.1.1 Wax Modeling

In terms of realism and craftsmanship, historical anatomical wax models are arguably the most impressive. Wax modeling, or *ceroplastics*, had been used widely in antiquity for artistic and religious purposes, with many wax figures still found in churches to this day. It is thought that the wax preservation technique was first introduced to the medical field in the early part of the fourteenth century (Marković and Marković Živković 2010). Alessandra Giliani (1307–1326) is believed to be the first woman recorded in historical documents to practice anatomy and prosection, and allegedly did so by disguising herself as a man during a period when women were barred from practicing medicine or conducting medical research. Giliani was an assistant to the Italian surgeon and anatomist, Mondino de Luzzi (1270–1326), whose text *Anathomia Mundini* revived the practice of anatomical dissection in medical curricula throughout Europe (Maraldi et al. 2000; Mavrodi and Paraskevas 2014). It has been suggested that Giliani herself pioneered the technique of wax injection directly into cadaveric blood vessels to form solid wax casts of the vessels that could then be extracted from the surrounding tissue as detailed models of the vasculature (Chen et al. 1999; Haviland and Parish 1970). Wax injection was also pursued by Leonardo da Vinci (1452–1519) who is thought to have created the first known wax neuroanatomical model by injecting wax into the ventricles of the ox brain to investigate their internal shape and relationship (Chen et al. 1999; da Mota 2020).

The oldest surviving wax anatomical specimen produced as an educational model is the "Anatomical Head," and was fabricated at the end of the seventeenth century by Sicilian artist Gaetano Giuliano Zumbo (1656–1701). Zumbo worked for the sixth Grand Duke of Tuscany, Cosimo III de'Medici (1642–1723) of the Medici dynasty, Florence, and was known for producing portraits and sculptures in wax, including gruesome artwork and scenes depicting plague and decomposing corpses (Chen et al. 1999; Talairach-Vielmas 2014). In 1695, Zumbo moved to Genoa to learn more about anatomy and began to collaborate with French surgeon and anatomist Guillaume Desnoues (1650–1735) (Maraldi et al. 2000). Desnoues proposed that wax anatomical models could potentially replace the horror of dissecting and examining rotting corpses (Riva et al. 2010), and also preserve the findings in any useful and interesting clinicopathological dissection specimens (Chen et al. 1999). Wax was considered to be a convenient and appropriate means of capturing the human form; it was malleable, obtained from several sources (including animal (beeswax), vegetable, and mineral), relatively durable, easy to work with, could be melted, poured and molded, and could be colored and decorated with various organic materials, including hair and teeth (Chen et al. 1999; Ballestriero 2010). Most notably, it provided a remarkable likeness to human flesh and tissue, far surpassing that of other available materials (Ballestriero 2010). Zumbo was commissioned to create exact duplicates of Desnoues' own dissections in colored wax. The steps involved in wax anatomical model preparation were complex and time consuming. Initially, a specific region of interest was carefully dissected for the subsequent creation of a plaster mold. Given the varying degrees of decomposition of the unpreserved bodies, it was sometimes necessary to dissect and harvest non-decayed tissue from the same region of multiple cadavers and combine these pieces together to create a single wax cast. The next step involved making a plaster mold by coating the surface of the reassembled dissected region with fat in order to prevent the liquid plaster from adhering to it (Olry 2000). All of the surface textures of the dissected tissue would be faithfully reflected into the plaster as it surrounded it and made contact with it. The plaster was then left to set and solidify into a hard mold. The tissue was then removed from the solidified plaster mold, leaving it hollow and ready to be replaced with wax. Before the wax was added, the mold was lined with liquid soap or oil to form a non-stick coating. This prevented the wax from sticking to the internal walls of the plaster mold, so that it could be easily released from it and retains essential textural details that were impressed into it from the inner plaster walls. The final steps involved removing the hardened wax from the mold, correcting defects, and adding detail using further tools specialized for this purpose and by painting the wax surface of the model (Olry 2000). Zumbo and Desnoues formed a partnership by making wax anatomical models, exhibiting them in museums to the paying public, and selling them, but ultimately quarreled over claims to the invention of the technique and associated financial compensation (Lightbown 1964; Haviland and Parish 1970; Talairach-Vielmas 2014). Only two of Zumbo's anatomical specimens have survived, including the "Anatomical Head" housed in Florence (Fig. 1.1) (Narang et al. 2021).

Two major workshops of wax modeling were established in Italy during the eighteenth century. The first was established at the Institute of Sciences and Arts in Bologna, which was earlier founded by Italian scholar Luigi Ferdinando Marsigli (1658–1730) under the auspices of Pope Benedict XIV. Dry mummified cadaveric specimens prepared by Antonio Maria Valsalva (1666–1723) and donated by his family had been used in Bologna for anatomical instruction, but they lacked morphological accuracy and deteriorated significantly with constant use (Maraldi et al. 2000). This led to the commissioning of a series of anatomical models using more durable materials. Pope Benedict XIV tasked the Bologna painter and sculptor, Ercole Lelli (1702–1766) with producing this collection of wax anatomical models, including eight life-sized statues for the newly founded Museum of Anatomy (1742–1747). These specimens were carefully constructed replicates of actual anatomical dissections and laid the foundations for the study of medicine at the University of Bologna (Maraldi et al. 2000). Two of the waxwork anatomical statues were fully intact nude models (a male and female referred to as Adam and Eve), while most of the remaining statues were posed as "skinned" to reveal increasingly deeper layers of the musculoskeletal system (Fig. 1.1) (Maraldi et al. 2000). Lelli's technique in creating these statues involved mounting muscle replicates constructed from colored wax onto actual human skeletons (Manzoli et al. 2022).

Over his program of work, Lelli apprenticed a young sculptor, Giovanni Manzolini (1700–1755)—along with Manzolini's wife and former pupil, Anna Morandi Manzolini (1714–1774) to help him complete these models, in addition to many more specimens that were

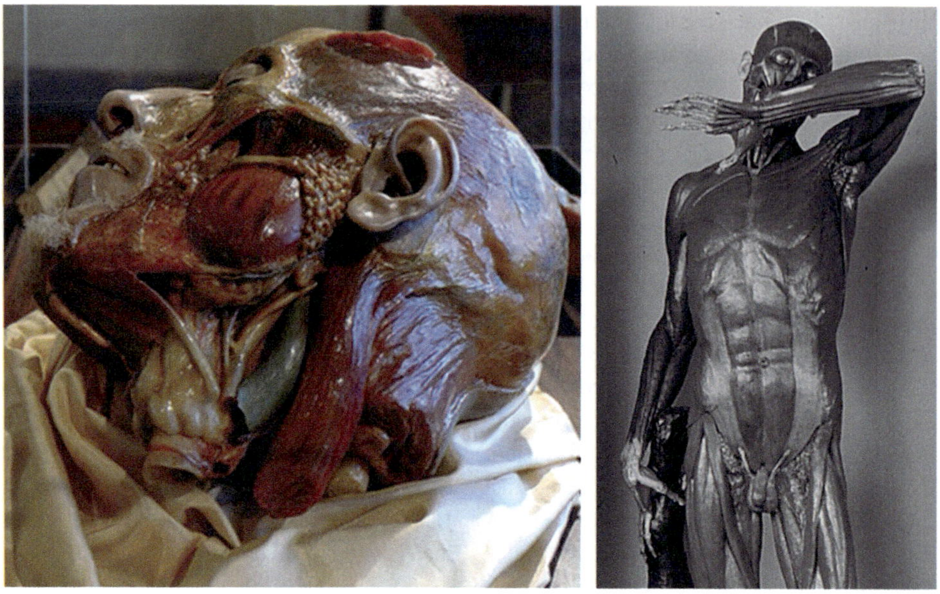

Fig. 1.1 (Left) "Anatomical Head" by Gaetano Giuliano Zumbo (Museum *La Specola*, Florence, Italy) (image from Ballestriero 2010, reproduced with permission); (right) "Skinned Man" by Ercole Lelli (Museo di Palazzo Poggi, Bologna, Italy) (image from Maraldi et al. 2000, reproduced with permission)

added to the collection. Lelli also created wooden statues, but the colors of the wax components made the waxworks far more lifelike. Anna Morandi Manzolini continued Lelli's work after his death, and after the death of her husband. Her collection had exquisite detail that was close to perfection and a testament to her vast anatomical knowledge (Haviland and Parish 1970). She achieved critical acclaim across Europe as the creator of anatomical masterpieces, including a series of models focused on individual body systems and those showing aspects of pregnancy and labor. Her most notable piece was a full female figure complete with internal organs which could all be removed and examined—considered to be one of the first educational innovations in wax model preparation (Haviland and Parish 1970).

Guiseppe Astorri (1795–1852) continued working on the Bologna collection and was appointed an official modeler of the Institute of Human Anatomy in 1816 (Marković and Marković Živković 2010). He created many wax models relating to anatomy, teratology, and pathology, including conjoined twins, infectious disease (one of his most well-known pieces illustrates a smallpox case), and detailed specimens depicting elements of the peripheral nervous system. Cesare Bettini (1814–1885)—an Italian painter, designer, illustrator, and model maker, was best known for his works in the field of neuroanatomy where his models were purposely scaled up for the benefit of the observer or user.

The art of wax anatomical modeling spread to Florence and the workshop of *La Specola* was established by Felice Fontana (1730–1805) in 1771 under the patronage of the Grand Duke of Tuscany, Peter Leopold (Chen et al. 1999). The workshop soon hired Clemente Susini (1754–1814), whose work would later be viewed as the peak of ceroplastics perfection (Ballestriero 2010). At *La Specola*, plaster molds would be created from dissected specimens into which wax was cast to create wax models. Wax figures were subsequently produced on a wooden scaffold; this differed from the Bologna waxes, which were initially handmade models of organs that were superimposed onto actual bone material (Riva et al. 2010; Manzoli et al. 2022). The use of molds also permitted the development of multiple model copies and slight variants of those copies in some cases. Individual members of the workshop team also became specialists at applying specific details to the models, such as blood vessels, lymphatics, and nerves, which were carefully added to specimens by using silk and cotton threads impregnated with wax. Hair was composed of actual hair dipped and embedded in the wax (Maraldi et al. 2000). The models produced at *La Specola* were received with such acclaim that the Austrian Emperor Joseph II commissioned an even larger collection from Florence for his newly founded Military Academy of Medicine and Surgery in Vienna (the Josephinum collection) used in the training of military surgeons. Over 1200 models were created at *La Specola* for transfer to Vienna and about 800 specimens remain there today at the Josephinum Museum of Medical History (Chen et al. 1999; Riva et al. 2010).

Fontana believed that these wax models would ultimately dispense with the need for dissection of decomposing cadavers and remove the sense of revulsion associated with that practice. Such was his conviction that the models were often created with an artificial beauty and came to be known as wax *Venuses*; for example, they may have been positioned in graceful poses on silks, all the while with no signs of death or decomposition of the cadavers from which they were modeled, and with a subtle removable anterior abdominal wall to reveal underlying tissues and organs (Fig. 1.2). These romanticized wax models were far removed from those produced by nineteenth-century English sculptor, Joseph Towne (1808–1879), who created around 1000 anatomical and pathological models at Guy's Hospital in London from 1826 until his death. Towne's models were depicted exactly as they were found in the dissection room—rugged and brutally realistic—a tendency evident in most of the wax casts of northern European countries from the late eighteenth and nineteenth centuries (Fig. 1.2) (Ballestriero 2010).

In the early nineteenth century, embryology began to develop as a modern science and a need grew for three-dimensional models to depict a process that could not be observed naturally. Early embryologists and anatomists like Wilhelm His (1831–1904) would create scaled-up free-hand designs of what they could observe under the microscope to provide a gross three-dimensionality of the staged growth and development of their organisms of interest (including humans). Eventually, these designs would be transferred to artists like Adolf Ziegler (1820–1899) who became the pre-eminent wax modeler in the field of development anatomy. Ziegler would construct oversized wax models and molds would then be made to reproduce many copies that could be sold to medical schools around the world (Fig. 1.2). These models would have been invaluable to students to allow them to comprehend the three-dimensionality of structures that they would not normally get an opportunity to see in the flesh.

These wax modelers made an indelible contribution to the education of anatomy during that period—allowing gross examination of the body with exquisitely detailed and lifelike three-dimensional specimens, in conjunction with anatomical textbooks and complementary lectures. Importantly, they addressed the need for alternative pedagogical resources due to the difficulties in sourcing cadavers and in particular, the absence of suitable embalming procedures. However, the process of wax modeling was highly time consuming, laborious, and expensive (Chen et al. 1999). The high costs and time required to construct these models made them accessible to only the wealthiest institutions. With the development and emergence of modern embalming techniques, the need for wax models to show both anatomy and embryology became somewhat redundant. Normal gross anatomy and pathology became routinely illustrated through formaldehyde fixation of cadavers and preparation of fixed specimens in glass jars.

1.1.1.2 Wood and Ivory Modeling

Early wood and ivory figures were developed in China during the fourteenth and fifteenth centuries for the apparent purpose of training acupuncturists by highlighting specific acupuncture points on the body using pierced holes (Marković and Marković Živković 2010). Up until the eighteenth century in Germany, France, and Italy, wooden and ivory figurines were developed with

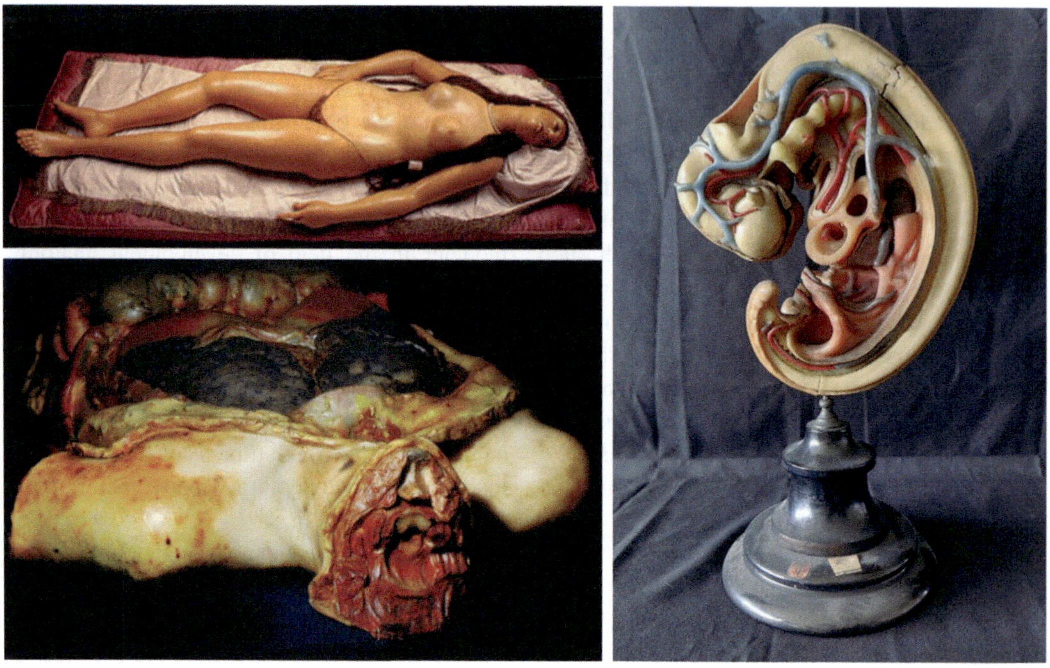

Fig. 1.2 (Upper left) A wax anatomical figure (The Medici Venus) by Clemente Susini (Museum *La Specola*, Florence, Italy). These models of the Florence workshop were idealized figures, and had detachable surfaces to reveal the underlying tissues and organs (image from Ballestriero 2010, reproduced with permission); (lower left) a wax anatomical figure by Joseph Towne showing the neck and open torso (Gordon Museum, King's College, London, UK). Towne's figures were exact, uncleansed replicas of the cadavers in front of him (image from Ballestriero 2010, reproduced with permission); (right) example of the embryological wax models manufactured by Adolf and Friedrich Ziegler (from the anatomical model collection, School of Medicine, University College Dublin)

some removable organs which provided an element of interactivity for the anatomy learner (Marković and Marković Živković 2010) (Fig. 1.3). The anterior torso could generally be removed from these models and reveal a basic representation of underlying viscera, often with a fetus in an advanced stage of development shown in the uterus, suggesting that they may have been used to educate the layperson in relation to pregnancy (the anatomical detail would have been inadequate for instruction of medical students) (Russell 1972; Olry 2000). Ivory figurines were considered a luxury and were generally much smaller than wooden models as they were constructed from single pieces of ivory (Marković and Marković Živković 2010).

Perhaps the most well-known wooden anatomical figures were created by Felice Fontana during the late eighteenth century (Olry 2000; Palouzié and Ducourau 2017). His most notable creation is that of a model of a standing male figure composed of approximately 3000 moveable wooden components—each component having up to 100 details (Fontana 1795). The model is believed to have taken about 10 years to complete and could also be converted to the female form (Olry 2000). It is understood that Napoléon Bonaporte (1769–1821) ordered a copy of the model, but changed his mind after the war due to the excessive costs (30,000 liras) (Sonolet 1977).

Model human skeletons have also been constructed from wood, most notably in Japan during the Edo era (1600–1867) when medical doctors and educators were prohibited from keeping human bones for teaching or research. The first of these wooden skeletons (*"mokkotsu"*) was created by a craftsman, Koji Harada at the request of a Hiroshima medical doctor, Ryoetsu Hoshino, to understand initially how to relocate a dislocated temporomandibular joint (Kataoka et al. 2007).

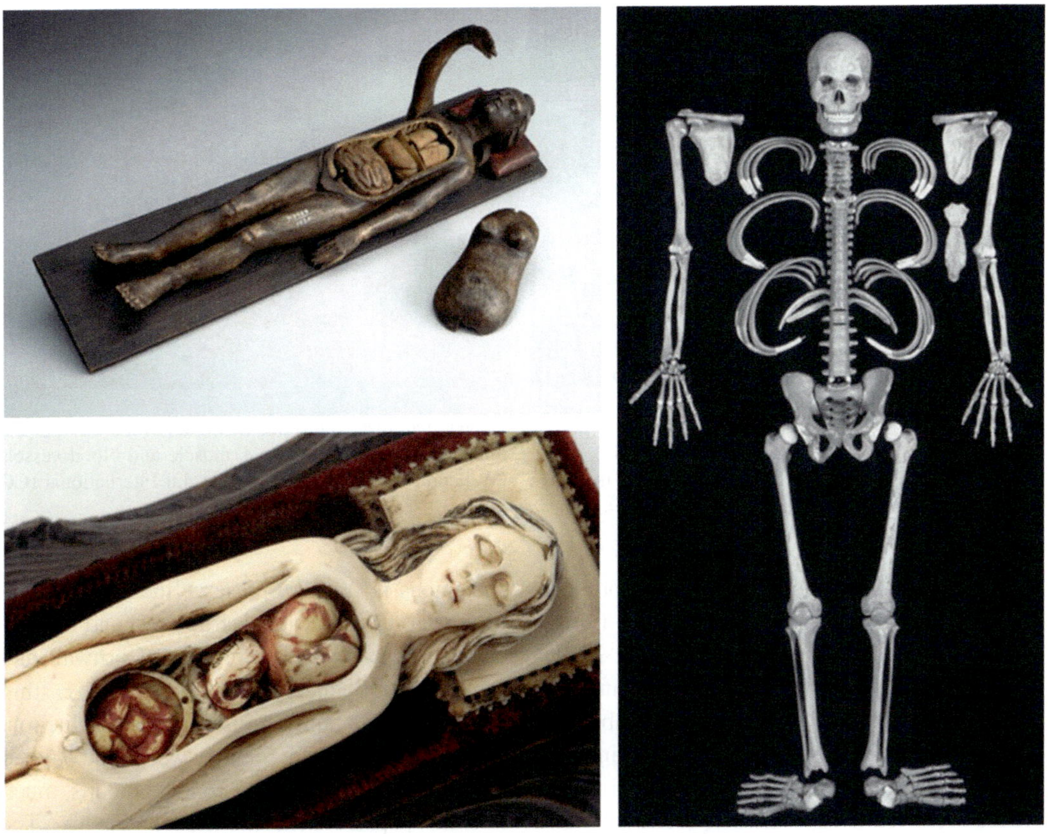

Fig. 1.3 (Upper left) Early wooden anatomical model with detachable components, seventeenth–eighteenth century (Wellcome Collection. Attribution 4.0 International (CC BY 4.0). Source: Wellcome Collection); (lower left) ivory anatomical model of a pregnant woman with detachable parts, seventeenth-eighteenth century (Wellcome Collection). Attribution 4.0 International (CC BY 4.0). Source: Wellcome Collection); (right) Hoshino wooden skeleton (image from Kataoka et al. 2007, reproduced with permission)

What later became known as the Hoshino wooden skeleton was crafted by replicating the skeletal bones of two executed criminals (all bones apart from the hyoid and ear ossicles), and the bones were generally copied with considerable accuracy (Fig. 1.3) (Kataoka et al. 2007).

1.1.2 Papier-mâché, Plaster and Commercial Plastic Models as Anatomy Teaching Resources

While wax, wooden, and ivory anatomical models continued to be produced into the nineteenth century, *papier-mâché* ("chewed paper") became a popular material for anatomical models due to its affordability, accessibility, and ease of use (Fig. 1.4). These models were popularized around the world by the French doctor Louis Thomas Jérôme Auzoux (1797–1880), who had recognized the major limitations associated with anatomical instruction during his own medical training. The limited supply of cadavers, poor cadaveric preservation, together with the expense and fragility of wax models inspired Auzoux's innovative spirit to come up with a more durable, inexpensive instruction tool (Olry 2000; Ortug and Yuzbasioglu 2019; Narang et al. 2021). He visited the workshop of Jean-Francois Ameline (1763–1835), a professor of anatomy at Caen, France, who had begun to create models from papier-mâché—a mixture of paper and gum—and found that he could refine the paste material

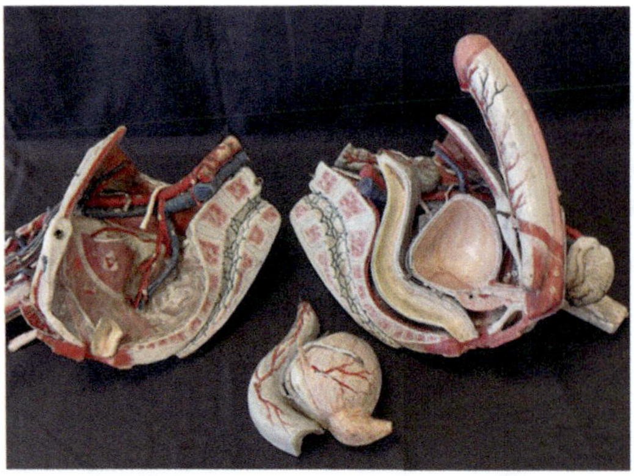

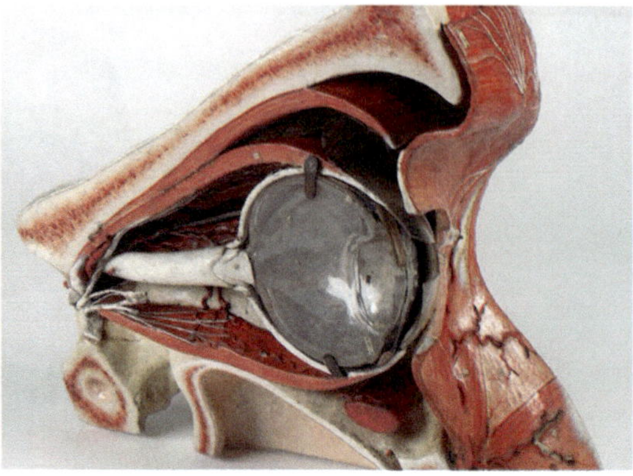

Fig. 1.4 (Left) *papier-mâché* model of the male pelvis which can be split into mid-sagittal section and is composed of several detachable components (from the anatomical model collection, School of Medicine, University College Dublin); (right) plaster model of a section of the human eye surrounding bone, muscle and blood vessels (Wellcome Collection. Attribution 4.0 International (CC BY 4.0). Source: Wellcome Collection)

and develop a molding process to mass-produce light, durable, weather-resistant models (Ortug and Yuzbasioglu 2019; Narang et al. 2021). Auzoux's models displayed additional functionality over wax models as they were movable and could be wholly disassembled to allow interactive engagement by students—an innovation that Auzoux referred to as "*clastic*" (Greek for *break into pieces*) (Valdecasas et al. 2009; Narang et al. 2021). Previous attempts to provide some interaction with wooden, ivory, and even wax models were quite limited, with a scarcity of removable components that lacked detail (Valdecasas et al. 2009). Each component of Auzoux's models was significantly detailed and all hand painted prior to receiving a final coat of varnish to enhance their durability and appearance. Such was the demand for Auzoux's models that he established a company in his home village of Saint-Aubin d'Ecrosville in France which went on to produce both human and animal anatomy models for medical and veterinary schools all over the world.

In the late nineteenth century, work was also beginning to produce anatomical models from plaster (of Paris) (Fig. 1.4), most notably by Franz Josef Steger (1845–1938), a German sculptor who collaborated with Carl Ernst Bock (1809–1874) and Wilhelm His, both at University of Leipzig (Cornwall and Smith 2014; Narang et al. 2021). At the suggestion of His, Steger produced his plaster casts on frozen anatomical specimens in order to preserve the natural form of the various tissues and structures, and thereby create highly realistic replicates (Cornwall and Smith 2014; Spencer 2008). The plaster casting process once again permitted mass production, and once models were set, they were hand painted to highlight the various anatomical details, much like the papier-mâché models of Auzoux. The Bock-Steger anatomical model company mass-produced anatomical models of limbs, torsos, embryos, heads, and other specific regions or organs for many decades up until the 1930s (Spencer 2008).

The emergence of the first fully synthetic plastic in the early twentieth century changed the course of anatomical model development, with plastic models replacing most other types of models from the 1920s onward. Notable plastic anatomical models include the Transparent Figures produced by the Deutsches Hygiene-Museum in Dresden, Germany during the 1920s and 1930s, including a Transparent Man, Transparent Woman, Transparent Pregnant Woman, and Transparent Cow (Bienholz-Radtke and Roeßiger 2022). These models are composed of a clear plastic shell or skin, with several of the body systems constructed from variable-colored plastics visible beneath the transparent skin. The figures have largely been used as exhibition pieces in various exhibition formats around the world. In the 1960s, American designer Richard

Rush developed a similar model called the Transparent Anatomical Manikin (TAM) (Fig. 1.5) which was more interactive in that it played an audio script to educate the audience about the various body systems, while the manikin rotated on a turntable and various parts of the body were illuminated as they were discussed (Marković and Marković Živković 2010; Narang et al. 2021). Rush produced 42 of these manikins, many of which are still on display in various museums and health centers. The model became immortalized on the cover of alternative rock band Nirvana's 1993 album, In Utero.

The use of plastic models in anatomy education took hold in the mid-twentieth century, coinciding with advances in plastic manufacturing technologies and establishment of several companies that mass-produced model replicas of whole bodies, systems, and/or individual organs. Commercial plastic replica models (Fig. 1.5) are still used in most medical programs today, either as complement to or replacement for anatomical dissection. Indeed, there has been a degree of cross-pollination of the two resources, with the invention of plastination by German anatomist, Gunther von Hagens (b. 1945) in 1977 (von Hagens 1979). Although beyond the scope of the

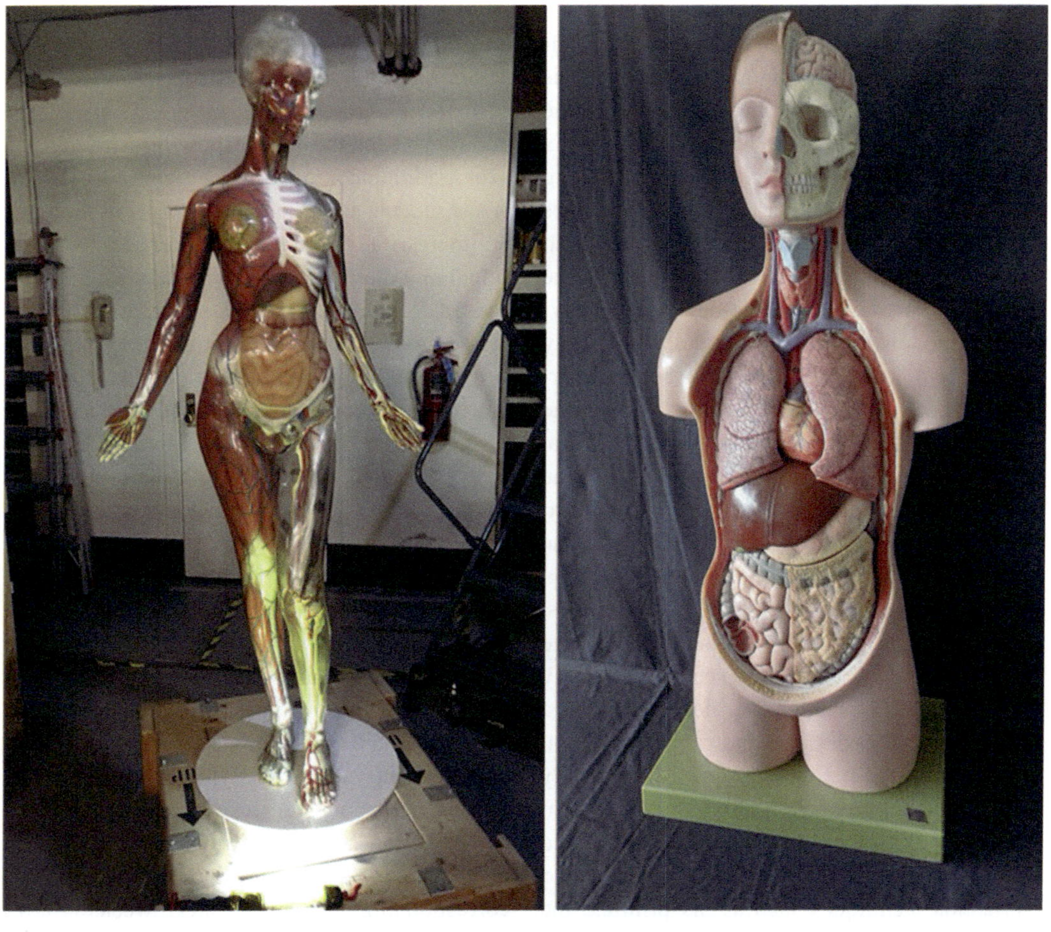

Fig. 1.5 (Left) Transparent Anatomical Manikin (TAM) created by Richard Rush (1968), with audio script, organ illumination, and a mechanical turntable (image by Nick Fujii, own work, Attribution-Sharealike 4.0 International (CC BY-SA 4.0), https://commons.wikimedia.org/wiki/File:Transparent_Anatomical_Manikin_(TAM).jpg); (right) an example of a commercial plastic model of the upper body with detachable organs - similar commercial models can be found in most medical schools today (from the anatomical model collection, School of Medicine, University College Dublin)

present chapter, given that the technique is one of tissue preservation, plastination has revolutionized how dissected anatomical specimens (prosections) are preserved and used for educational and display purposes, primarily by replacing the water and fat in body tissues with polymers, which hardens to create durable and odorless prosections (Sora et al. 2019). These preserved anatomical tissues, which feel like plastic replicas after processing, retain their intricate structure and pose no health risks from chemical fixatives like formaldehyde, making them safer to handle and store. In addition, they do not require refrigeration and can be kept indefinitely with minimal care, making them highly practical and economical for educational institutions.

From the descriptions of the historical development of models for anatomy education, it is clear that the majority of the trailblazers in the field were either artists or master craftspeople, and a creative mind was essential in developing these beautiful pieces. In a sense, this continues today in that an element of creativity is required in developing anatomical models—whether these are highly anatomically accurate, or even low-fidelity models where the detail is entirely stripped back to allow students to focus on a particular aspect of functional anatomy, and at the same time reduce the cognitive demands (Chan and Cheng 2011; Chan 2015). To support the teaching and visualization of more complex, dynamic anatomical processes, and to fill the gaps in the available commercial model catalog, many anatomical educators, including ourselves, have resorted to developing low-tech, bespoke models in the classroom, which can be extremely effective at explaining difficult concepts. However, such models are, more often than not, single-use items, perishable items, or generally have a limited life span without any significant reusability. With the technological advances and increasing accessibility associated with rapid fabrication techniques such as 3D printing, the development of more advanced, durable, and reusable bespoke anatomical resources is now within reach for most anatomy educators (Brumpt et al. 2023; Shahrubudin and Lee 2019; Torres et al. 2011). In addition, as we hope to demonstrate in the next section, one does not necessarily need to be a master craftsperson to develop their own effective 3D-printed teaching and learning models for anatomy education.

1.2 Case Studies

1.2.1 3D Printing to Visualize "Impossible Anatomy"

One of the most important applications of anatomical models generally is to allow the visualization of anatomical structures that cannot be seen in cadaveric material, for example, due to scale (structures too small to be seen easily), lack of availability (e.g., developing human embryos) or simply by nature of the structure being defined as a space (for example, the ventricles of the brain). One of the advantages 3D printing has over traditional model making is the ability to create models not purely on the basis of the artistic impression of the model maker, but based on actual and realistic anatomical data. The confluence of 3D printing and high-resolution 3D imaging technologies means that actual anatomical structures can be reproduced repeatedly anywhere in the world, requiring nothing more than a functional 3D printer. This advance has also been facilitated by increasing availability of 3D anatomical data as supplementary to the peer-reviewed literature.

An excellent example of the application of these resources is the developing human embryo. Representing embryonic structure presents two problems, specifically the size of the structures which make them difficult to see at the earliest stages, and the availability of embryonic tissue; human embryos are not typically available for anatomy teaching due to ethical and logistical considerations. The 3D structure of developing human embryos from Carnegie stage 12 to 23, corresponding to approximately weeks 4–8 of development was recently published (Azkue 2021). The 3D structures themselves were generated from across a range of 3D imaging modalities, including optical projection tomography for the smallest embryos through to high-resolution episcopic microscopy and magnetic resonance imaging for progressively larger embryos, and subsequently manually processed into versions ready for 3D printing. As part of this publication, the actual models themselves have also been made available for download as supplementary material to the paper (Azkue 2021).

Importantly, because these models are generated from real 3D scan data of actual human embryos, they can be 3D printed and made available to students as accurate representations of the actual anatomical structure during this complex and dynamic phase of development. Even the smallest of these models can be printed with high-resolution resin 3D printing at actual size. This means that as well as creating enlarged 3D prints of these models allowing the external structure of these embryos to be visualized at a scale where details are perceptible, the models can also be printed at actual size, giving learners an opportunity to appreciate the true size of the developing embryo and the how the size changes during this phase of development (Fig. 1.6). This is a distinct advantage that physical models offer over digital 3D models (such as de Bakker et al. 2012), which can demonstrate 3D structure well, but realistic size and scale can be difficult to convey. This has obvious uses in embryology education in a traditional academic setting, but also as a visually appealing and accessible demonstration aid in other settings, such as patient education—in addition to using these models in the classroom, we have provided them to obstetricians for use in their practice.

Another challenging set of structures for realistic visualization on account of their size, location, and accessibility are the structures of the inner ear associated with the auditory and vestibular system. The

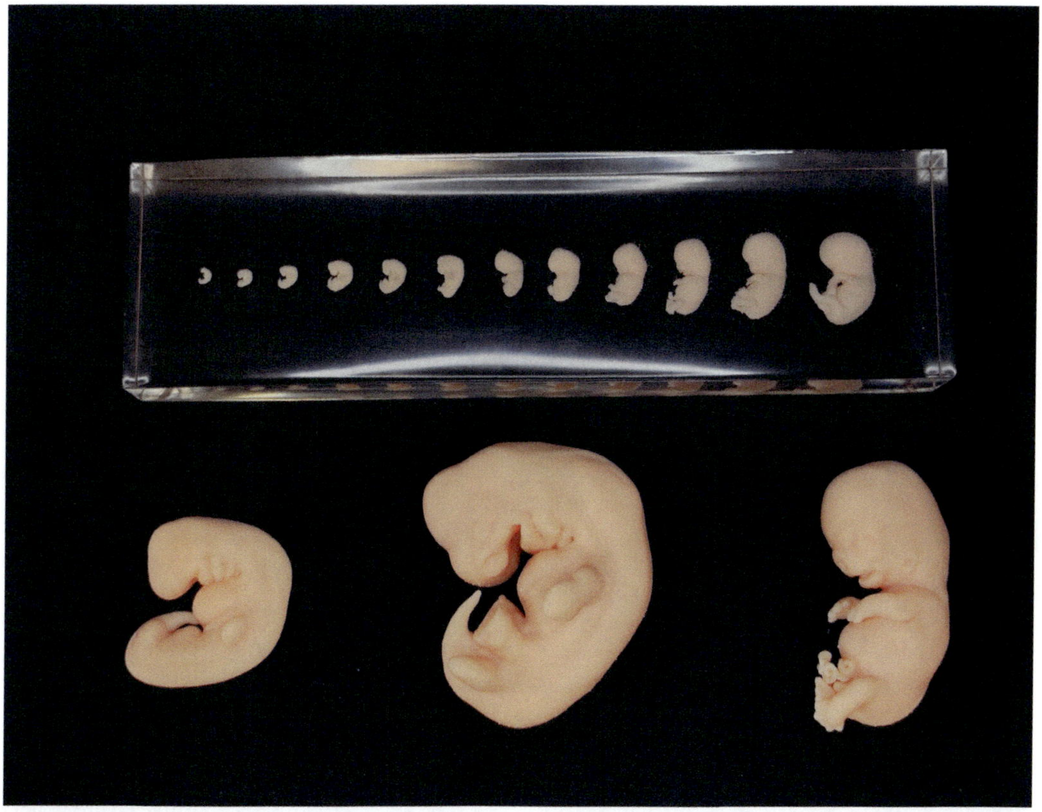

Fig. 1.6 3D-printed human embryos representing Carnegie stages 12–23. Models printed at actual size are embedded in an acrylic resin block for handling and display and give a realistic sense of scale, while enlarged 3D prints allow the appreciation of fine structural details

bony labyrinth itself is challenging to dissect, and the functional part of the system, specifically the fluid-filled space enclosed within the walls of the bony labyrinth, is inherently impossible to render directly visible through cadaveric dissection. However, these structures can be resolved with 3D imaging instruments (for example, CT scanning; Gunz et al. 2012). 3D printing then allows the additional step of rendering this 3D data accessible to non-expert users. Similar to the embryo models mentioned previously, 3D imaging data (specifically MRI data in this case) has been processed and converted into a publicly available file describing the structure of the inner ear, which can be 3D printed (https://www.thingiverse.com/thing:27340). Once printed, this creates a model which can enlarge the structure for ease of interpretation, and allow the complex three-dimensionality to be appreciated—the ability to hold the model in the hand and observe it from multiple angles would greatly facilitate an understanding of the relationships between, for example, the cochlea and the semicircular canals (Fig. 1.7).

3D printing can also be used to integrate functional information with structural information, something which is not particularly easy using cadaveric anatomy alone. An example of such an application is the functional anatomy of the cerebral cortex. We have created a model of the human cerebral cortex that includes both the gray matter of the surface of the left cortex, as well as the white matter of both cortices and the connecting structures (Fig. 1.8). As the gray matter is only included for one hemisphere, the resulting model reveals the underlying white matter with a clarity and integrity that would require significant technical skill to achieve through cadaveric dissection. This model was created using human data from BodyParts3D (Mitsuhashi et al. 2009), and can be printed at actual size to produce a

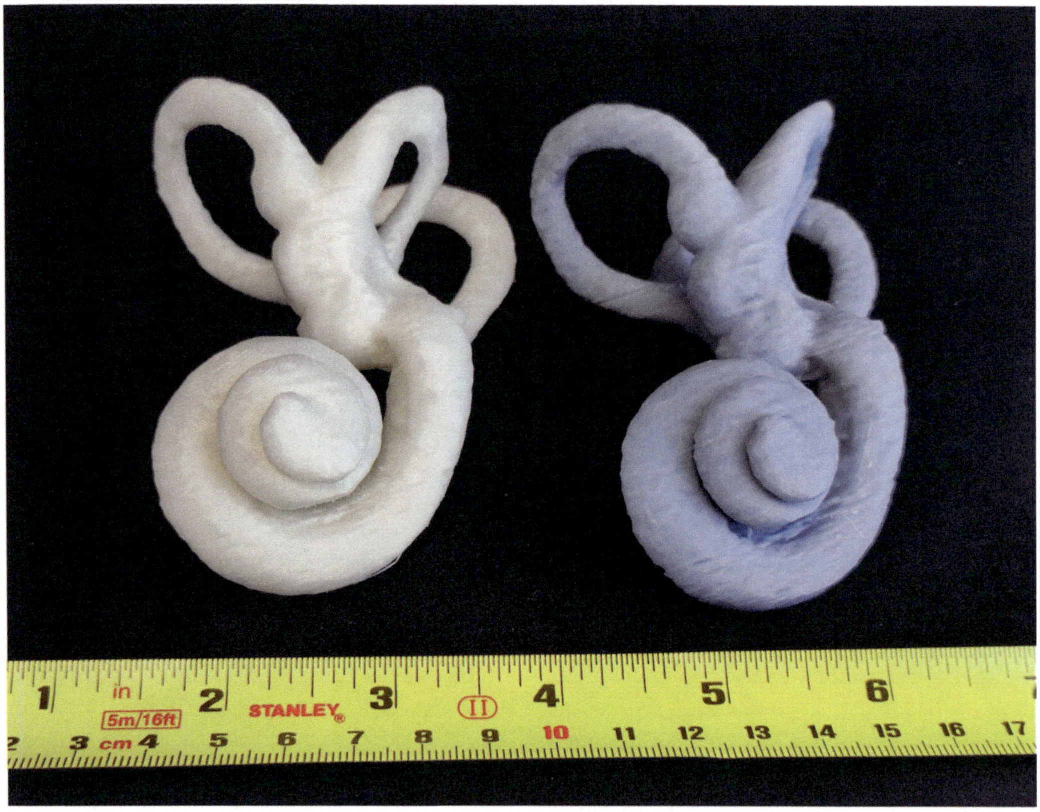

Fig. 1.7 Enlarged 3D-printed models of the inner ear structures enclosed by the bony labyrinth, printed using both filament printing (left) and resin printing (right). The model is derived from human MRI data, the ridges in the model (most visible in the resin print) are an artifact of the 3D MRI processing, rather than the printing process

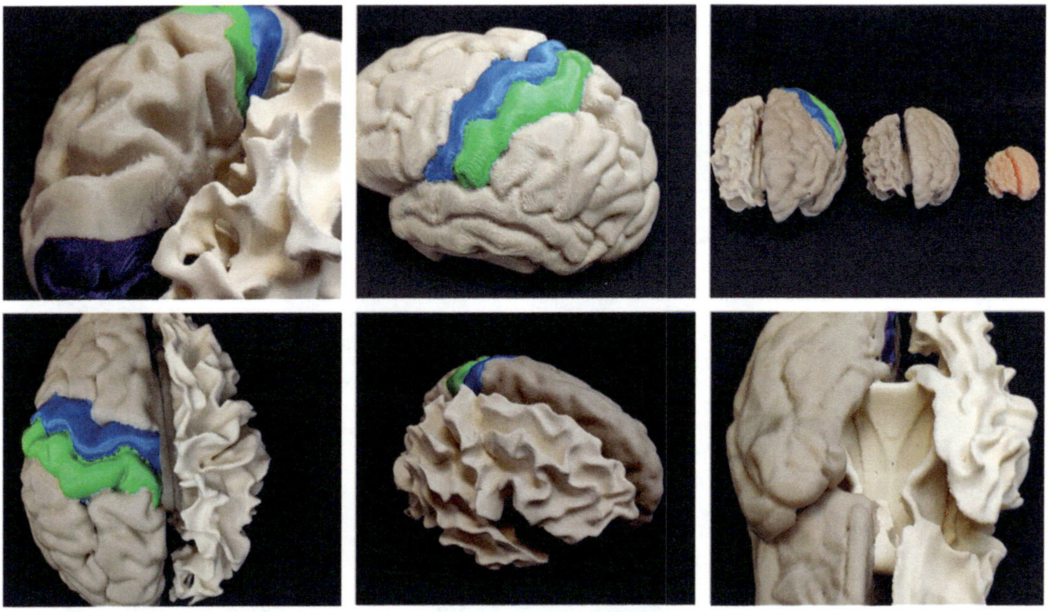

Fig. 1.8 Multi-color printing of the cerebral cortex allows clear delineation of structural and functional regions. In this model, the gray and white matter of the cortex are printed in darker and lighter materials respectively, the pre and post-central gyrus in blue and green, respectively, and the primary visual cortex in purple. The model can also be printed in a single color; the upper right panel shows the model printed full size in multi-color and at a reduced scale in a single color, as well as the smallest version, which is printed in resin

high-fidelity representation of the true anatomical structure. While the model can be printed in a single color, the use of multi-color printing allows the model to be created with gray and white matter printed in contrasting colors as an additional way to visually represent the differences between the two regions.

However, multi-color printing can also be used to convey additional information, highlighting specific structures or functional regions that may not be readily apparent in a cadaveric specimen. In this specific example (Fig. 1.8), the precentral gyrus, postcentral gyrus, and primary visual cortex have been printed in different colors, highlighting their location relative to the rest of the cortex. This approach can be used to create focal points of the user's choosing on a model and can easily be customized to fit the user's needs. We have made this model publicly available through an online 3D print-sharing repository (https://www.printables.com/model/345477-brain-with-grey-and-white-matter).

A final example of the use of 3D printing to represent "impossible" anatomy—anatomy which is impossible to observe directly by itself—is using 3D printing to replicate historically important but unique anatomical specimens. A specific case where this approach has been applied is that of Phineas Gage. While working as a railroad construction worker in 1848, an accident caused a penetrating injury to Gage's left frontal lobe as a tamping rod entered the cranium, passing through the back of the orbit before exiting through the top of the cranium. Gage famously (and surprisingly) survived this injury, and went on to become a classic textbook example used to illustrate the impact of traumatic injury to the frontal lobe, and by inference the function of the frontal lobe. While more recent critical reappraisals of the evidence surrounding the case of Phineas Gage, and the inferences that can be drawn from this case, have called much of what is frequently stated about Gage into question (Macmillan and Lena 2010), the case is one of persistent interest to this day. Six years after his eventual death in 1860, Gage's head was exhumed and his skull was displayed in Harvard Medical School's Warren Anatomical Museum, and this skull is the only persistent biological evidence surrounding his case. The importance of having this artifact became apparent in

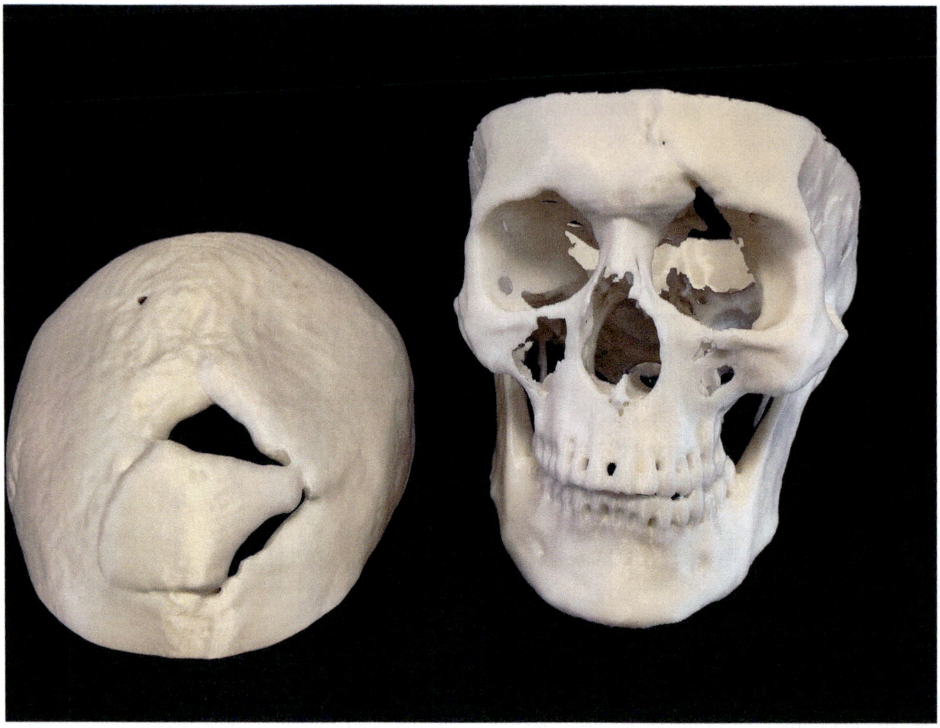

Fig. 1.9 3D-printed copy of the skull of Phineas Gage

recent decades when 3D imaging tools allowed detailed analysis and modeling of the injury using the digital models that were thus created (Ratiu et al. 2004; Van Horn et al. 2012). These digital scans are now available in a format processed to allow 3D printing (https://3d.nih.gov/entries/3DPX-003118) (Fig. 1.9). This has allowed an unlimited number of copies of this historically important specimen to be reproduced globally, allowing the model to be accessed on a scale that was previously impossible; the only way to directly examine the skull was to visit it where it was displayed in the Warren Anatomical Museum in Harvard Medical School.

1.2.2 3D Printing for Active and Interactive Exploration of Anatomy

An alternative way to use 3D printing in anatomy, particularly for educational purposes, is to create interactive models or models that are designed for active engagement and activity-based learning. Frequently, this will involve simplification or abstractions of realistic anatomical structures, for instance, as 2D representations of structures (like textbook diagrams) rendered in 3D. This might involve, for example, rendering lines and boundaries between structures in relief, or printing flattened 2D representations as structures that can be positioned relative to each other to show 3D relationships. An interesting and illustrative example of this "2.5D" approach is a simplified model of the inner ear, available for download from a public 3D model repository (https://3d.nih.gov/entries/3DPX-011385). In contrast to the anatomically accurate model discussed earlier, this is a simple, quick 3D print in the form of a kit card where individual parts are slotted together in 3D (Fig. 1.10). The model itself is assembled into a simplified 3D structure by slotting together flat parts that are effectively rigid 2D representations of the structure. While the structure is obviously less accurate than the aforementioned 3D print of the structure, it does encourage active engagement with structures and their relationships through its assembly, as well as being much eas-

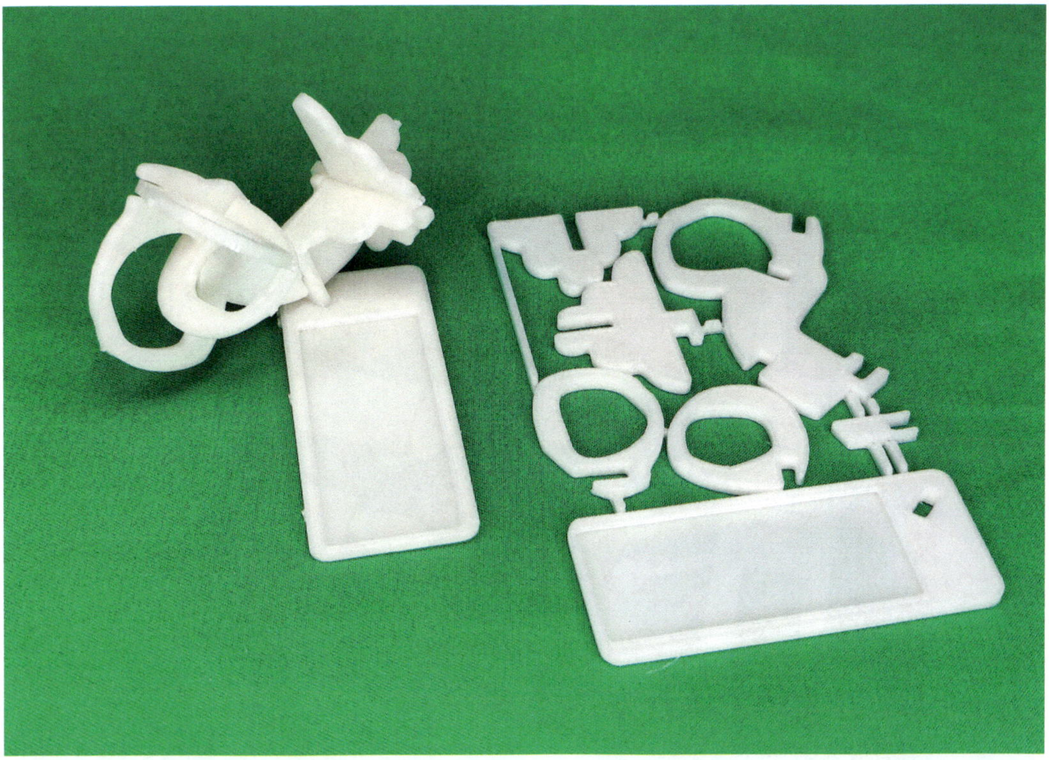

Fig. 1.10 In this simple model of the structures of the inner ear, the structures are produced as a flat "kit" (right) where pieces are snapped off and assembled into a simplified version of the 3D structure including the cochlea and semicircular canals

ier, quicker, and cheaper to produce in large quantities, meaning it may be feasible to produce one of these models for every student in a class.

One of the simplest 3D-printed models we use in our own teaching also employs this "2.5D" approach. This model is used in an activity designed to help students understand the organization of the autonomic nervous system (Fig. 1.11). The model is very basic, consisting of a simplified outline of the central nervous system and sympathetic ganglia, augmented with small pegs integrated into the 3D print. This print is then attached to a board along with a set of simple shapes with pegs, labeled with organs and systems regulated by the autonomic nervous system. Students then use rubber bands to connect elements with the pegs, using different colors to represent different neurotransmitters. The 3D-printed elements are incredibly simple; the three-dimensional nature of this model is not used to convey any structural, anatomical information that could not be conveyed with a two-dimensional diagram on a page, but rather to add interactivity and active engagement with the information, and is therefore designed with that in mind.

A similar approach was used to create an activity-based "2.5D" model to help students learn and understand the sensory pathways in the spinal cord (Fig. 1.12). This model consists of 3D-printed representations of transverse sections of the spinal cord and brainstem and a coronal section of the cerebrum. 3D-printed parts are mounted onto a large board (clear acrylic in this case) in the correct anatomical sequence, and the board can be drawn on with dry-erase markers (the type of erasable marker typically used on whiteboards). Students are then provided with colored wires, each corresponding to a different sensory modality from a specific region (e.g., left foot and fine touch) which they then feed through the model from the periphery through the spinal

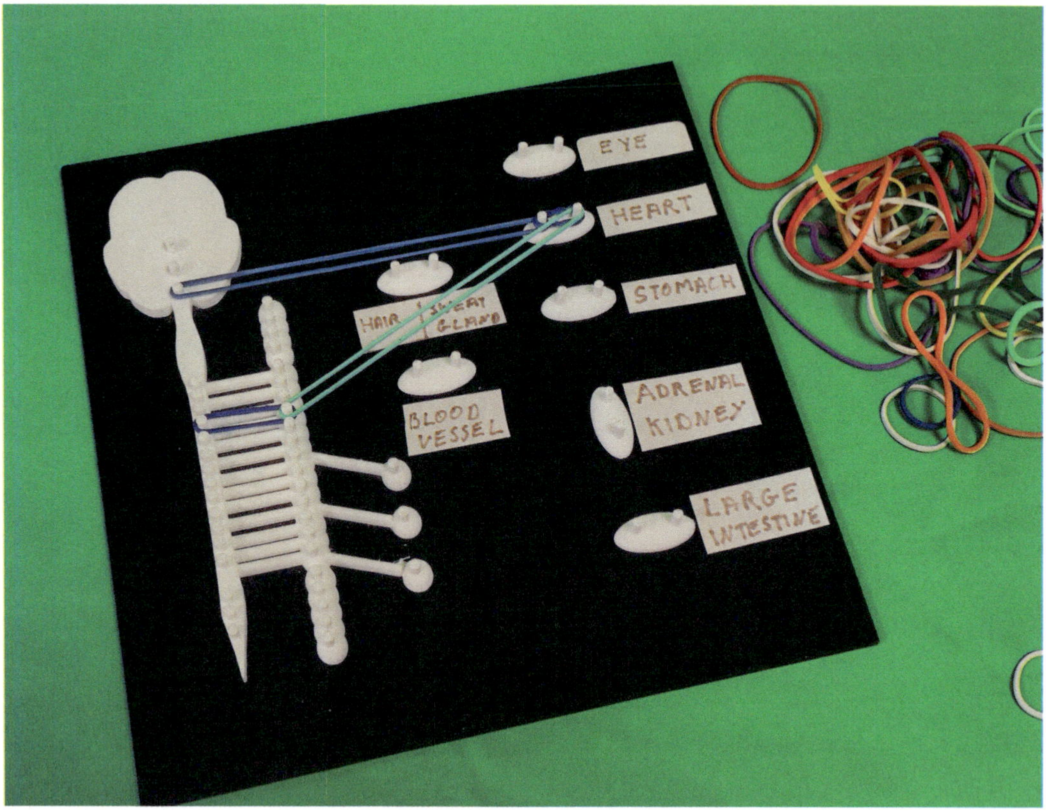

Fig. 1.11 This simplified schematic of the autonomic nervous system using 3D-printed components forms the basis of an activity where students connect organ systems to the autonomic nervous system using rubber bands, with different colors representing different neurotransmitters

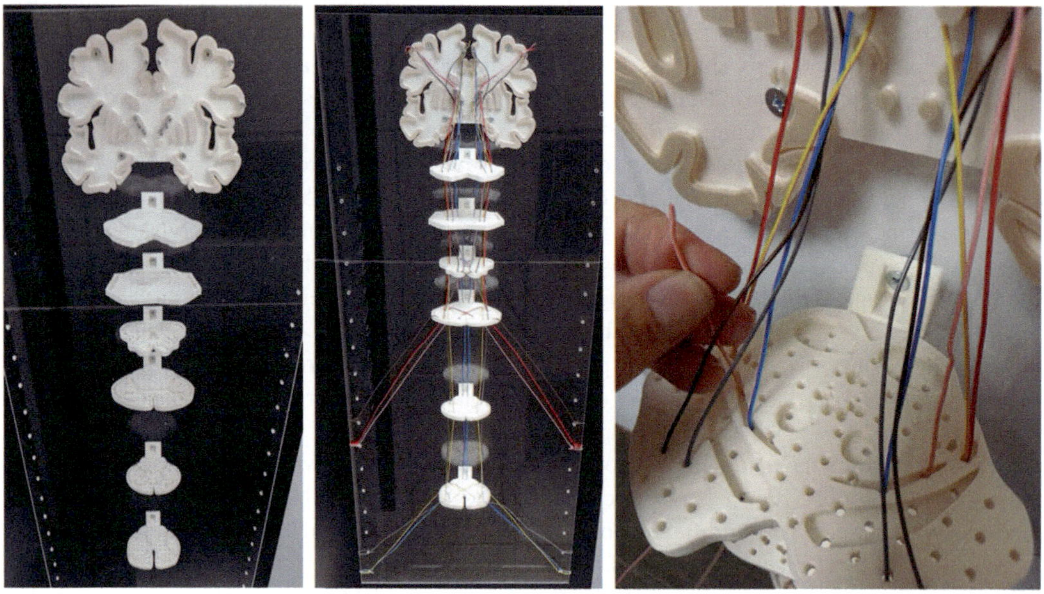

Fig. 1.12 3D-printed model used to demonstrate the sensory pathways in the spinal cord. Students using this model trace the spinal pathways using colored wires from the periphery through the spinal cord and brainstem to the cerebral cortex, recreating the 3D structure of the spinal pathways in the process

cord, brainstem, thalamus, and finally to the cortex. This approach places two-dimensional textbook sections in a 3D context with the aim of helping students appreciate the relationship between 2D and 3D structures while actively engaging with the activity. This activity can be built upon and expanded to descending motor pathways or for simulating the effect of spinal cord lesions. As with our previous example, this model is shared publicly and the 3D print files can be freely downloaded (https://www.printables.com/model/127895-spinal-cord-pathways-model).

Not all activity-based or interactive 3D-printed models require an abstraction or simplification of the structure to be effective. While the multi-color model of the brain (Fig. 1.8) may be effective for demonstrating various functional regions, it is complex and time consuming to print and requires more specialized 3D printers to fabricate. One consequence of this is that scaling up production of this model to create many copies becomes difficult, primarily in terms of the time required to produce each model (which may be several days) and the additional cost of multi-material capable printers. Simpler 3D prints of a single hemisphere of the cerebral cortex can be produced in a single color at a reduced size, meaning many prints can be produced to allow individual students or small groups to receive their own hemisphere that still retains realistic anatomical structure. We have used these types of models in the past and built mini active learning activities around these models by encouraging students to "color in" specific structures in these cortical models, requiring them to engage in understanding the relationship between functional regions of the brain and 3D structure (Fig. 1.13).

Fig. 1.13 Functional neuroanatomy can be taught using simple 3D prints without the need for multicolor printing. In this case, small cerebral hemisphere models are given to students, who are encouraged to "color in" the functional regions in the 3D structure

1.2.3 Combining 3D Printing with Other Tools and Techniques: 3D Printing+

Another approach we have developed to explore functional neuroanatomy borrows from the Phineas Gage case, but requires combining 3D printing with silicone molding to construct the final working model (Fig. 1.14). The goal of this model is to create a system which allows students to simulate a penetrating brain injury (similar to Gage's), and consider the relationship between the anatomical structures that have been damaged and what the functional impacts might be.

To create this, we printed a model of a lateral half of the brain (including cortex, cerebellum, and brainstem) using the same data source used for the previous multi-colored cerebral cortex model. The brain was used to create a mold with soft silicone. In addition, we 3D printed a model of the corresponding lateral half of the cranium from the same data set, modified with an opening below the orbit and a corresponding opening at the top of the cranium. The silicone mold is then filled with a gel substance (we have used gelatin and carrageenan gels in the past) to create a soft brain model with a structure reflecting the original brain. This cranium model is then placed over the brain model, with the half-brain fitting into its normal anatomical position in the corresponding half-cranium. A metal rod can then be passed through the opening, through the underlying brain, creating a simulated injury. Brain models can be reproduced over and over using the silicone mold; we have created over a hundred soft brains from a single mold. While we have used this to create a frontal lobe injury, the same approach could be used to create injuries in any brain region by simply modifying the cranium model.

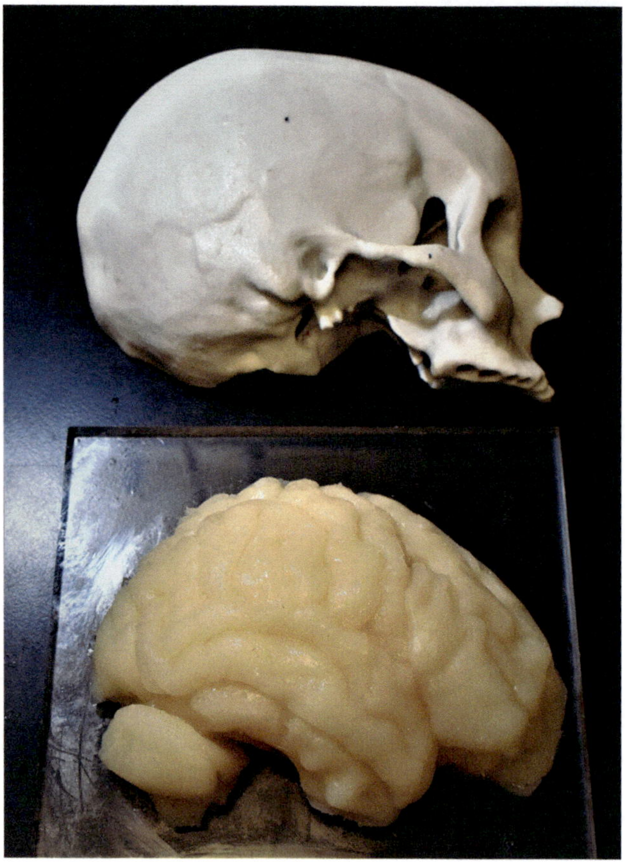

Fig. 1.14 Gelatinous brain models created using molds formed around 3D prints can be paired with modified cranial models from the same human dataset. In this case, the skull has been modified with openings, the brain placed inside it and a penetrating injury is simulated. The gelatinous brain model is destructible and can be remade over and over using the same mold. Note the skull in this photo was printed using a specialized 3D printer (using binder jet technology), but standard 3D printing methods would be just as effective here

Adding complementary fabrication techniques to 3D printing can be effective in improving the usability and visual impact of 3D-printed models. A simple example of this is the 3D-printed embryo models discussed earlier (Fig. 1.6). For the realistic scale models, the small size makes handling difficult, so the entire series was cast in a block of resin, creating a coherent display. Embedding the 3D print in clear resin not only improves robustness of the piece but also improves the visual appeal. We have also used a combination of mold making and resin casting to create a model of the brainstem in resin which shows the relationship between brainstem structure and the cranial nerve nuclei within (Fig. 1.15).

To achieve this, a large-scale model of the brain stem was 3D printed and used to create an impression of it in a silicone mold. The mold was then filled with clear acrylic resin in layers, with cranial nerve nuclei placed in their approximate brainstem position during this process. The end product is a model of the brainstem that demonstrates both the structure of the brainstem itself and the relative structure and location of the nuclei within it. The clarity of the internal structure was enhanced by using resin with a refractive index matched to water. Out of water, the brainstem structure itself is most evident, but when submerged in water, the internal structures become the clearest feature (Fig. 1.15).

In addition to mold-making and casting, further supplementary components can be added to 3D prints as part of this "3D printing +" approach, to add functionality and usability. This is particularly effective when used to demonstrate dynamic, functional anatomy, such as the action of mus-

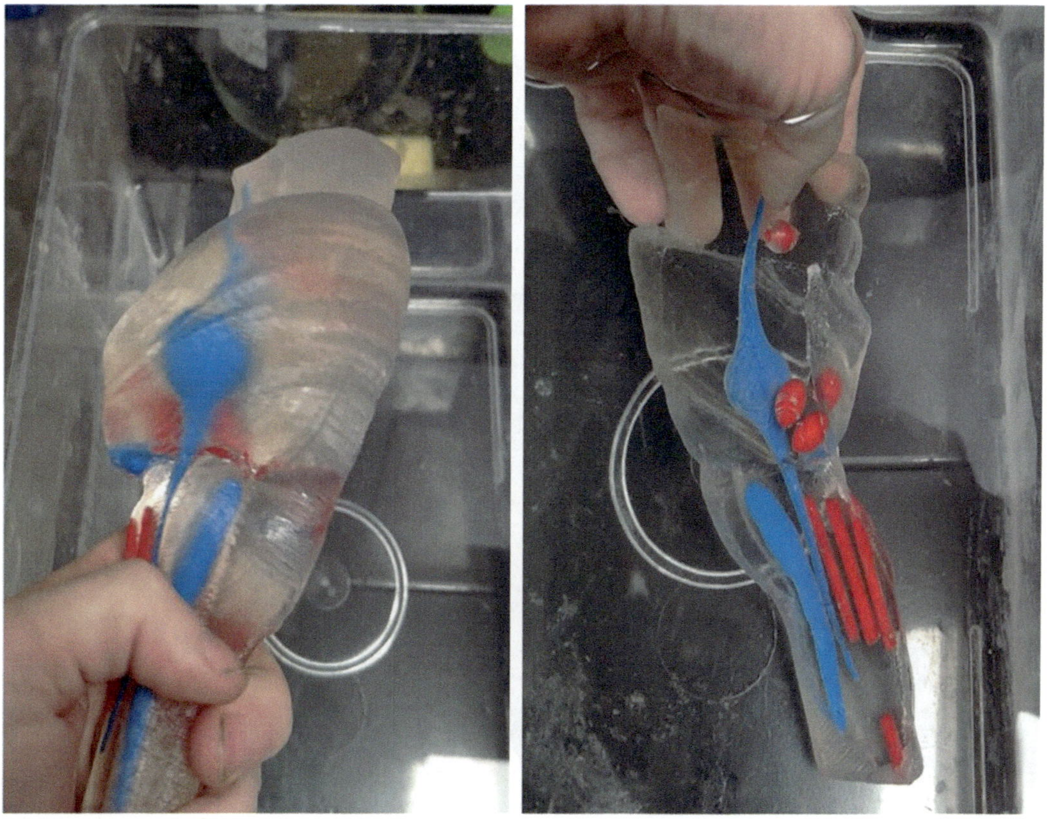

Fig. 1.15 In this model, motor and sensory nuclei of the brainstem were 3D printed, and the printed parts embedded in an acrylic resin brainstem model, itself made using a mold created using 3D printing. The acrylic has the same refractive index as water; when in the air, the surface of the brainstem is visible, but submerging the model in water renders the interior structure more visible

cles. A particularly good application of this approach is for the demonstration of the actions of the extraocular muscles on the position of the eye in the orbit. While this is a system which has been modeled many times, one of the more interesting versions is largely 3D printed and publicly available for download (https://www.printables.com/model/336284-extraocular-muscles-human-eye-model). This design integrates a 3D-printed eye and supporting structures with more rigid mechanical components for structural stability as well as elastic band components representing the six extraocular muscles. The muscles are "contracted" by pulling on the specific elastic bands, generating accurate eye movements in response. This model is particularly effective because it reduces the anatomical accuracy to the functionally essential components, i.e., the axis of the orbit and the direction of action of the muscles, while abstracting the remaining anatomy (e.g., the actual shape of the muscles and the structure of the orbit), making it easy for learners to focus specifically on what the model is designed to demonstrate.

We have modified the original design with a simple addition of a crosshair laser inserted into the center of the eye (Fig. 1.16). This projects a cross-hair pattern in front of the eye, which allows a user manipulating the "muscles" from behind in order to see the effect of the muscle action on eye position, showing not only changes in where the eye is pointed, but also how the eye is rotated. This clear demonstration of the actual movement makes what is traditionally a difficult concept to illustrate (especially with traditional, static textbook diagrams) much more obvious to

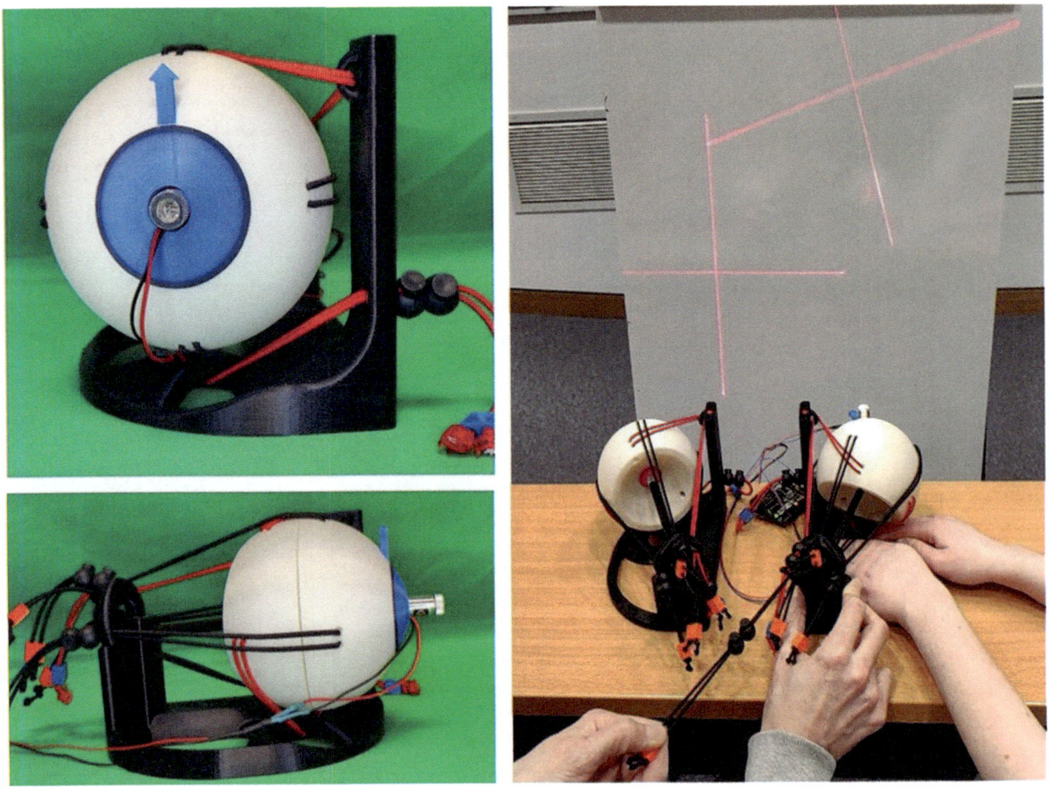

Fig. 1.16 The actions of the extraocular muscles can be effectively demonstrated by this mostly 3D-printed model, where muscles are represented by elastic cords. The addition of a simple cross-pattern laser to the front of the eye allows the position of the eye to be visualized during movement. The image on the right shows the model in use, in this case showing that contraction of the superior rectus results in not only elevation of the eye, but also adduction and intorsion

learners, and encourages them to experiment with the different muscle actions and combinations to explore their functions.

Our final example of "3D printing+" takes the abstraction of actual anatomy even further to focus almost exclusively on functional representation. We created a model which demonstrates the movement of the tympanic membrane of the ear in response to sound, by designing a device which renders these movements visible using a reflected laser beam (Fig. 1.17). The 3D-printed part consists of a sound cone with either a speaker mount or open back to allow the device to demonstrate the response to live human voice, and an attached mount for a laser diode. A rubber sheet (i.e., a balloon) is stretched over the opening of the sound cone, and a small laser diode is directed toward the sheet. By gluing a small mirror to the point where the laser strikes the balloon, the laser is reflected off the mirror and onto a wall in front of the device. As the balloon (and mirror) move, the laser will then also move. This means that as the balloon vibrates in response to sound, just like the tympanic membrane does, the laser point will also trace these vibrations, rendering the complex frequency and amplitude components of the sound into visible shapes and patterns. As well as demonstrating normal function, this model can be used to demonstrate pathological states, such as a perforated tympanic membrane

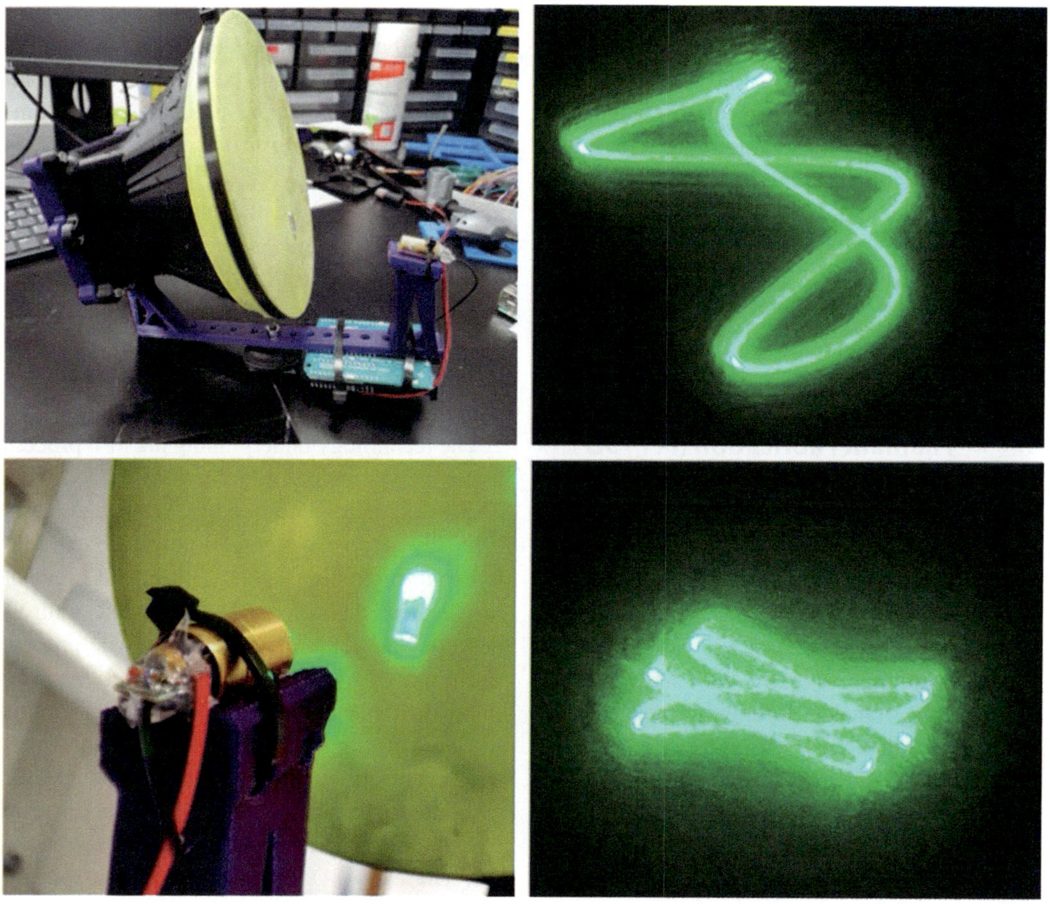

Fig. 1.17 This 3D-printed part allows visualization of the movement of a membrane (representing the tympanic membrane) in response to sound. The 3D-printed parts consist of a mount for a small laser diode, and a holder for a stretched balloon. A small mirror glued to the membrane reflects the laser, creating complex patterns in response to the sound. The two examples are single tones with different waveforms, producing different movements of the membrane

by placing a tear in the balloon which will result in visibly reduced laser movements, reflecting reduced sound transmission. Once again, this model is shared publicly and available to be reproduced easily by anyone who would find this model of use (https://www.printables.com/model/293269-sound-reactive-laser-reflector-tympanic-membrane-m).

1.3 Beginners Guide to 3D Printing and Model Making

While the advantages of 3D printing for anatomy education may be clear to most, it can still seem technically daunting for those with no prior 3D printing experience. However, modern 3D printing has become increasingly accessible both in terms of the cost of the 3D printers and their ease of use. To help demystify the 3D printing process for those unfamiliar with it, the following beginner's guide has been written to provide even complete novices with advice and guidance on each step in the process, from selecting a 3D printer to post-processing the final model.

1.3.1 Printing with Filaments: Fused Filament Fabrication (FFF) 3D Printing

The first step when starting out in the world of 3D printing is deciding what type of printer to invest in. While many exotic forms of 3D printing have been developed over the last 30 years, including binder jet printing, selective laser sintering, and paste printing of materials ranging from ceramics and concrete, to biopolymers and even chocolate, most of these techniques are either too specialized in purpose or too expensive in terms of materials and infrastructure to be broadly useful and accessible for anatomical model creation. The two dominant technologies in the consumer market are also the two easiest and most cost-effective techniques that would be most appropriate for those with an interest in embarking on anatomical 3D printing, and so this guide will focus on consumer 3D printers based on these technologies (Fig. 1.18).

Fused filament fabrication (FFF) operates by passing filaments of thermoplastic material into a heated nozzle where the material is melted and extruded from the tip. Moving the nozzle along a flat surface (the print bed, or build plate) during this extrusion results in a line of the thermoplastic material being deposited on the bed, and continuing this process across the bed fuses adjoining lines together into a solid flat layer (slice) of material. The extrusion nozzle is then elevated by a fixed amount and the process is repeated, with the next layer of thermoplastic fusing to the previous layer to form a 3D volume. This dynamic process continues in this manner, with each layer composed of fused individual lines of thermoplastic, and adjacent layers fused together until a completed 3D structure has been constructed. This simple process can result in incredibly complex 3D shapes as long as one important factor is considered—each layer (or region of a layer) must be deposited on the layer below or supported in some way. As each 3D print is constructed from single thin lines of thermoplastic, the time taken to print a model is a function of the volume of material in that model, regardless of the individual model dimensions (as will be noted later, this is not true for all 3D printer types).

The technology underpinning the FFF process is relatively simple. The minimum needed for a 3D printer is a heater to melt the filament, motors (typically stepper motors) to move the extruder nozzle in three dimensions and to feed filament into the heated nozzle, and an electronic controller to operate these elements. The basic process was invented by S. Scott Crump in 1989, who founded Stratasys Ltd. and held a patent on this process (Crump 1992). FFF is also known as fused deposition manufacturing, although the abbreviation FDM is actually a trademarked term (FDM®), with the trademark owned by Stratasys Ltd. The early 3D printers were developed for, and marketed toward, the commercial manufacturing and engineering sectors, meaning that these printers were prohibitively expensive and complex for use outside those sectors.

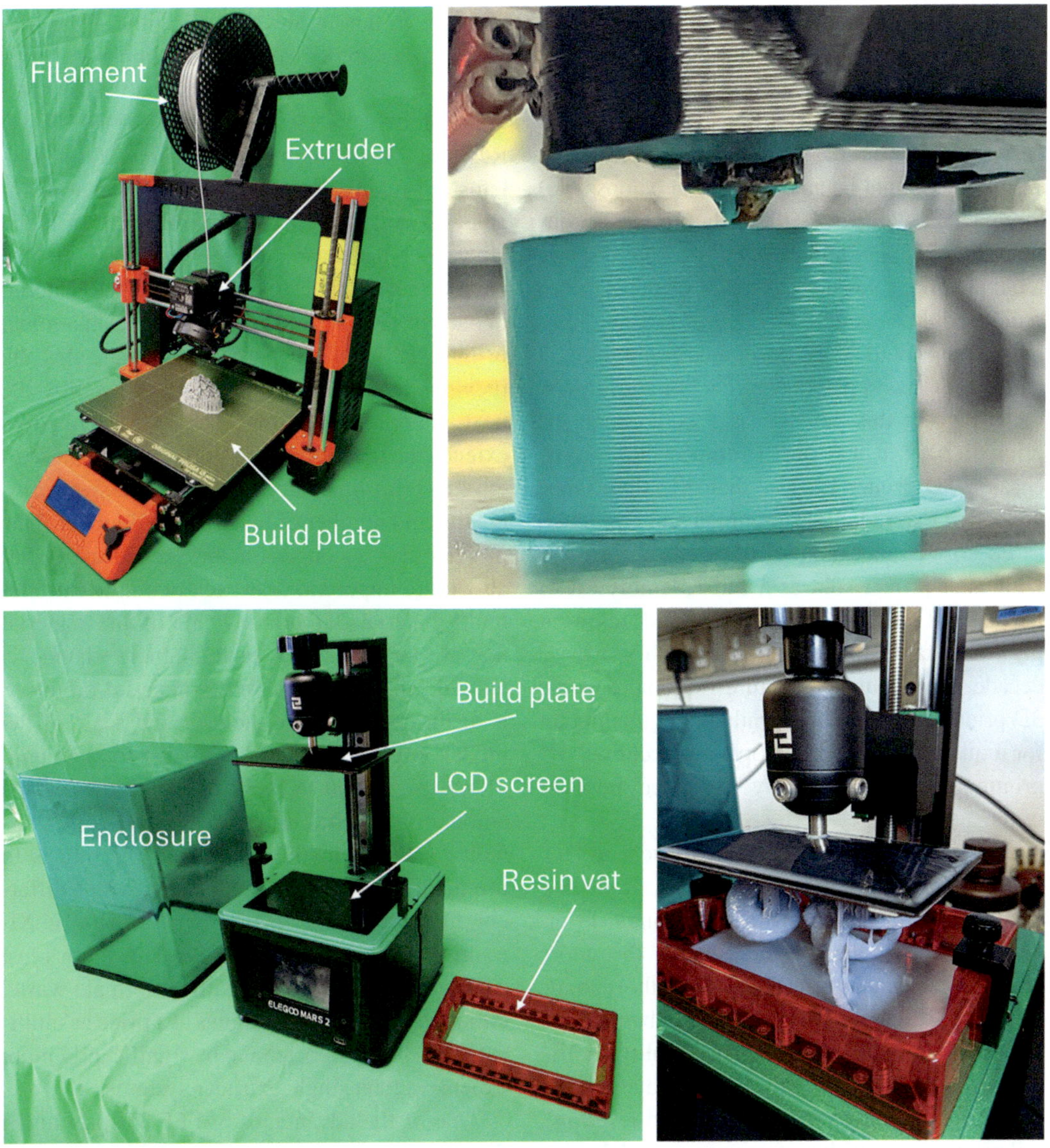

Fig. 1.18 Anatomy of a 3D printer. The two most common types of 3D printers used for anatomical model making are FFF or filament printers, represented by the Prusa i3 mk3 on the upper panels, and the SLA or resin printers, represented by the Elegoo Mars 2 on the lower panels. Key components of FFF printers common to all systems are a build plate where the model is constructed, and an extruder which pushes filament into a heated nozzle; the filament is then extruded in single lines as the nozzle moves in 3D (upper right panel). Resin printers also contain a mobile build plate, a vat with a transparent bottom to contain the resin, and a screen that exposes an entire layer of the model at once. As each layer is cured by UV light, the build plate rises and a solid print emerges from the liquid resin (lower right panel)

The expiry of the Stratasys FDM patent in 2009 was arguably the most important milestone in the history of 3D printing, as it allowed new manufacturers to enter the 3D printer market. However, this would not have mattered without the ongoing efforts to make 3D printing more accessible both in the academic environment and in the burgeoning network of hackers, hobbyists, and enthusiasts connected globally through various internet communities. Specifically, the other key milestone around this time was the creation of the first RepRap 3D printer design at the University of Bath (Jones et al. 2011). The RepRap project is focused on developing a low-cost 3D printer built around the philosophy that a 3D printer should be able to (insofar as possible) fabricate its own parts—the name itself is a contraction of *Replicating Rapid Prototyper*. While the innovations developed as part of the RepRap project could be adopted by enthusiasts and hobbyists, they could not be developed commercially until the Stratasys patent expired. For consumers, the expiry of this patent was a game changer; prior to 2009, it was next to impossible to purchase a 3D printer for less than €15,000. Once the patent expired, prices tumbled and continued to do so. As an example, in 2011 the original Ultimaker printer sold as a self-assembly kit of mechanical and electronic parts and laser-cut plywood frame for around €1000. Just over a decade later, countless feature rich and capable 3D printers are currently available on the market for a quarter of that price, opening access to 3D printing to an enormously expanded pool of users.

Entry-level FFF 3D printers such as the Creality Ender series are capable of producing excellent quality prints, with the primary differences between these printers and the more advanced consumer printers (such as those made by Prusa Research or Bambu Labs) being print speed and reliability, as well as quality of life improvements that simplify or automate parts of the user experience (such as print bed leveling). All printers are capable of printing a model using a single material, although some printers also have capabilities which allow more than one material (or more than one color of the same material) to be used in a single print. The ability to print in multiple materials can be valuable in the context of anatomical models, allowing multiple structures to be represented in a single model, thereby facilitating a demonstration of the relations between those structures. The brain model demonstrating gray and white matter and functional regions (Fig. 1.8) was generated using this approach. The most common way to implement 3D printing in multiple materials uses a mechanism that allows the changing of the filament in a single extruder. Examples of this approach are seen in the Bambu Labs AMS (automatic material system) and the Prusa Research MMU (multi-material unit). These systems are relatively inexpensive and often available as add-ons for single material printers. However, there are two drawbacks to this approach: the first is that every time the material in the extruder is exchanged, a significant amount of material must be "purged" to ensure that the previous material is completely replaced by the next material. The time taken for each purge extends the total print time, and can add hours-to-days to print times depending on the size, complexity and number of materials in a model. This process also accounts for the second and arguably more important drawback of single extruder multi-material 3D printing; all of the material that is purged during each filament change is wasted. The volume of waste material produced is again dependent on the size and scale of the model, but it is not unusual for the total amount of waste material to be several multiples of the amount of material in the model itself. An alternative solution for multi-material printing is to use multiple extruders, each loaded with a different filament, as permissible by the Prusa XL multi-tool printers, for example. This allows fast multi-material printing with minimal waste, although this efficiency comes at a price as multi-extruder 3D printers tend to be significantly more expensive than single extruder systems.

One of the advantages of FFF 3D printing is the wide range of materials available for model manufacture. This provides the user with an enormous range of choices and control over both the appearance and mechanical properties of the final model. The choice of filament material is informed by the desired mechanical properties and visual appearance of the final model, but also the ease of use of the material. Some thermoplastic materials, such as acrylonitrile butadiene styrene (ABS) and nylon, which are commonly used in other manufacturing techniques like injection molding are challenging materials to work with in 3D printing and are usually avoided unless there are specific reasons for their use. ABS and nylon, along with materials like polycarbonate and acrylonitrile styrene acrylate (ASA) need to be extruded at a higher temperature than other materials (~280 °C), and also require the print bed to be heated to a high temperature (100 °C) to ensure that the model adheres to the bed. As a result, these materials are often printed in sealed enclosures to maintain this warmer environment.

On the other hand, materials like polyethylene terephthalate glycol (PETG) and polylactic acid (PLA) can be printed at lower extrusion temperatures (~210 °C for PLA and ~235 °C for PETG), and while PETG works best with a print bed heated to 85 °C, PLA can be printed successfully without any bed heating (although heating to around 60 °C does produce more reliable prints). PLA is arguably the easiest material to work with and is suitable for printing the finest details that FFF printing is capable of, although it does have some drawbacks. Mechanically, PLA can be brittle, and it has low resistance to UV light, making it unsuitable for outdoor use. However, the primary weakness of PLA is its thermal resistance; while the melting point of PLA is ~160 °C, it begins to soften and deform around 70 °C.

As a material, PETG can struggle with very fine details in prints but will produce excellent quality parts in most cases and the prints themselves are very robust. The thermal and UV stability of PETG is better than PLA, and the bonding between layers is particularly strong. A consequence of this is that objects printed in PETG can be watertight, which can be useful for functional prints.

Both PLA and PETG are relatively inexpensive, typically costing between €20 and 40 per kilo of material, and both are available in an extensive range of colors and finishes from matt to high gloss. There are even commercial services (such as Color On Demand from colorFabb) that can produce filaments in any specific color that the user requires.

In addition to the basic materials like PLA and PETG, filaments with embedded additives such as glass or carbon fiber can also be used in 3D printing. These additives change the mechanical properties of the final print (altering hardness, wear resistance, and tensile strength, for example) and also the visual appearance. More unusual additives, such as metal or wood particles, glow-in-the-dark, and fluorescent materials are also available, although these are less likely to have utility in anatomical model making. While many of these more specialized filaments are likely to have limited utility in the creation of 3D-printed anatomical models, there are some specialized filaments that may have specific applications in anatomical 3D printing.

Soluble thermoplastics are a type of filament that has a specific utility in FFF 3D printing. Polyvinyl alcohol (PVA) and butenediol vinyl alcohol copolymer (BVOH) which are water soluble, and high-impact polystyrene (HIPS) which is soluble in limonene, can be used in multi-material printing as support materials. In these cases, the model itself is printed in an insoluble material, but the model is supported by a structure that is printed in a soluble material. After printing, immersion of the model in solvent will remove the support material while leaving the model itself intact and clear of extraneous supports. While a relatively complex process, this is the easiest way to 3D print hollow structures that require support in the interior, which in the anatomical context, is a useful technique for creating hollow organs, cranium, etc. without needing to print in multiple parts and assemble afterward.

Polyvinyl butyrate (PVB) is soluble in alcohol and ABS in acetone, and one of the uses of this property is a technique called vapor smoothing. Here, the printed model is either immersed briefly in the solvent, or more often enclosed in a sealed chamber with the solvent at the bottom, which is heated to fill the chamber with solvent vapors. The effect of this is to gently and evenly dissolve a very thin layer of the surface of the model, which smooths out the printing layer lines, producing a smooth, glossy finish similar to what would be seen in an injection molded part.

The other filament type which may be particularly of interest in an anatomical context is a flexible filament, such as thermoplastic polyurethane (TPU). Flexible filaments hold the promise of being able to produce softer structures, and potentially create models that replicate not only the look but also the feel of soft biological tissues. However, while the majority of these materials are softer than rigid materials like PLA and PETG, they are still much stiffer than biological tissues, with hardnesses closer to hard rubbers. Additionally, these materials can be difficult to work with, and the softer the material, the harder it is to use. One of the reasons for this relates to the FFF process itself—extrusion of filament needs mechanical means to push the filament into the heated nozzle, and flexible filaments are more likely to buckle and tangle during this process. If very soft materials are required in a model, combining 3D printing with molding soft silicones may be a better option than direct 3D printing.

1.3.2 Printing with Resin: Stereolithography (SLA) 3D Printing

While the consumer 3D printer market has been dominated by FFF printers for over a decade, a second 3D printer type has become more readily available in the consumer market in more recent years—stereolithography (SLA) 3D printers. SLA printers are also known as resin printers, and this name derives from the basic process that underpins SLA printing. Unlike FFF printing, the raw material for SLA printing is a photocurable liquid resin, typically cured by UV light. In an SLA printer, a movable print platform is lowered to the bottom of a vat of printing resin with a transparent film bottom. A pattern of light representing a single slice through the model is projected through the vat after being formed by a Liquid Crystal Display (LCD) screen below the vat of resin. This light cures and hardens the liquid resin into a solid slice attached to the build plate. The plate then rises by the thickness of a print layer, and liquid resin flows into the space below before the curing process is repeated with the next layer fusing to the first. As with FFF printing, repeating this process over and over, layer by layer, allows the gradual construction of a complex 3D structure. Similar to FFF printers, each layer must be attached to the previous layer. A consequence of this process is that the print time is a function of the total build height, as the entire layer is formed simultaneously in most SLA printers (an exception being scanned laser SLA, although this is relatively unusual in the consumer printer sector).

SLA printers typically print in layers of 50 μm or less, significantly smaller than FFF printers, and these layers typically have cleaner edges than those of FFF prints. As a result, visible layer lines are much less prominent in resin prints compared to filament prints where fine ridges along the surface are a characteristic feature of those prints. In general, resin-printed models have much higher fidelity and overall detail than filament-printed models.

SLA printers are simple devices, consisting essentially of a UV light source, an LCD screen, and a motor to move the build platform as the printed object is formed. While early SLA printers were expensive, the entry of manufacturers such as Elegoo and Anycubic into the market has seen an explosion of SLA printers available for <€300. Most of these printers (such as the Elegoo Mars series) have limited build volumes, with build plates typically measuring 150–180 mm by 70–90 mm, with the build plate size limited by the screen size. While specialized larger-scale SLA printers do exist, they typically cost many times the price of the smaller-scale consumer printers. The surge in availability of these low-cost printers is related to the build size, as manufacturers have been able to tap into large-scale production processes for LCD screens; it is no coincidence that the size of LCD displays in consumer SLA printers is similar to the LCD display size in mass-produced tablets and phones.

Prints from SLA printers always require an element of post-processing. At a minimum, the model must be washed directly after printing to remove the excess uncured resin, and a final post-curing in UV light is also essential. While some resins can be washed in water, most require alcohol for washing. As a consequence, resin printing, unlike FFF printing, usually requires supplemental infrastructure in addition to the printers themselves, and many printer manufacturers also sell washing and curing systems to automate and simplify the post-processing part of the resin printing workflow. The majority of SLA prints are also printed with some element of support scaffolding which must be removed. This is usually relatively easy to do, and support structures will often detach effortlessly by hand, especially when performed before the post-curing process when the attachment points are softer; however, this needs to be conducted carefully to prevent scarring or bumps from remaining at the attachment points which can mar the surface finish. These surface imperfections are typically more noticeable in SLA prints compared to FFF prints since the surface quality is otherwise so favorable. These surface imperfections can be dealt with through sanding and polishing as most SLA resins are hard and will stand up well to such treatment, although this can add considerable post-processing time and work in preparing a final model.

Multi-material printing is not realistically possible with SLA 3D printers, and the range of photocurable resins is much more limited in terms of both color and mechanical properties compared to that of filaments for FFF printing. However, resin-printed models are very suitable for painting (Fig. 1.19), as evidenced by the popularity of SLA 3D printing in the miniature gaming community. Therefore, while resin printing has excellent fidelity in terms of shape and detail, the visual aesthetic options are limited without additional (and potentially skilled) craft, making this technique most suited to representations of morphology specifically, rather than more general visual representation. As an example, the SLA print of the brain model with both gray and white matter visible (Fig. 1.20) shows excellent smooth continuous detail across the entire surface compared to the FFF print, but the use of color to clearly indicate the different functional and structural regions is only realistically possible with the multi-material printing used to create the FFF version of this model.

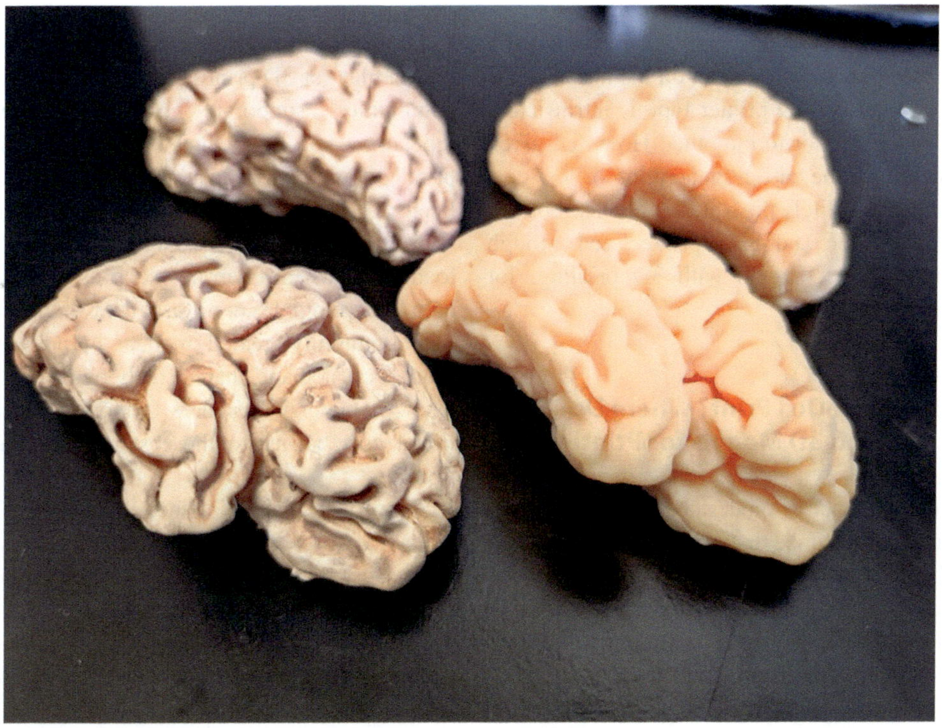

Fig. 1.19 Resin printing is restricted to a single color, but resin prints are suitable for painting. While some paint finishes may require a skilled hand, simple techniques can produce useful effects. In this case, the brains on the left were sprayed with white primer before being covered in a thin acrylic wash which accumulates in the hollows of the model, enhancing shadows and detail relative to the unpainted models on the right

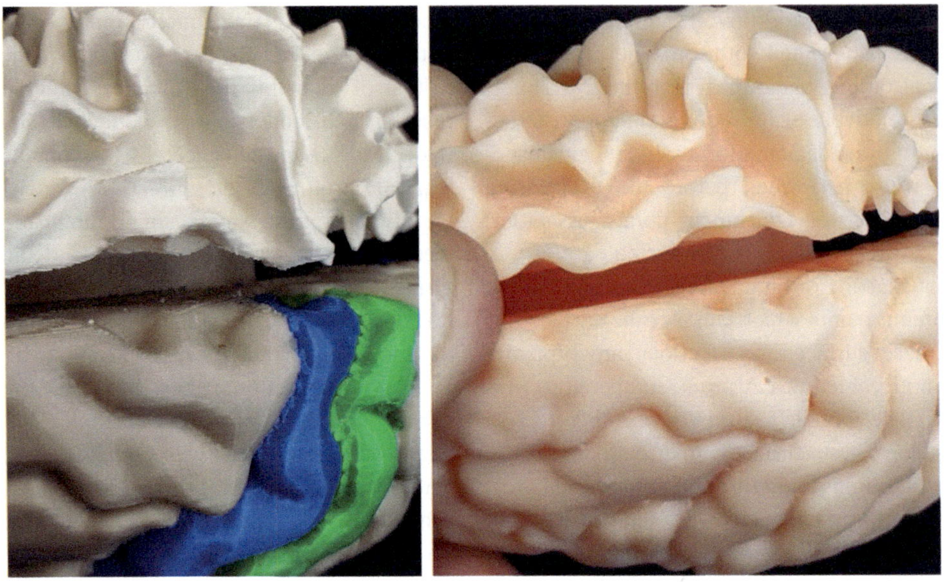

Fig. 1.20 A side-by-side comparison of the same structures printed with FFF (left) and SLA (right) printing shows the increased resolution and fidelity of SLA printing, although the multi-color printing seen in the FFF print is not possible with SLA printing. Note also that the two models are not printed at the same scale—see Fig. 1.8 for relative sizes of these models

Ultimately, the choice between FFF and SLA printers is entirely dependent on the needs of the user, and neither 3D printer type is intrinsically superior or inferior to the other. However, given the quality of the current generation of low-cost printers, it is entirely possible to purchase both an entry-level FFF and SLA printer, and enough materials to get started with 3D printing, for under €500.

1.3.3 What Do I Print? Where Do I Find 3D Models for Printing?

One consequence of modern consumer 3D printing having a heritage that originates in enthusiast and hobbyist communities is the persistence of a culture of open sharing with other community members. The open-source origins of the RepRap project are reflected today in open-source hardware designs from manufacturers like Prusa Research, and both of the most commonly used "slicer" programs (namely PrusaSLicer and Cura), which are essential to go from digital design files to FFF 3D printer commands, are distributed freely under open-source licenses.

However, this culture also extends to digital design files for 3D printing, with the 3D printing community supported by online public repositories of freely downloadable "ready-to-print" 3D model files, usually shared under open licenses (e.g., creative commons or similar). While paid models do exist and can be downloaded elsewhere, websites such as thingiverse.com and printables.com distribute thousands of 3D community-created models in addition to models created and shared freely by professional and commercial designers, removing an additional cost barrier to those starting out in 3D printing.

Due to the nature of these repositories as a "one-stop-shop" for 3D models, it can be somewhat overwhelming for those looking for specific purpose 3D models (e.g., anatomical model making). However, in addition to general-purpose online repositories, more specialized public repositories are also available, some of which are particularly good sources of anatomical models.

At the time of writing, NIH 3D (formerly the NIH 3D print exchange, https://3d.nih.gov/) contains 651 models tagged as "anatomy." Many of these models are based on real-world human anatomical data, and include both normal anatomical structure as well as examples of specific structural changes associated with pathological states, such as the lateral ventricles of the brain in hydrocephalus, including specific anatomical structures from the National Library of Medicine visible human project (https://www.nlm.nih.gov/research/visible/visible_human.html). While the subset of structures available is not comprehensive or complete, the material available is downloadable in a file type ready for 3D printing, eliminating the need for the technical skills and software required to otherwise extract this information from the raw data available elsewhere.

Resources available on NIH 3D also include 3D scans of historic artisan models, such as the eighteenth-century model of the placental vessels and fetus attributed to Felice Fontana (https://3d.nih.gov/entries/3DPX-020529). This allows historic models to be reproduced repeatedly anywhere in the world, making the skills and artistry of anatomical model makers of the past accessible to modern audiences globally. However, it should be remembered that 3D printing can only reproduce the shape of these models; other visual details of color and texture are lost during this process. NIH 3D also includes examples of printable anatomical structures from famous textbook cases, such as the 3D scanned cranium of Phineas Gage (Ratiu et al. 2004), making classic educational case studies available in a tactile, realistic way and in a manner that can only be achieved by a combination of 3D printing technologies and a culture of open sharing of resources.

For comparative anatomy, many natural history museums internationally have made models generated from 3D scans of their specimens available publicly, such as those made available on Sketchfab (https://sketchfab.com/NHM_Imaging) by the Imaging and Analysis unit of the Natural History Museum, London. These initiatives provide access to unique specimens which might only be otherwise available by actually visiting the museums where they are stored and also allow handling and examination in a way that may not be otherwise possible due to the fragile nature of the specimens themselves.

Another useful resource is BodyParts3D (Mitsuhashi et al. 2009)—a comprehensive database of 3D anatomy for an adult human male, which is accessible online (https://lifesciencedb.jp/bp3d/) and shared under a CC-BY-SA license. The entire body of this single individual has been segmented and users can select anatomical structures, filtered by anatomical region and structure size, to download in a format suitable for 3D printing (.obj files). Importantly, these models retain not only the details of 3D structure but also accurate scale and relative position to each other. This means that users can not only be confident that structures printed on a one-to-one scale are accurate in size, but multiple structures can be combined and aligned to create bespoke, complex models to suit the user's needs. This process was used to fabricate the gray and white cortical matter brain model, which was created using individual structural components from BodyParts3D (Fig. 1.5). However, it should be recognized that the resolution of the 3D data describing anatomical structure in this database is limited, and some structures, especially smaller structures, may reveal a loss of detail (Fig. 1.21).

1.3.4 Creating Your Own Digital Printing Files

If the desired 3D printing models do not already exist, or do not exist in a form that meets the user's needs, a 3D printing user will inevitably need to venture into some 3D design and modeling of their own. 3D design software includes both traditional computer-aided design (CAD) software as well as 3D sculpting software, and in both categories open-source software (e.g., OpenSCAD and Blender) exists alongside commercial software (e.g., AutoCAD and ZBrush). However, in almost all cases, these particular 3D design applications have a significant learning curve.

The simple to use, yet surprisingly capable solution for 3D design is Tinkercad. This is a browser-based, free-to-use program which employs a simple graphical interface allowing users to assemble 3D structures from simple

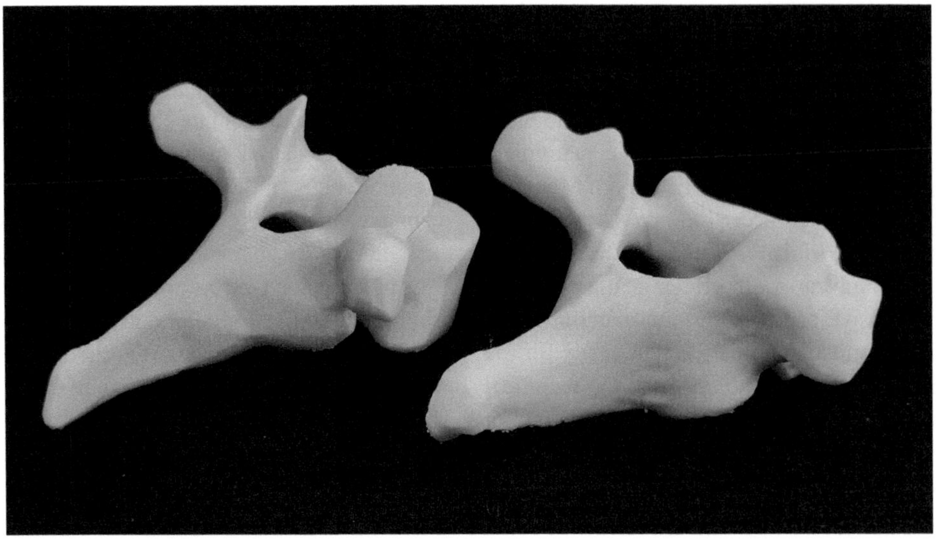

Fig. 1.21 The processing of 3D imaging data can result in loss of detail and accuracy in models derived from real human imaging data. The vertebra on the left was printed using data from BodyParts3D, and the simplification of the structure is evident in the flat faceted surface relative to the more accurate model on the right, a hallmark of reduced polygon count in a 3D model. While these slightly simplified models may be completely adequate for many purposes, it is important to consider how much detail and precision are required when sourcing 3D models for printing

primitive geometric shapes, and models that are created using Tinkercad can be exported in a format ready for 3D printing. While this approach is obviously difficult to use for creating anatomically realistic structures from scratch, there are a number of ways in which Tinkercad may be useful for a user interested in 3D-printed anatomy. Firstly, Tinkercad allows importing of other 3D model files, which can then be manipulated using simple computer modeling tools. For example, this approach can be used to subtract areas from a more complex model, creating cutaways or openings that reveal internal structures that would not be visible if the 3D model was printed intact. Other modifications could include adding text or markers to the structure, allowing labels to be 3D printed in place on the model.

Tinkercad also allows 2D vector graphic images to be imported, which it then renders as 3D structures with adjustable "thickness." This technique is particularly useful for creating "2.5D" models, i.e., 3D representations of 2D information, such as sectional anatomy, with specific details printed in relief. An example of this approach is seen in the spinal pathways model (Fig. 1.12), and the 3D-printed parts used in the autonomic nervous system model (Fig. 1.11) were created with a combination of simple basic shapes and imported 2D drawings rendered into relief (specifically for the spinal cord and brain part of the model).

1.3.5 How to Print: Slicer Software

While many different file types exist to represent 3D structural data, the file types most important for 3D printing are .obj, .stl, and .3mf files. Each of these file types contains 3D geometry data which describes the vertices of polygons which in turn describes the surface of the 3D object. The simplest file type is .stl, which contains only the 3D structure, while .3mf and .obj contain additional information, not all of which is relevant for 3D printing. For example, .obj files can contain data about textures and materials relevant to displaying the model on a screen, while .3mf files can also contain additional information about material and color which is useful for 3D printing, e.g., multi-material printing.

However, 3D data in these file types cannot be directly parsed by a 3D printer. The instructions for printing a 3D model in a consumer FFF 3D printer are encoded in a gcode file, a specific file type that describes the sequence of motion and heating that is required to produce the final model layer by layer. Gcode instructions are generated by slicer software, so called because it renders the 3D shape into individual stacked slices which are then 3D printed. The two most commonly used slicers are PrusaSlicer and Cura. These programs allow the user to load a 3D model, control the printing parameters, and then generate a gcode file which can be transferred to the printer (either on a USB stick or SD card, or remotely for network-enabled printers). Both Cura and PrusaSlicer have very similar feature sets and capabilities, so the choice between these slicers largely comes down to personal preference. Both PrusaSlicer and Cura are open-source software distributed free of charge.

Once a 3D model file has been imported into the slicer, the user then has control over how that model will be sliced, and hence the properties of the final product. Almost every aspect of the slicing process can be adjusted to tweak parameters and optimize settings for the model and the materials being used, but preset configurations are usually adequate for a novice user. Some key parameters must be decided on, however, and can have a significant impact on the properties (both physical and visual) of the final model. One of these parameters is the density and structure of the interior fill of a printed object. It is not usually necessary or desirable to print models as completely solid objects—most FFF 3D prints have solid walls, top and bottom layers, but the interior is mostly hollow. The purpose of any infill is to add structural rigidity and strength, as well as support the top layers.

The specific geometric pattern and density of the infill can be defined in the slicer. Geometric patterns can affect the strength and rigidity of the model, although the difference between patterns is mostly relevant to mechanical, load-bearing parts, and therefore not so critical for most anatomical applications. The infill can be effective at quite low densities; for 3D prints made using PLA or PETG, typically infill density of 10–15% in any pattern will produce robust models for most applications.

Arguably the most important parameter in FFF printing is the height of each printed layer. The exact range of layer heights usable in 3D printing is defined by the size of the extrusion nozzle—in simple terms, you cannot print lines of filament that are taller than they are wide. The most common size for an extrusion nozzle used in FFF printing has an internal diameter of 0.4 mm, so typically this means layer thickness ranging from 0.05 mm to 0.3 mm. The most obvious consequence of changing layer thickness is the appearance of layer lines in the final print. The resolution of the 3D print in the Z-dimension is defined by the layer height, so increasing layer height can cause loss or obscuring of fine details. These layer lines can reduce the visual impact of a model, with thicker layer lines distracting from the overall appearance. This is most prominent on sloped or curved top surfaces, where the regular stepping between layers becomes most obvious (Fig. 1.22).

It may be tempting to print at the smallest layer height possible to improve the appearance of a 3D print. However, this improved appearance comes at a cost. Adjusting layer height may have (relatively subtle) effects on the mechanical strength of the print, but this effect is small and not significant for most anatomical applications. However, arguably the most significant impact of adjusting layer thickness is how it affects print times. Printing in thinner layers improves appearance, but takes a lot longer as the number of layers needed to fill a volume increases. As an example, a 25-mm cube printed in PLA on a Prusa i3 mk3s printer with 15% infill will take 3 h to print at 0.05 mm layer height. The same model printed at 0.3 mm layer height can be printed in 20 min on the same printer. This difference in printing time with a layer height makes printing large and complex multi-material models at the smallest layer heights prohibitive, as print times would extend out to several days. Slicers like PrusaSlicer and Cura both allow for variable layer heights, which allows different layer heights in different parts of the model. This

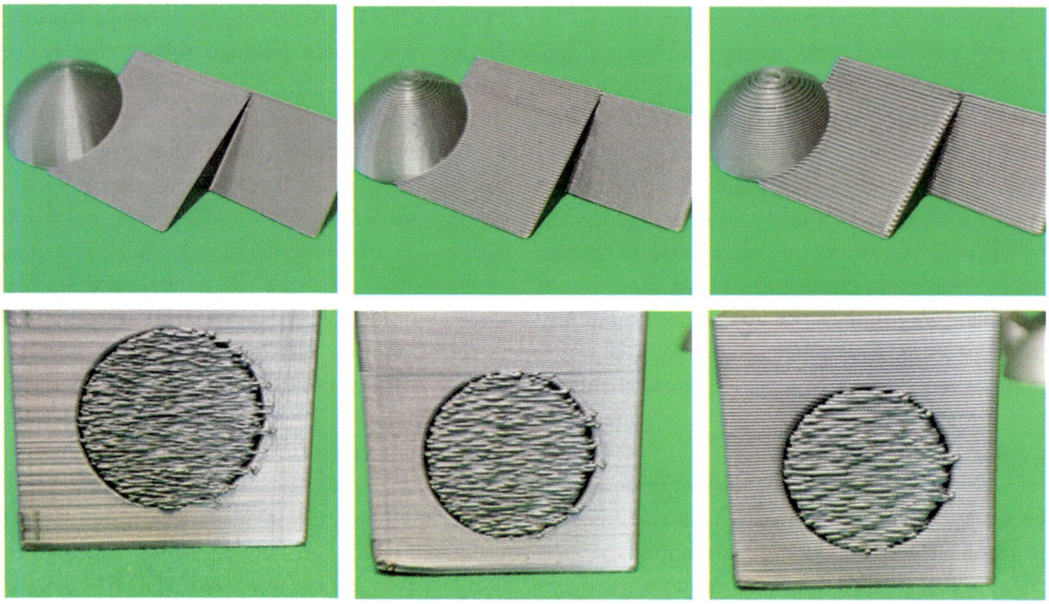

Fig. 1.22 Layer height in FFF printing has an impact on appearance of the final print, as well as print speed. Layer lines are least prominent when printed at 0.1 mm (left panels), becoming more visible at 0.2 mm (middle panels) and most obvious at 0.3 mm (right panels). The effect is seen most clearly on sloped surfaces; in the small prints on the top row it is more prominent in the 22.5° slope than the 45° slope, and toward the top of the dome. The effect of layer lines is much less obvious on vertical surfaces (bottom row). The circles in the bottom rows show the effect of a "fuzzy skin" modifier in the slicer in a specific region of an otherwise basic model (the model itself is a basic cube). This surface effect can be used to change both the look and feel of a 3D print

permits the use of reduced layer heights on top surfaces, while increased layer heights can be used on other parts of the model where their visual impact is less obvious, although the effectiveness of this approach is dependent on the geometry of the actual model.

Slicer software also allows other ways to modify the model during the printing process. An example of this is the "fuzzy skin" modifier, which can be applied to an entire model, or selectively to specific regions. This modifier adds a random jitter to the path of the extrusion along the exterior surface of a model. When this is repeated layer after layer, it creates a textured finish to an otherwise smooth surface (Fig. 1.21). This can be used to create a more organic appearance, but also it can be used to create different tactile properties on the model surface. An interesting application could be to convey information and highlight different regions in a model through texture and haptic response, which may be a useful alternative to conveying information with color when multi-material printing is not available.

Another modifier that can be added through the slicing process is to introduce pauses or filament changes during the print. Pausing mid-print can allow additional, non-printed elements to be added to a 3D-printed model. As an example of how this could be used, a model could be created with a hollow space internally that is specifically designed to fit a magnet. If the print is paused before the hollow space is completely enclosed, a magnet can be glued into the 3D print in situ before resuming the print. This would result in the magnet being completely enclosed within the print. Careful alignment of magnets in a second printed part could then produce a pair of prints that snap together magnetically in a specific alignment, which would be useful for disassembly and reassembly of multi-part anatomical structures.

Alternatively, introducing filament changes during the print can allow different layers in a model to be printed with different materials (as distinct from true multi-material printing, where different materials are used within the same layer). This can be used to add color contrast to raised text, for example, or to more generally apply color highlights to structures printed at different heights within the model. This technique is particularly appropriate for the "2.5D"-style 3D models, and was used to create the clear delineation of gray and white matter in the coronal brain section of the spinal pathways model (Fig. 1.12). Such a technique can be a useful workaround if using printers that are not capable of multi-material printing.

One of the biggest challenges in 3D printing realistic anatomical structures is that natural organic structures are not the ideal shape for 3D printing. Specifically, FFF printing works best with flat bottom surfaces which will facilitate good adhesion to the build plate, and a shape that ensures each printed layer can be supported by the previous (underlying) layer. Overhanging structures are difficult and prone to collapse during printing, with steeper overhangs most prone to issues. The solution to 3D models that do not match this ideal is to use support structures. These are generated by the slicer software automatically or can be defined by the user by selecting the placement of the support structures using the software. Modern slicers allow for different patterns of support material, including rectilinear and honeycomb patterns, as well as organic supports, which are sometimes termed "tree supports" (Fig. 1.22). Different support patterns have advantages and disadvantages, primarily in terms of the amount of material used (generally lowest for organic supports) and ease of removal; the optimal support varies from model to model, and may require some trial and error to determine the best option.

Soluble supports printed in soluble thermoplastics are theoretically the optimal way to support a complex 3D model. However, this is only possible with multi-material printing. For single material printing, support must be broken away by hand. In FFF printing, this can be a difficult and time-consuming process, and the quality of the surface finish where support is removed is typically inferior to the finish on top surfaces. The general principle is to orient the model on the print bed in a manner that minimizes the requirement for support, as well as the complexity of that support. The gray and white matter brain model shows a good example of this (Fig. 1.23). The model was printed with the gray matter (being a relatively smoother and simpler surface) facing down toward the print bed and taking the bulk of the support, and organic support was used to ease support removal. In the region between the hemispheres, the support structures were attached to the model itself (rather than the print bed). This creates particular challenges for removal of supports, and an earlier version of this model shows where rectilinear support was used instead of organic, resulting in damage to the model surface when the support was removed.

Slicers for SLA printers produce a different type of file, which needs to describe timing of the switching of the UV light source and the motion of the print platform, but also the image data which corresponds to each slice of the model to be displayed on the screen. Therefore, SLA slicer software needs to be compatible with the specific printer being used. Two popular slicers that are used in SLA printing are Chitubox and Lychee Slicer, and these slicers generally have good compatibility with the SLA printers from the major consumer printer manufacturers. Unlike PrusaSlicer and Cura, Chitubox and Lychee Slicer are not open source, although they do have fully featured free versions (in addition to subscription versions with additional features).

SLA slicers offer fewer adjustable printing parameters and therefore less control over the final 3D print, although this is due largely to the relative simplicity of the SLA process (i.e., there are fewer parameters to adjust). Layer height can be adjusted, although in general SLA prints are created at much smaller layer heights than FFF printers (typically 50 μm or less), which allows for the high-resolution, high-fidelity prints characteristic of resin printing. The most important parameter that needs to be adjusted with SLA slicers is the generation of support material. Like FFF slicers, SLA slicers will generate appropriate support scaffolds automatically, with the option of user-defined supports as required.

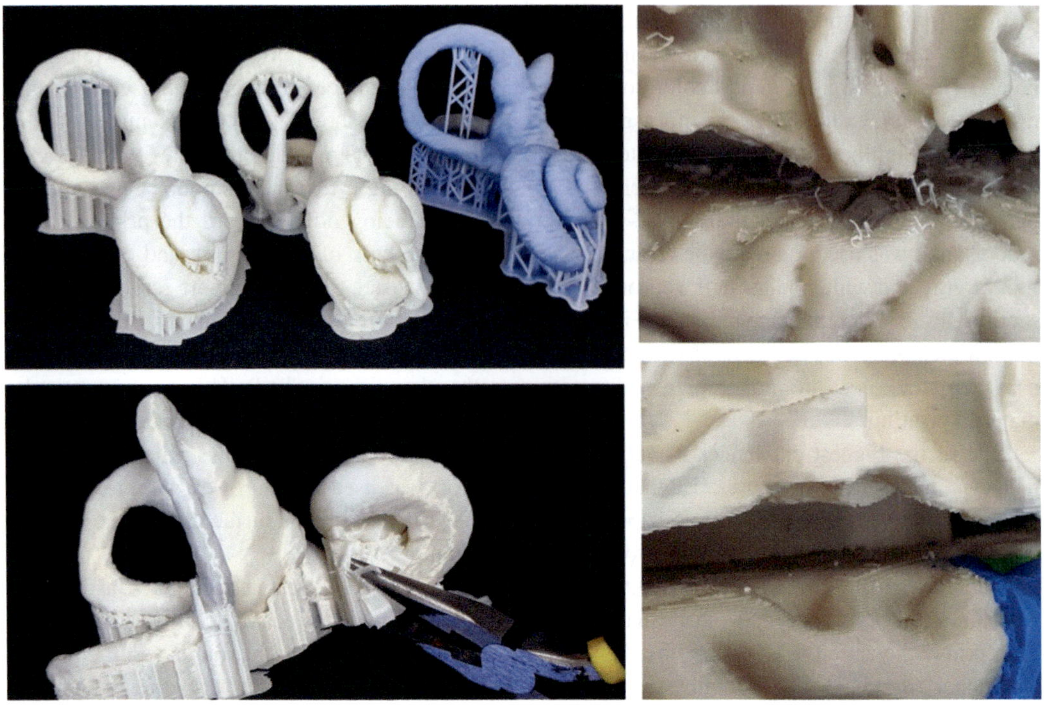

Fig. 1.23 Complex 3D structures, such as the inner ear, require support structures in order to be 3D printable. Support structures will be in place at the end of printing (top left). The beige models were printed using an FFF printer, the model on the left uses rectilinear supports, and the right uses organic tree supports. The gray model is SLA printed and shows the fine scaffold generated by SLA printing. Unless soluble support material is used, support must be broken away, which often requires tools and can be difficult when support is within the model itself. The upper right panel shows the poor surface quality when the rectilinear support was removed between the hemispheres in the brain model shown in Fig. 1.8. The surface is much better in the lower right image due to the increased gap in the larger model and the use of more sparse organic support in this case

Unlike FFF printing, almost all SLA prints require support. This is because the optimal orientation of a model for SLA printing is not the same as for FFF printing. If you take a simple shape like a rectangular slab as an example, this would be optimally created using FFF printing by aligning the largest flat surface on the print bed. In SLA printing, this orientation would create issues. Since the model lifts between layers, liquid resin must be able to flow into the space between the printed part and the transparent film at the bottom of the resin vat. Minimizing the surface area needed for this will result in more successful printing, so in this case, the flat slab is optimally printed at an angle to the bed, with support structures generated to attach the model to the build platform. SLA slicers also allow models to be hollowed out, typically with drain holes incorporated into the model, which will reduce the surface area of each printed layer and can reduce this suction effect resulting in better print performance.

In addition to slicer software, some users will also use printer management software which can allow remote uploading of gcode files and remote monitoring of 3D prints (e.g., using live video feeds from cameras mounted on the printer). These systems can be useful for monitoring long prints—the largest and most complex 3D prints can take several days, meaning that it is inevitable that the printer will need to operate overnight. While printer management systems may offer some usability and quality of life benefits, they are "optional extras."

1.4 Summary

While cadaveric anatomy is the gold standard in anatomy education, the use of anatomical models has always played an important role in anatomy education in the past, continues to do so in the present, and will continue to do so in the future. What is different now is that 3D printing, for the first time, not only gives people access to a wider range of anatomical models than ever before but more importantly gives educators the means of production of their own anatomical models. Any educator can now become a producer of anatomical models, not just a consumer, with less investment of time and money than at any time in the past.

After many years (if not decades) of promise, 3D printing is now truly available to everyone. The tools and resources for 3D printing are more accessible than ever, and becoming more accessible year after year, with fewer barriers to entry, both in terms of economic cost and technical expertise. 3D printers are not only cheaper than ever before but increasingly reliable and easy to use. Open-source software allows equitable access to 3D design tools, as well as the best tools to create the exact 3D prints you need. Public repositories of freely shared models to print in 3D, as well as online communities where expertise, experiences, and advice are also freely shared, are also growing year after year.

For the creative educator, 3D printing opens a world of easy experimentation with new ideas; 3D printers are optimized for rapid prototyping, allowing novel ideas to be realized, tested, and revised quickly and repeatedly. However, the future of 3D printing in education may not be centered around the educator as the focus of the design and fabrication process; 3D printers are becoming increasingly common in secondary schools and public libraries, meaning students arriving in tertiary education are increasingly familiar with and potentially experienced in 3D design and 3D printing. The expectations of these students may shift as a result, and the limitations of commercial 3D models will become apparent to them. However, more importantly, these learners will be more capable of creating their own models and learning resources according to their own needs. In that regard, 3D printing may be a truly disruptive technology capable of bringing about a genuinely equitable and accessible educational ecosystem with learners and educators as partners and collaborators.

References

Azkue JJ (2021) External surface anatomy of the postfolding human embryo: computer-aided, three-dimensional reconstruction of printable digital specimens. J Anat 239(6):1438–1451

Ballestriero R (2010) Anatomical models and wax Venuses: art masterpieces or scientific craft works? J Anat 216(2):223–234

Bienholz-Radtke J, Roeßiger S (eds) (2022) Transparent figures: researching and preserving objects of plastic, Deutsches Hygiene-Museum. Sandstein Verlag, Dresden

Brumpt E, Bertin E, Tatu L, Louvrier A (2023) 3D printing as a pedagogical tool for teaching normal human anatomy: a systematic review. BMC Med Educ 23(1):783

Burrows R (2019) The anatomy act of 1832: the story of bodysnatching, dissections, and the rise of anatomy. Tenor Our Times 9(1):8

Chan LK (2015) The use of low-tech models to enhance the learning of anatomy. In: Chan LK, Pawlina W (eds) Teaching anatomy: a practical guide. Springer, Cham, pp 259–266

Chan LK, Cheng MM (2011) An analysis of the educational value of low-fidelity anatomy models as external representations. Anat Soc Edu 4(5):256–263

Chen JC, Amar AP, Levy ML, Apuzzo ML (1999) The development of anatomic art and sciences: the ceroplastica anatomic models of La Specola. Neurosurgery 45(4):883

Cornwall J, Smith C (2014) Anatomical models by FJ Steger (1845-1938): the University of Otago Collection. Eur J Anat 18(3):209–211

Crump SS (1992) Apparatus and method for creating three-dimensional objects. US Patent 5,121,329, 9 June 1992

da Mota GM (2020) From the wax cast of brain ventricles (1508-9) by Leonardo da Vinci to air cast ventriculography (1918) by Walter E. Dandy. Rev Neurol 176(5):393–396

de Bakker BS, de Jong KH, Hagoort J, Oostra RJ, Moorman AF (2012) Towards a 3-dimensional atlas of the developing human embryo: the Amsterdam experience. Reprod Toxicol 34(2):225–236

Fontana F (1795) Courte description d'une statue anatomique qui se démonte en plus de 3,000 parties. Magasin Encycl 2:35–36

Gunz P, Ramsier M, Kuhrig M, Hublin JJ, Spoor F (2012) The mammalian bony labyrinth reconsidered, introducing a comprehensive geometric morphometric approach. J Anat 220(6):529–543

Haviland TN, Parish LC (1970) A brief account of the use of wax models in the study of medicine. J Hist Med Allied Sci 25(1):52–75

Jones R, Haufe P, Sells E, Iravani P, Olliver V, Palmer C, Bowyer A (2011) RepRap–the replicating rapid prototyper. Robotica 29(1):177–191

Kataoka K, Suzaki E, Ajima N (2007) The Hoshino wooden skeleton, the first wooden model of a human skeleton, made during the Edo era in Japan. Anat Sci Int 82:38–45

Lightbown RW (1964) Gaetano Giulio Zumbo-I: the Florentine period. Burlingt Mag 106(740):486–496

Macmillan M, Lena ML (2010) Rehabilitating Phineas Gage. Neuropsychol Rehabil 20(5):641–658

Manzoli L, Ratti S, Cocco L (2022) Anatomy and ceroplastic School in Bologna: a heritage with unexpected perspective. Ital J Anat Embryol 126(1):93–101

Maraldi NM, Mazzotti G, Cocco L, Manzoli FA (2000) Anatomical waxwork modeling: the history of the Bologna anatomy museum. Anat Rec 261(1):5–10

Marković D, Marković Živković B (2010) Development of anatomical models—chronology. Acta Medica Medianae 49(2):56–62

Mavrodi A, Paraskevas G (2014) Mondino de Luzzi: a luminous figure in the darkness of the middle ages. Croat Med J 55(1):50

Mitsuhashi N, Fujieda K, Tamura T, Kawamoto S, Takagi T, Okubo K (2009) BodyParts3D: 3D structure database for anatomical concepts. Nucleic Acids Res 37(suppl_1):D782–D785

Narang P, Raju B, Jumah F, Konar SK, Nagaraj A, Gupta G, Nanda A (2021) The evolution of 3D anatomical models: a brief historical overview. World Neurosurg 155:135–143

Olry R (2000) Wax, wooden, ivory, cardboard, bronze, fabric, plaster, rubber and plastic anatomical models: praiseworthy precursors of plastinated specimens. J Plast 15(1):30–35

Ortug A, Yuzbasioglu N (2019) Tracing the papier mache anatomical models of Ottoman Turkish medicine and Louis Thomas Jerôme Auzoux. Surg Radiol Anat 41(10):1147–1154

Palouzié H, Ducourau C (2017) De la collection Fontana à la collection Spitzner, l'aventure des cires anatomiques de Paris à Montpellier. In Situ Revue des patrimoines

Ratiu P, Talos IF, Haker S, Lieberman D, Everett P (2004) The tale of Phineas Gage, digitally remastered. J Neurotrauma 21(5):637–643

Riva A, Conti G, Solinas P, Loy F (2010) The evolution of anatomical illustration and wax modelling in Italy from the 16th to early 19th centuries. J Anat 216(2):209–222

Russell KF (1972) Ivory anatomical manikins. Med Hist 16(2):131–142

Shahrubudin N, Lee TC, Ramlan (2019) An overview on 3D printing technology: technological, materials, and applications. Procedia Manuf 35:1286–1296

Sora MC, Latorre R, Baptista C, López-Albors O (2019) Plastination - a scientific method for teaching and research. Anat Histol Embryol 48(6):526–531

Sonolet J (1977) A propos d'un mannequin anatomique en bois: Napoléon Bonaparte et Felice Fontana. Atti del I Congresso Internazionale, La ceroplastica nella scienza e nell'arte / Proceedings of the International Congress on Wax Modelling in Science and Art: Florence, Italy 1975. Olschki LS (Ed.) pp. 443–458

Spencer L (2008) Chance, circumstance and folly. Richard Berry and the plaster anatomical collection of the Harry Brookes Allen Museum of Anatomy and Pathology. University of Melbourne Collections vol 2, pp 3–10

Talairach-Vielmas L (2014) Anatomical models: a history of disappearance? Hist Méd Santé 5:9–20

Torres K, Staśkiewicz G, Śnieżyński M, Drop A, Maciejewski R (2011) Application of rapid prototyping techniques for modelling of anatomical structures in medical training and education. Folia Morphol 70(1):1–4

Valdecasas AG, Correas AM, Guerrero CR, Juez J (2009) Understanding complex systems: lessons from Auzoux's and von Hagens's anatomical models. J Biosci 34:835–843

Van Horn JD, Irimia A, Torgerson CM, Chambers MC, Kikinis R, Toga AW (2012) Mapping connectivity damage in the case of Phineas Gage. PLoS One 7(5):e37454

Von Hagens G (1979) Impregnation of soft biological specimens with thermosetting resins and elastomers. Anat Rec 194(2):247–255

2 Evolution of Veterinary Anatomy Education: A Paradigm Shift from Dissection to Digitalization via 3D Visualizations, 3D Reconstructions, and 3D Models

Kritima Kapoor, Nirmal Singh, and Anuradha Gupta

Abstract

Veterinary anatomy education is crucial in shaping the future of budding veterinary professionals. It has evolved significantly in terms of its teaching pedagogies with a shift from routine traditional methods to a new era of teaching. This has been possible with the upsurge of technological interventions in anatomy, especially in virtual reality (VR) and augmented reality (AR). The inclusion of AR and VR has transformed the teaching and learning in veterinary anatomy into an era of hybrid education. The intervention and application of these innovative technologies have enabled learners to visualize and perform dissections via the virtual interface, thereby substituting actual dissections with digital dissections. These technologies provide an alternative to animal dissections and thus safeguard ethical considerations. Due to the scarcity of animals to be euthanized for dissection and the associated ethical considerations, cadaveric dissections are being supplemented or replaced with alternative methods of three-dimensional (3D) visualizations such as computed tomography (CT) or Magnetic Resonance Imaging (MRI). These images can be utilized to reconstruct a complete 3D virtual model of anatomical parts and structures, including the less accessible ones. These 3D reconstructions from 3D visualizations have been made possible by the use of various software programs with the expansion of the use of artificial intelligence (AI) in anatomy. The use of the digital interface in anatomy and its amalgamation with 3D printing technology has revolutionized veterinary anatomy teaching and learning. 3D printing technology is a boon for learners in this era as it can convert any organ or bone structure into its replica 3D model. By utilizing 3D reconstruction data, a 3D model of any anatomical structure can be developed from a live animal without actually euthanizing the animal for studying anatomy. Veterinary anatomy teaching has also evolved in terms of fixative-free cadaver dissections using various methods such as plastination and the glycerin method of preservation. The integration of clay modeling in veterinary anatomy is being increasingly

K. Kapoor (✉)
Guru Angad Dev Veterinary and Animal Sciences University, Ludhiana, India

Department of Veterinary Anatomy, College of Veterinary Science, Guru Angad Dev Veterinary and Animal Sciences University, Ludhiana, India

N. Singh · A. Gupta
Guru Angad Dev Veterinary and Animal Sciences University, Ludhiana, India

L. Shapiro (ed.), *How to use 3D Printing Innovations and Digital Storage to Democratize Anatomy Education*, Biomedical Visualization 6, https://doi.org/10.1007/978-3-031-68501-9_2

explored as a novel way to enhance creativity, critical skills, and the hand-on learning experience by students.

Keywords

Augmented reality · 3D visualization · 3D printed model · Clay modeling · Virtual reality · Veterinary anatomy

2.1 Introduction

Anatomy is an essential discipline which is illustrative and vivid with a strong focus on its scientific and practical significance in understanding the diversity of animal bodies in veterinary medicine. This makes it essential for veterinarians to become acquainted with anatomical knowledge in order to build a strong foundation for future exposure to clinical cases as a practitioner. Teaching veterinary anatomy by traditional methods includes the use of a variety of routine teaching methods such as the use of slides, books, and the anatomy atlas, for theoretical information. However, the use of formalin-fixed animal specimens and the dissection of cadavers are essential for practical learning experience. Animal dissection in veterinary medicine has always been interpreted as the benchmark prototype of educational anatomy. Conventionally, animal dissections and museum specimens have therefore played an important role in the teaching and learning of anatomical details within animals (Kapoor and Singh 2022). However, due to ethical constraints, the dissections of healthy animals have been prohibited in most veterinary institutions in India. Several other educational methods including the use of plastinated specimens, glycerin specimens, and organ casts of lungs, have been developed to preserve specimens without the use of toxic fixatives like formaldehyde. This has generated odorless, tangible, and durable specimens, making learning more student-friendly (von Horst et al. 2019). As such, anatomy education has evolved in terms of minimal use of animals for dissection, while at the same time making learning more immersive and interactive by shifting the focus toward digital alternatives like VR and AR alongside the hands-on dissection experience. Additionally, different visualization techniques are being used to demonstrate regional anatomy in living animals as well and utilizing this information for the creation of exact 3D replicas of that particular region of animal anatomy (Li et al. 2018; Alcântara Leite dos Reis et al. 2019). This has revolutionized teaching veterinary anatomy entirely. Innovative 3D visualizations and imaging techniques have enhanced the modeling of internal anatomical structures which are crucial, especially when implemented in teaching gross anatomy to facilitate efficient diagnostics in veterinary medicine. In the past few decades, a wide-range of imaging techniques such as CT scans, MRI scans, micro-computed tomography (micro-CT), scanning electron microscopy (SEM), conventional light microscopy (LM), and confocal laser scanning microscopy (CLSM) have been developed and are being used in biological studies including veterinary medicine (Kapoor 2024). By utilizing information from MRI scans, CT scans, and X-rays to create images of the animal's body, it provides a 3D image of morphological details of normal and damaged anatomical structures to educators and clinicians. Using data acquired during clinical imaging studies, these technologies can be incorporated into clinical practice to improve the diagnosis and treatment of disease, along with improved understanding of each case and an increase in pet owner satisfaction (Wulster 2022). This software-generated information is then utilized by another novel technology which is the latest industry revolution being adopted swiftly in veterinary medicine, i.e., 3D printing. The use of 3D printing has transformed teaching and learning, particularly during this new era of hybrid veterinary anatomy education. 3D printers can create any desired structure such as bones or organs, which can be produced within minutes to hours (for larger products) depending on size and complexity.

The chapter is focused on discussing the most recent interactive pedagogy and methodologies adopted for improved learning in veterinary anatomy (Fig. 2.1). The main aim is to document

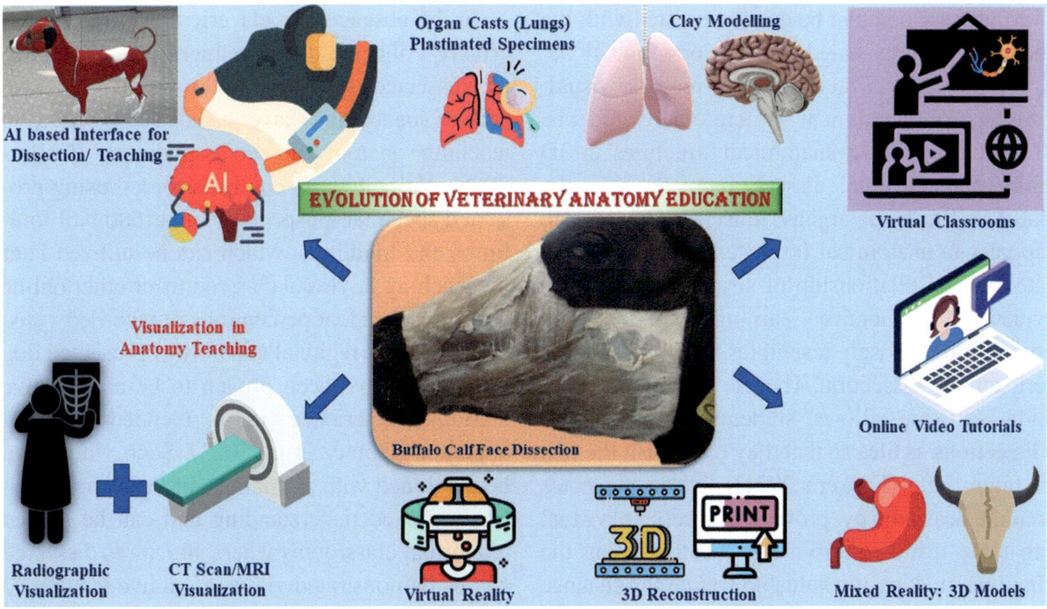

Fig. 2.1 Evolution of veterinary anatomy education to new era methods like digital dissections, 3D visualizations, 3D printing, virtual reality, augmented reality, virtual classrooms, use of plastinated specimens, organ casts, and clay modeling

newer innovative alternatives being adopted to visualize and demonstrate the anatomy of animals and to elevate the teaching of veterinary anatomy to students, researchers, and clinicians.

2.2 Advances in Digitalization in Veterinary Anatomy

2.2.1 Veterinary Anatomy Teaching via Digital Dissections, Virtual Reality, and Augmented Reality

Virtual dissection or digital dissection has become increasingly popular as it provides students with advanced and innovative learning opportunities in veterinary anatomy by simulating the process of dissecting anatomical structures without the need for physical specimens. However, cadaveric dissection is not replaceable as it is an essential practice for enabling students to master physical skills such as using surgical instruments coupled with the simultaneous tactile exposure of tissues and organs (Kapoor and Singh 2022). These virtual dissections in the form of pre-recorded videos of actual dissections were utilized in anatomy education before the COVID-19 pandemic, but their potential use gathered momentum during the COVID-19 pandemic itself (Sadeghinezhad 2023a). In addition, due to ethical constraints about euthanizing animals for dissection, virtual dissections have been proven to be of great value where students can operate a software-based computerized dissection through touch screens to understand complex anatomical relationships (Darras et al. 2019). Currently, virtual dissection is performed on anatomy visualization tables or virtual dissection tables with the aid of animated anatomical structures. These enable students to dissect life-size digital cadavers of a variety of animals with virtually available scalpels that can be rotated, magnified, and labeled and includes the ability to zoom in and zoom out (Youssef 2021). Several virtual dissection software programs have been created by software professionals in collaboration with anatomists. This collaboration has allowed for the retention of anatomical details of different species. These programs are available for purchase and can be easily operated with the help of computer-aided interface or applications.

Digital dissection tools offer realistic visualizations of anatomical structures to students. They can visualize anatomical structures in 3D form and separate, dissect, recreate, magnify, label and transect organs, muscles, and bone to appreciate anatomical form, texture, and associated spatial relationship of structures. This provides an immersive learning experience to students that closely resembles traditional cadaveric animal dissection. The use of digital dissection software allows students to save their dissections as files so that they can revisit them at a future date (Karkera 2021). It also promotes remote learning by providing access to virtual anatomy resources from any location using the Internet, making it a valuable source for distance education programs or in pandemic situations where access to physical anatomy laboratories is limited (Evans et al. 2020). Several online web-based platforms like Vet-Anatomy and EasyAR (Shanghai, China) are also available that offer interactive 3D models of animal anatomy, including veterinary-specific content such as real-time canine dissections. Users can explore different anatomical systems by performing virtual dissections, and access additional learning resources such as quizzes and tutorials. Recently, this virtual anatomy experience has been rated by several anatomy professionals in human medical schools to be significantly more likely to stimulate student interest (Niedermair et al. 2022; Zhang et al. 2019). It is used with increasing frequency in veterinary medical and anatomical education (De Bose 2020; Kapoor and Singh 2022). Based on qualitative and quantitative analyses of survey results by Martin et al. (2023) it is suggested that most students (92.5%) agreed that the use of virtual animal anatomy helped them learn anatomy easily and 82.2% indicated that it is a valuable resource of online learning. Although, these software systems require initial expenditure, they are highly cost-effective over time. Digitalization of dissections is a safe, ethical, and humane way of learning as animals are not harmed or euthanized and this eliminates the need for physical specimens.

In the absence of cadaveric dissection in veterinary colleges, due to ethical considerations, pre-dissected specimens (also known as prosected specimens) have been widely used in teaching anatomy to undergraduates (Collins 2008). Video lessons can be recorded using prosected specimens, especially of buffalo bull forelimbs and hindlimbs which can be utilized later for teaching in virtual classrooms or other online platforms. Student perceptions of recorded video lessons on the regional anatomy of prosected dog specimens have been proven to be an effective learning experience in a blended learning approach (Gomez et al. 2023). Such videos can be combined with animations of the same region for improved understanding and can be played via virtual classrooms where structures dissected can be demonstrated via video in live lectures to students. The teacher can play and pause the video dissections at specific junctures according to their lecture requirement, and students can have real-time interactions and discussions on the same online platforms (Fig. 2.2). This method is also highly regarded and proves that recording videos is an effective strategy to improve the performance of students for their active learning in veterinary anatomy (Senos et al. 2022). Therefore, the video demonstrations of limb dissections, visceral organs comprising splanchnology have proven to be of great help in facilitating learning (Senos et al. 2022; Gomez et al. 2023). These can also be uploaded on social media platforms like YouTube for future reference (Sadeghinezhad 2023b).

Digital dissections can be further integrated with several other evolving technologies such as VR or AR to further enhance the learning experience in veterinary anatomy. AR and VR platforms allow students to visualize anatomical structures in 3D providing an innovative, immersive 3D visualization experience for a deeper understanding of spatial topographical relationships of anatomical structures, in contrast to traditional 2D images. AR and VR applications incorporate topographic anatomy related to clinical cases and scenarios to illustrate the practical relevance of anatomical knowledge in veterinary medicine. VR influences the capacity of students

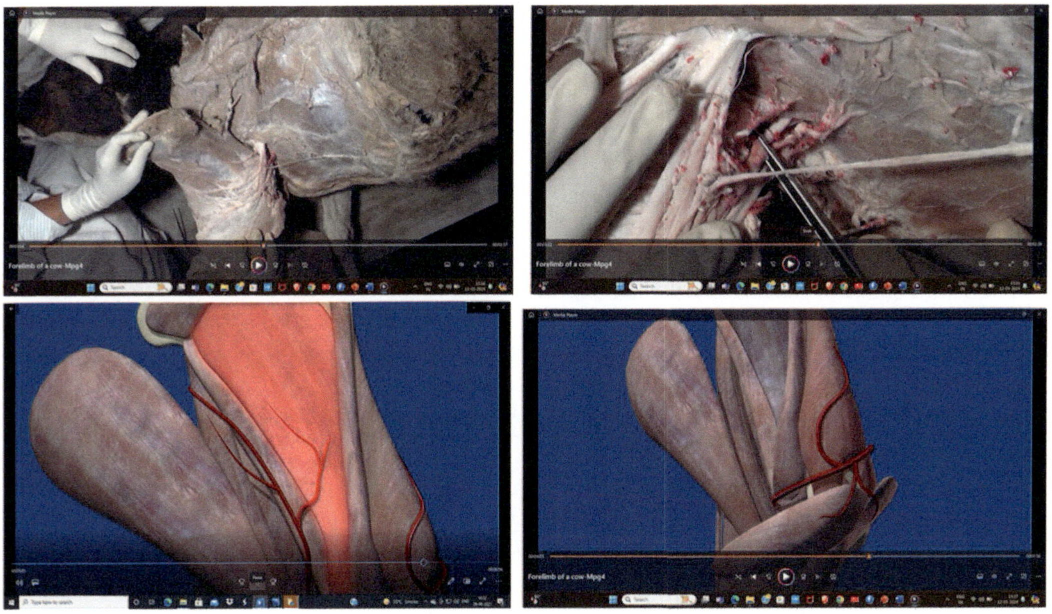

Fig. 2.2 Pre-recorded videos of dissections, prosected specimens blended with anatomical 3D animations to represent same area muscles, blood vessels that can be utilized for better understanding

to replicate the real-world environment via simulations in clinical situations by exploring virtual patient cases and practicing diagnostic procedures in a realistic clinical setting (Rachmatullah and Sukihananto 2020; Fitria 2023).

Learning anatomy through VR offers realistic sensations with audio–visual as well as haptic and visual perceptions of organ topography while students enjoy the comprehension of content being taught. For this purpose, separate virtual anatomy laboratories are increasingly being set up in premium veterinary colleges (e.g., GADVASU, Ludhiana) where students can explore and dissect anatomical structures of different species in a simulated environment (Kapoor and Singh 2022). These laboratories often include high-resolution images or 3D models of anatomical specimens for hassle-free, digital dissection of any area of an animal's body, thus allowing students to interact with tissues and organs digitally. For instance, students can enhance their VR experience by using AR headsets to add audio–visual effects to virtual anatomical models to mimic real-world objects for a more advanced and immersive experience (Fitria 2023).

The anatomical contour of the animal body can be viewed in a 3D virtual form using smartphones and iPads in combination with AR technology via applications (Küçük et al. 2016). AR model programs like Unity3D software (San Francisco, USA) and EasyAnatomy AR (Shanghai, China) are being used efficiently by anatomists to teach the complex anatomy of canine skull (Jiang et al. 2024) and anatomy of the heart by the IVALA® augmented reality (AR) program (Little et al. 2021). Several other mobile applications are also available, such as Vet-Anatomy and Biodigital Veterinary, which provide free access to gross anatomy and imaging anatomy (MRI, CT) of animals to learners worldwide. An AR mobile application has been developed to teach detailed anatomy of domestic animals to veterinary students at our institute (Guru Angad Dev Veterinary and Animal Sciences University, Ludhiana, Punjab, India). This is being utilized via students' mobile devices as well as in the VR laboratory facility created at the institute for learning anatomy. In this AR application, the students have access to the anatomy of the skeletal and muscular systems, the blood supply of the head, neck, and hindlimbs, the digestive system and its organs, and the anatomy of the heart in bovines and canines. It offers a user-friendly interface with easy navigation and search functionalities to study bones, organs, and muscles depicting canine anatomy. In this application, the species to be studied, e.g., a dog, can be rotated through 360 degrees and in all directions to view a 3D view of its axial skeleton and dissection of a lateral and medial group of muscles of the whole body (Fig. 2.3).

This comprehensive collection of animal anatomy in one application enables targeted learning objectives and curriculum alignment with the veterinary anatomy course. It was observed that students had a clearer understanding of anatomical topography from studying AR dog models as compared to their study of the living dog only (De Bose 2020). Several surveys conducted on first year veterinary medical students who learned the anatomy of the heart, canine skull, via AR applications have confirmed that AR tools can improve students' motivation for learning and enhance the comprehension of anatomical structures in 3D (Little et al. 2021; Jiang et al. 2024). The digital veterinary anatomy applications cater to diverse learning styles and preferences, providing users with convenient tools to learn from and reinforce their understanding of animal anatomy. Additionally, the VR and AR technologies facilitate remote learning by providing access to virtual anatomy resources worldwide via the internet, thereby improving personalized learning experiences and comprehension and knowledge retention by students. Another area of digital learning exploration in anatomy is by a combination of both VR and AR, known as mixed reality (MR) which allows for interaction of both real-world and digital objects through a head-mounted, transparent display. In a MR environment, students can interact with and manipulate both the physical forms and virtual items, using next-generation sensing and imaging technologies. Objects can be projected as holograms in front of the user's eyes in 3D space, reducing the need for holding a screen or camera and the hologram can be manipulated using gesture controls with hand and fingers or with a traditional keyboard or handset. This has great

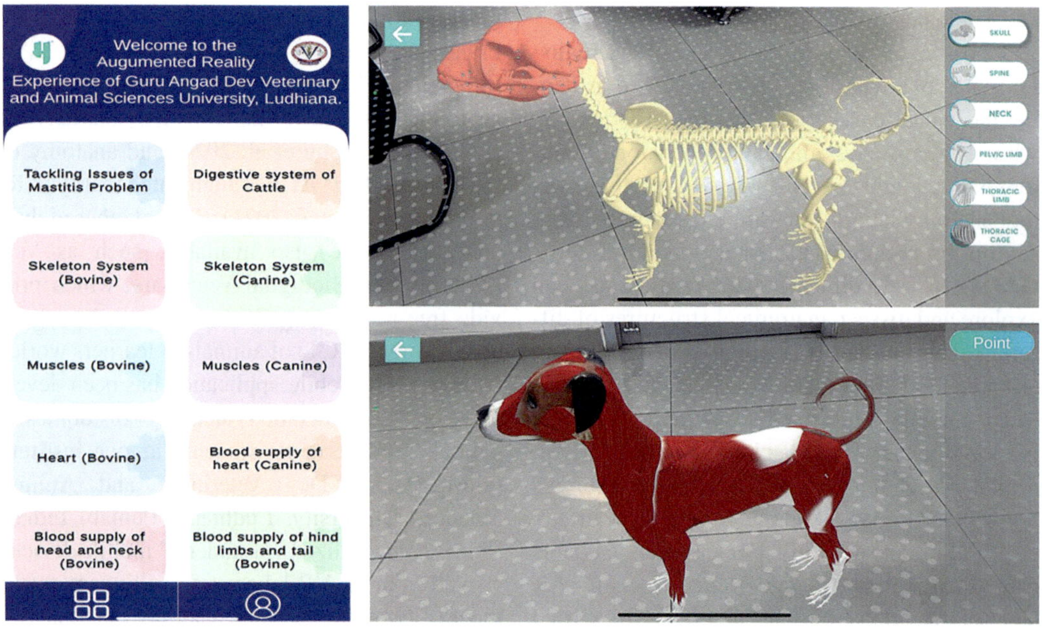

Fig. 2.3 Augmented Reality application developed by Guru Angad Dev Veterinary and Animal Sciences University for students to have access to anatomy of musculoskeletal system, blood supply to head, neck, and hindlimbs, and visceral organs in bovines and canines

potential for allowing multiple users to visualize the same structure simultaneously if the headsets are linked in communication with each other (Stirling and Moro 2020). Mixed reality (MR) is still a developing technology which can enhance learning outcomes if adapted in routine teaching in veterinary curriculum (Xu et al. 2023).

2.2.2 Veterinary Anatomy via Animations

Animation is a powerful tool in teaching veterinary anatomy and offers visualization of complex anatomical structures in a dynamic and visually engaging manner, enhancing students' understanding of anatomical relationships. The relevant anatomical knowledge is extracted from textbooks along with practical knowledge by experienced anatomists before conveying it to first year veterinary students. However, the process of conveying theoretical knowledge sometimes becomes monotonous. This is overcome by lecturers by including 3D anatomical animations in their pre-class anatomy lectures before practical demonstrations or dissections. This helps the students to visualize the anatomy of a particular region in advance (such as the muscles of forelimb or hindlimb) with the ease of practical, hands-on dissections and bridges the gap between theoretical knowledge and its practical application. The 3D animations are constructed by converting anatomical knowledge using computer software programs, such as clipart media in PowerPoint or photogrammetry tools, to provide an actual representation of the anatomy which is close to the real animal specimen experience as these animations are 3D and can rotate through all possible directions. This animation-based learning clarifies difficult-to-understand theoretical anatomical details as a sequential depiction of anatomical structures by providing images in a step-by-step sequence, helping students to grasp concepts more easily than would otherwise be possible with static illustrations or text alone. Animated anatomy lectures can adequately depict the anatomy of forelimb muscles, blood supply to the muscles of the forelimb, and nerve supply to that region, with 360 degrees of rotation showing all the anatomical structures of that particular region (Kapoor and Singh 2022). Teaching through this mode simulates audio–visual perception of students for a better understanding of spatial orientation and the course of blood vessels, nerves, and muscles. Some animated anatomy lectures can also be prepared as a combination of 3D animated organs and drawings through PowerPoint which can be animated to depict the course of these blood vessels and nerves (Fig. 2.4).

Some veterinary educators have developed a full web-based e-learning platform, denoted as an Interactive Veterinary Education Tool (IVET), which is designed to support teaching and learning in veterinary science by making content creation easier (Xiberta and Boada 2019). IVET provides theory, exercise, and image editors with functionalities to prepare exercises and theoretical content including 2D images, 3D models, and Digital Imaging and Communications in Medicine (DICOM) files which can be further manipulated by the users for teaching purposes (Xiberta and Boada 2019). Animated anatomical teaching engages multiple senses, including sight and sound, facilitating a deeper understanding of anatomical concepts, which ultimately enhances learning and knowledge retention. Therefore, incorporating animations into veterinary anatomy teaching enhances innovation and creativity in both learners and educators.

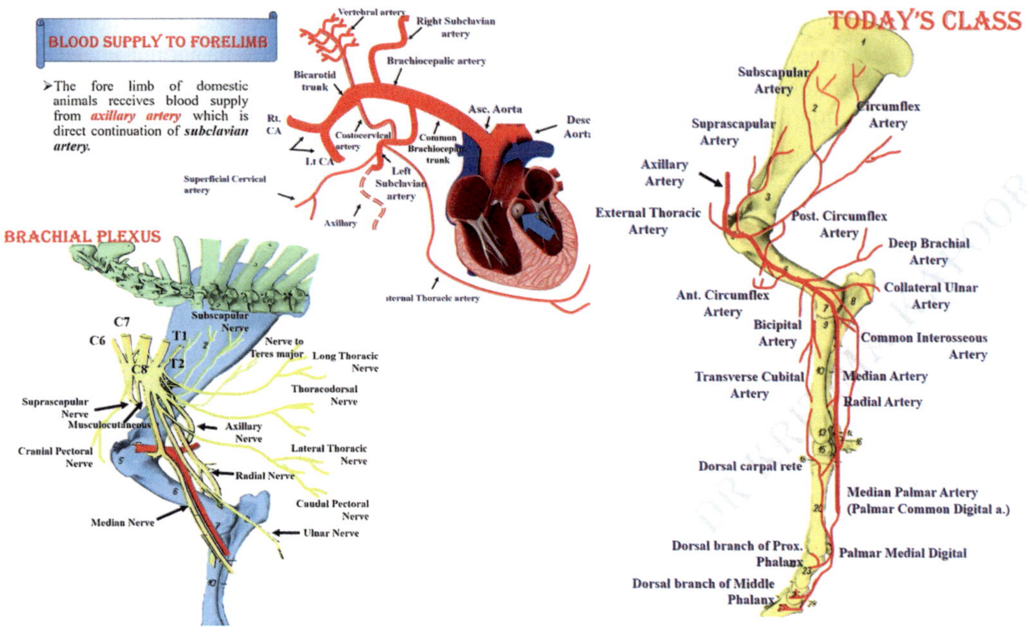

Fig. 2.4 3D animated drawings in PowerPoints combined with animated organs to depict the course of blood vessels and nerves in limbs (*created by the first author)

2.3 3D Imaging and Visualization for Teaching Anatomical Structures

Integrating 3D visualizations with veterinary anatomy teaching has gained momentum recently. It reveals the organization of organs within whole animal topography through imaging modalities like CT and MRI scans. 3D visualization provides fast and precise 3D image data with comprehensive, high resolution, and detailed views of anatomical structures, internal organs, tissues, and systems. By incorporating these diagnostic imaging visualization methods into anatomy teaching, students gain pre-clinical exposure to these modalities and learn how to interpret diagnostic images in the context of anatomical structures. It provides cross-sectional views of anatomical topography, which can be challenging to visualize using routine anatomical teaching methods, thus developing a deeper understanding of cross-sectional anatomy and the spatial relationships between different anatomical structures. These modalities which generate precise 3D image data in high resolution can be further converted to computer-assisted 3D reconstructions (Edelmers et al. 2022). Analyzing 3D visualizations from CT or MRI images enhances problem-solving skills and critical thinking among students by correlating anatomical structures with their appearance in those images. It helps them to develop abilities to solve clinical cases relating them to anatomical knowledge through problem-based learning which is essential for veterinary practice (Mizer et al. 2002; Borroni et al. 2021). In addition to CT and MRI imaging, other frequently used diagnostic imaging modalities, such as ultrasound or radiography, can also be integrated with osteology teaching combined with radiographic anatomy.

2.3.1 CT and MRI Scans in Anatomy Teaching

CT scanning is an imaging modality that facilitates assessment of soft tissues, orthopedic imaging, and cardiac structure, among many others in both normal and diseased conditions in living animals. It also aids scientists to compare the topography of anatomical structures in different animal species in cross-sectional views to study comparative anatomy of organs which are difficult to understand using traditional methods (Kapoor 2024). Anatomical structures are displayed as individual shades of gray on a screen and these shades allow differentiation between tissues and structures according to their radiodensity (Lauridsen et al. 2011). Apart from bones, muscles, and the nervous structures of the brain and eye, the most routinely assessed anatomical structures based on CT imaging are the nasal cavity, paranasal sinuses, retrobulbar region, jaw, dental arcade, diagnostic lameness, skull, temporomandibular joints, ear ossicles, and tympanic bullae in different animal species, which are more accurately assessed by CT than conventional radiographs (Gielen et al. 2012). These CT images have been used in recent anatomical studies to provide descriptive information about the normal cross-sectional anatomy of paranasal sinuses in Holstein cattle (Turgut et al. 2022), nuchal perforations in cattle skulls (Onar et al. 2022), coelomic organs in bamboo shark (Ito et al. 2023) and comparative studies of normal stifle menisci of donkeys, goats, and dogs, by comparing CT images with gross anatomical cross-sections (Abumandour et al. 2019). The transverse or cross-sectional anatomical studies of these complex structures that constantly change throughout life can be helpful to clinicians for easy interpretation of CT studies related to clinical conditions. These anatomical studies via CT images which provide a sound knowledge of cross-sectional anatomy, pave the way for a better clinical approach toward the affected region.

The morphological studies based on CT imaging are very useful for exploring wildlife anatomy for the study of animals in captivity (Lombardero and Yllera 2023). The CT studies on the heart and closely aligned lungs and liver of the alligator, microvasculature of coronary arteries in the heart of the yellow anaconda, coronary arteriography of the giraffe heart, and whole-body angiography of the domestic pig were conducted to display the mammalian vasculature (Lauridsen et al. 2011). Anatomists are utilizing this latest CT visualizing technique for 3D reconstruction in ontogenic studies of the brain and its relative position which are difficult to visualize in routine anatomical studies (Lessner and Holliday 2022). Cross-sectional CT images of skeletal elements, organ structures, and vasculature are further analyzed by using volume rendering software to construct 3D representations of anatomical structures, thereby creating 3D digital models which are movable in all possible directions. Colored 3D reconstructions via CT scans are an excellent, recently used medium to study complex bones in the skull of wild alligators, lesser accessible bones like inner ear bones, and membranes (Lessner and Holliday 2022; Wilhite and Wölfel 2019; Smith et al. 2022). This can help further in studying vertebrate evolution including patterns of neurological diversity in crocodiles. In several recent anatomical studies, researchers have used CT scans to detect the localization of one os cordis dextrum in ovines (Massari et al. 2021) and three in the camel, i.e., ossa cordis dextrum, ossa cordis sinistrum, and os aorta (El-Gendy et al. 2022). This finding in camels is novel as it reported three ossa cordis for the first time and also depicted that the wider cranial part of ossa cordis dextrum was composed of bone, whereas the caudal aspect contained both bone and cartilage.

MRI is another type of advanced imaging modality that provides detailed images of the body with the aid of a strong external magnetic field which translates these magnetic waves into gray-white signal intensities. This makes MRI particularly sensitive to delineate all anatomical structures other than air-filled regions and calcified compounds (skeleton, bones) (Haacke et al. 1999). Because of its higher soft-tissue contrast and multi-planar and multi-slice imaging capabilities, it reveals excellent anatomical information of canine and feline eye, orbit, and optic nerves in very fine detail (Yitbarek and Dagnaw 2022). Both MRI and CT are non-invasive techniques, keeping the animal intact so that both of these modalities can be used on live animals. As such, MRI and CT allow for repeated analysis on the same animal, which calls for longitudinal investigations of particular anatomical phenomena such as investigations on tissue regeneration and volumetric changes in digestive system during digestion (Lauridsen et al. 2011). CT and MRI pose a major disadvantage as they require the animal to be completely immobilized during the entire scan procedure.

Data acquired by both MRI and CT export the acquired biological images as cross-sectional images in DICOM, a standard data format to store, exchange, and transmit medical images. These DICOM images are then further used for creating 3D reconstructions or 3D models using free DICOM-viewer software and therefore, there is a critical connection between 3D printing technologies and receiver-specific medical imaging records (Lauridsen et al. 2011; Memarian et al. 2022). Some of the comparative imaging studies have also been achieved by merging these two diagnostic imaging modalities (CT, MRI) with cryosectioned images and 3D models of the canine brain (Czeibert et al. 2019).

2.4 3D Printing via 3D Reconstructions into 3D Models

3D printing is a novel innovative technology that allows for the creation of highly customized models of almost any anatomical structure such as bones, organs, or entire body systems with intricate details and very high accuracy, tailored to provide students with realistic representations of anatomical structures. It creates 3D-printed models by converting digital images obtained by 3D visualizations like CT and MRI into physical objects and thus has a great potential to reinforce student learning with better practical exposure to various anatomical structures, such as bones, organs, and blood vessels of animals, without any need for cadavers. 3D Slicer is a freely available software that is used for visualization, processing, segmentation, and reconstruction of images from CT or MRI, as DICOM files. Data is converted into a series of thin slices and these slices are then transmitted to the 3D printer for fabrication (Kathirvel et al. 2022). Some other popular 3D slicing software includes Ultimaker Cura (ultimaker.com/software/ultimaker-cura), Simplify3D (www.simplify3d.com), PrusaSlicer (www.prusa3d.com/en/page/prusaslicer_424), and Slic3r (slic3r.org). Each slicer has its own features and user interface, but they all serve the same fundamental purpose of preparing 3D reconstructions for printing into models. These 3D reconstructions can also be used as virtual 3D models to explain and display dynamic 3D anatomical models via computer screens which can be rotated and manipulated virtually (Fig. 2.5).

3D printing involves a sequence of steps to materialize images into physical objects. Firstly, a virtual 3D design of the item to be printed is developed either using computer-aided design (CAD) or sculpturing software programs like ZBrush. The 3D design is prepared by scanning the real bones, museum specimens, or freshly collected organs with specialized scanners such as 3D laser scanners (e.g., SIMSCAN 3D Scanners), through a phone camera, or by converting CT or MRI files to a 3D printable format. The 3D design is saved in a

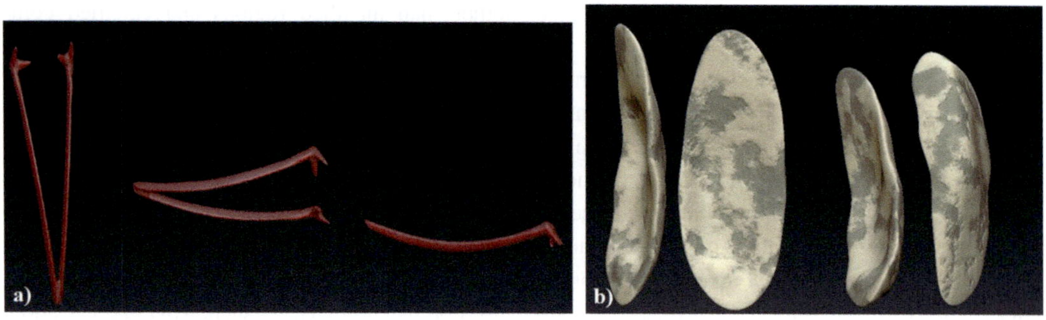

Fig. 2.5 Virtual 3D reconstructions by Ultimaker Cura software generated from (**a**) real mandible of fowl. (**b**) Spleen of goat (* at Guru Angad Dev Veterinary and Animal Sciences University; GADVASU)

file format supported by slicing software such as .stl, .obj, which can be read by 3D printer software (Wilhite and Wölfel, 2019). The output of CAD sculpturing software or .stl/.obj file generated is then processed in slicing software, determining various parameters including print quality, percentage of infill, size of print, quantity, support to print, and its percentage. An open-source slicing application, i.e., Ultimaker Cura is being used for this purpose at our institute (Singh et al. 2024). After determining the parameters, the file is then saved as a .gcode file. The .gcode file is then transmitted to the 3D printer either through networking or through an external storage device or WiFi, as supported by the printer (Fig. 2.6).

After the modeling step, the structure is sliced into printable layers and the final step is printing. During the printing step, an object is constructed by adding another layer of main materials on top of layers that usually overlap to create a structure that is prototyped. It is a unique process of depositing material layer by layer, known as "Additive Manufacturing" (AM) and "Layered Manufacturing," which means creating an object by adding printing material to the object layer by layer material until a new part is formed (Bozkurt and Karayel 2021). This differentiates it from the rest of the traditional methods of modeling. 3D printing is based on different techniques such as selective laser sintering (SLS) material jetting, stereolithography (SLA), material extrusion and binder jetting, etc., which can be used for different materials and areas. At our institute, we are using a Fused Deposition Modelling (FDM) printer, which is based on the principle of material extrusion as its 3D printing process. A spool feeds thermoplastic filament through an extruder that heats and melts this filament, and then extrudes the liquid thermoplastic through a thin nozzle to create a model of desired structure which is deposited layer by layer by AM method (Fig. 2.7).

The 3D printer develops physical models based on the .gcode file, using filament. The printed physical model reflects all angles, forms, and features from the digital 3D CAD model. Usually, the model printed is processed further in the finishing of the final physical print by removing the supports—which are extra filament material added to support the model while printing it. These 3D-printed models are sometimes painted to make them more realistic and appealing. The time and effort involved in printing varies from minutes to

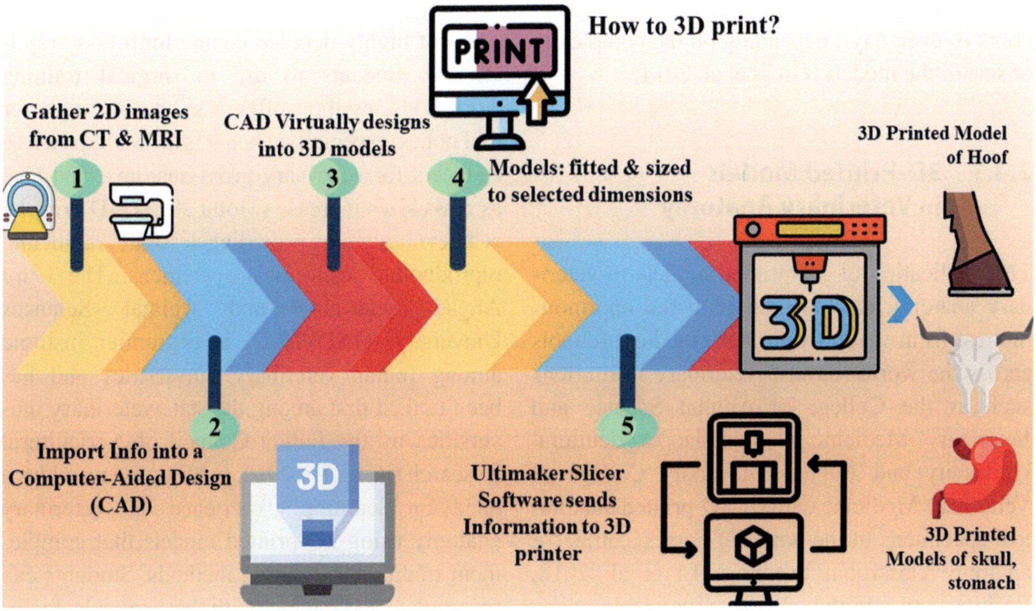

Fig. 2.6 Schematic flow showing steps involved in the process of 3D printing technology, i.e., from 3D imaging to 3D modeling of organs, bones, hoof, etc. (*created by the author)

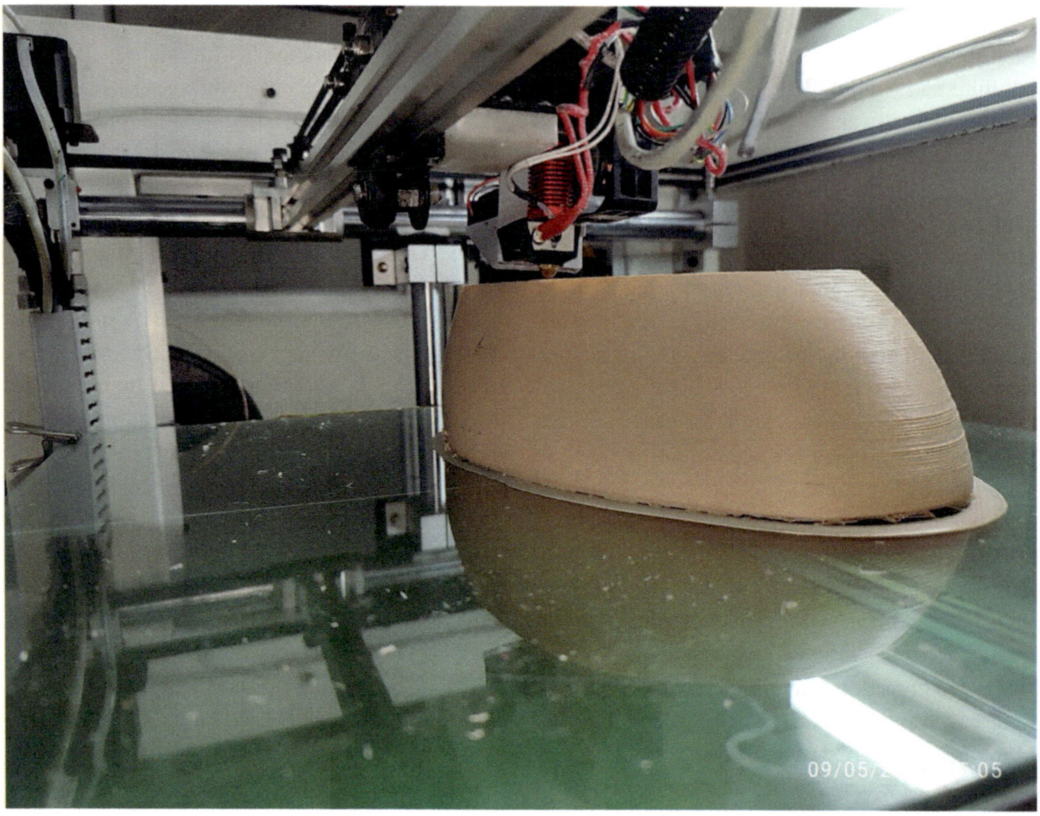

Fig. 2.7 Fused Deposition modeling (FDM) 3D printer printing the 3D model of bovine mammary gland at Guru Angad Dev Veterinary and Animal Sciences University

hours or even days, depending on the complexity or size of the models (Gross et al. 2014).

2.4.1 3D-Printed Models in Veterinary Anatomy

The application of 3D-printed models in veterinary anatomy teaching has gathered enormous interest from several veterinary medical schools across the world. Several veterinary institutions such as the College of Animal Science and Veterinary Medicine of Qingdao Agricultural University and Auburn University College of Veterinary Medicine employ 3D-printed anatomical specimens of the bones of horses, cattle, for anatomy classroom teaching (Li et al. 2018; Wilhite and Wölfel 2019). The Ohio State University College of Veterinary Medicine is using 3D printers to create educational prototypes of highly detailed canine limb bone replicas for students to use in surgical training workshops, as they offer a good substitute for real bones, especially when it is difficult to source cadavers for veterinary gross anatomy, and also as class sizes increase (Motta 2023). 3D printing is an evolving area providing accurate, rapid, and reproducible anatomical specimens. The Guru Angad Veterinary and Animal Sciences University (GADVASU) is a premier institute among Indian veterinary universities and has been ranked first among the state veterinary universities by the Indian Council of Agricultural Research (ICAR) in 2023. GADVASU provides a hands-on learning experience in veterinary anatomy using 3D-printed models that complement traditional teaching methods. Students can manipulate and interact with these models, facilitating enhanced visualization in three dimensions, tactile learning, and improving spatial understanding for better retention of anatomical knowledge.

3D-printed models are useful predominantly for anatomical structures that are considered small, inaccessible, and complex such as mouse cochlea (Urata and Okabe 2023), inner ear ossicles, guttural pouches in horses (Wilhite and Wölfel 2019), and sesamoid bones in equine hooves (Preece et al. 2013). Similarly, 3D models of the hyoid bone in domestic animals printed using an FDM printer are beneficial for studying its comparative anatomy as the hyoid bone is very fragile and difficult to macerate (Bakıcı et al. 2019). However, it has been widely used for printing the bones and visceral organs of both small as well as large animals. The 3D-printed radius-ulna and tibia of an ox, and the humerus and scapula of a dog are almost exact replicas of real bones and are being incorporated for teaching anatomy in an interactive way for students at our institute (Fig. 2.8).

Recent studies in veterinary teaching through 3D models of the equine foot have established that students find the use of 3D-printed models more engaging and interesting than other traditional learning methods, thus improving the learning experience of students (Preece et al. 2013; Hespel et al. 2014; Alcântara Leite dos Reis et al. 2019). Diverse studies on the comparison of thoracic limb bones in equines, cats, and dog skulls using their 3D-printed models revealed that anatomical characteristics of 3D-printed models were not dissimilar from those of the real bone specimens as the anatomical characteristics were successfully identified in printed ones (Preece et al. 2013; Alcântara Leite dos Reis et al. 2019; Elias' et al. 2021; da Silveira et al. 2021). Similar findings were revealed by 3D-printed models of cattle ribs and cervical vertebrae, precisely depicting all anatomical features (Li et al. 2018).

The 3D-printed models of visceral organs such as the liver and spleen are also being used to supplement anatomy teaching with greater accuracy as all the structural details of organs are faithfully replicated in the printed organs. The convenience of supplementing teaching using models of visceral organs is due to the fact they provide hassle-free manipulation of the models which are formalin-free, in contrast to preserved real organ specimens (Fig. 2.9).

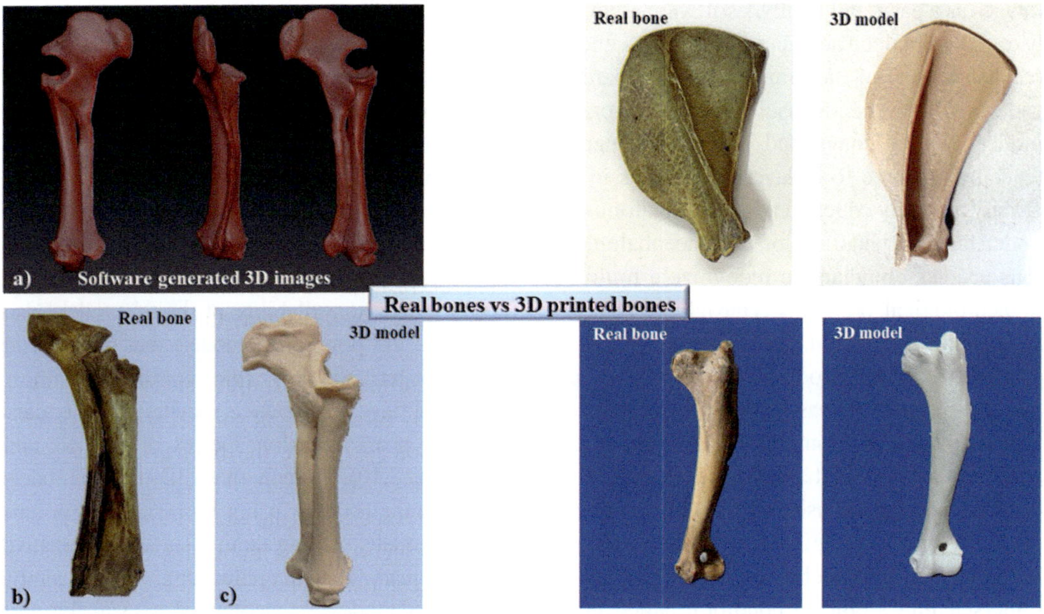

Fig. 2.8 3D-printed models generated from real bones via a) 3D images created with Ultimaker Cura, b) from real radius-ulna bone of an ox, c) 3D-printed radius-ulna; scapula of a dog (top right; real and 3D printed) and humerus of a dog (bottom right) (printed at GADVASU)

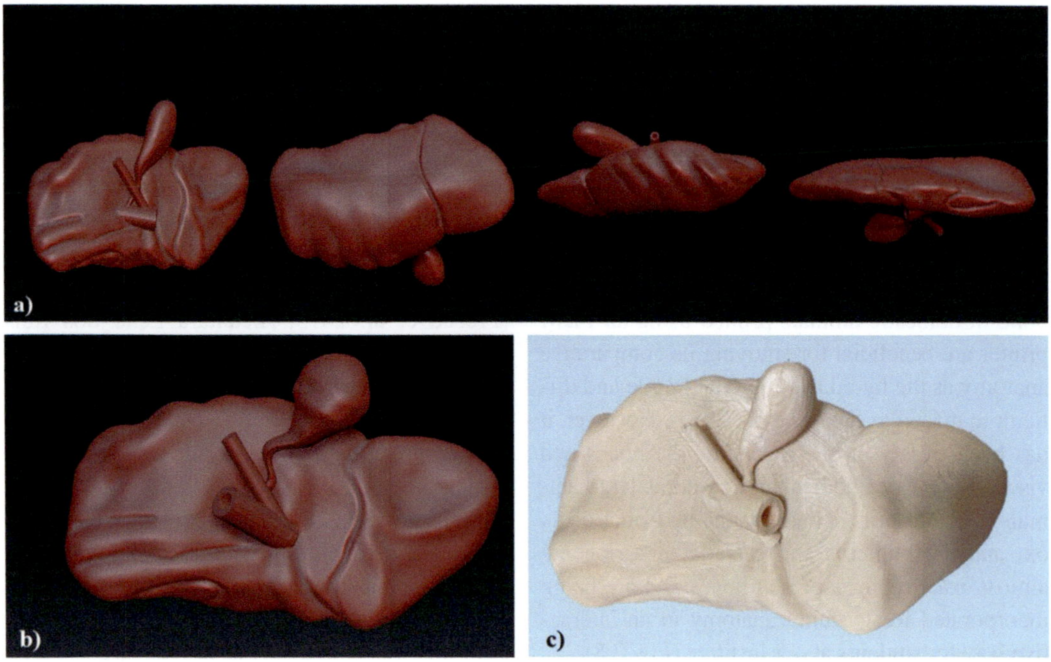

Fig. 2.9 Software (Ultimaker Cura) generated images of the liver of an ox showing intact fine anatomical details (**a**), (**b**); creating its exact replica as a 3D-printed model (**c**) *at GADVASU

Neuroanatomical studies were conducted on the effectiveness of 3D-printed sheep and dog brain models obtained by using different MRI images as a basis and EspINA software as an analyzer. The obtained segmented structures and 3D reconstructions of the brain, when compared to cadaveric brain specimens, found that 3D-printed models can be considered as a supplementary teaching resource to cadaveric materials in veterinary anatomy education. It helps anatomists to understand the anatomy of the encephalon, and thus enables clinicians to recognize a multitude of neurological problems (Haroglu et al. 2022; Blázquez-Llorca et al. 2023). The anatomical structures can be reproduced into extremely precise 3D-printed replicas from living animals and thus used for educational purposes in the field of anatomy without euthanizing an animal. Such models can also be used as working models to present clinical conditions, e.g., a model of the mammary gland model in a cow can be used for understanding a clinical condition called mastitis. Additionally, the digital versions of these models generated via software can also be analyzed, studied, and modified prior to them being printed (Fig. 2.10).

Although the initial setup cost for 3D printing may be substantial, once established the production of multiple copies of anatomical models can be more cost-effective than purchasing traditional ready-made anatomical models or cadaveric specimens. The cost of 3D printing technologies is further reduced and it becomes more cost-effective when producing anatomical models from digital models which are available at public sources in stl. or obj. format. This enables one to skip the step of modeling in CAD which makes it more expensive due to personnel operating costs (Bejdić et al. 2021). This pre-availability of digital models also facilitates the printing of models that depict wild animal or bird anatomy which is otherwise a near impossible task to reproduce, specifically for endangered or near extinction species of animals and birds (Fig. 2.10). As such, these 3D-printed models eliminate the need for using fresh cadavers or animal specimens making them a suitable alternative for veterinary anatomy education. Consequently, 3D-printed models are recognized as a valuable tool in veterinary anatomy, offering a hands-on learning experience, accuracy, customization, and technological integration to enhance student learning and retention of anatomical knowledge.

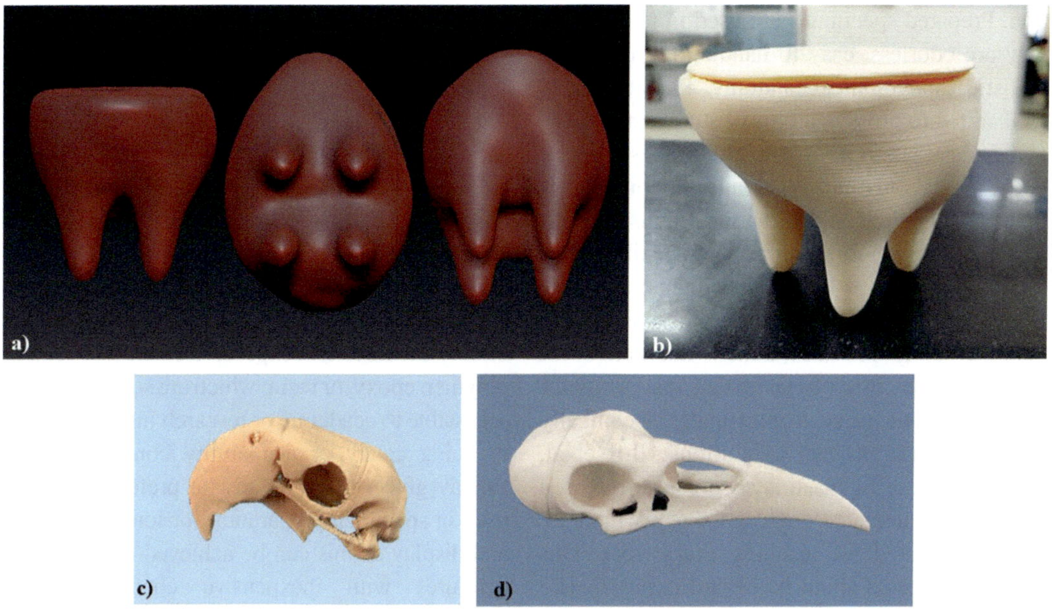

Fig. 2.10 (**a**) Digital 3D version of the mammary gland of a cow generated via software from scanning udder of an alive cow, (**b**) converted into a physical 3D model, (**c**) 3D model of parrot head, and (**d**) 3D model of crow skull (Printed at GADVASU)

2.5 Preparation of Fixative-Free Real Anatomical Organ Models

The organs and specimens in anatomy are commonly fixed with traditional preservation methods using formaldehyde as fixative in the form of a working solution made with diluted concentrations designated as 10–20% formalin. The fixative has its own disadvantages as its exposure leads to toxic and carcinogenic effects on handlers. Therefore, the need for innovative inventions for fixative-free preservation led to the discovery of new techniques to prepare real and dry anatomical specimens without using chemical preservatives or fixatives in veterinary anatomy. The intervention of revolutionizing methods like plastination and glycerin preservation for gross anatomical specimen preparation has made preservation possible without the use of fixatives, yet retaining a more natural appearance and texture, closely resembling living tissues. This preserves the realism of anatomical specimens and provides for more accurate representation, flexibility, and manipulation, thus facilitating a more realistic hands-on experience as well as better shelf-life. Other fixative-free anatomical models can be prepared by making silicone molds from organs to create clay models. This aids in more practical and action-oriented learning of veterinary anatomy. Plastinated and glycerin method specimens may not entirely replace the preserved specimens but offer a more practical, non-toxic, and cost-effective solution with realistic representations of anatomical structures for teaching and learning.

2.5.1 Plastinated Specimens

Plastination is a technique used to preserve biological tissues by forced impregnation of polymers like silicone (S10), epoxy E12, polyester (P40), or polymerizing emulsion PEM under vacuum into anatomical specimens followed by curing for long-term preservation and display (von Horst et al. 2019). The plastination procedure consists of four basic steps, i.e., fixation, dehydration, forced impregnation, and tissue curing or hardening. The S10 silicone technique is the most commonly used technique in plastination. Prepared specimens are used in teaching as they are odorless, easy to handle, and present a realistic topographical anatomy. Plastination is well suited to gross anatomical studies and well-made plastinates can last over decades without decay or degradation, especially silicone specimens. The method of plastination retains the realistic natural color, texture, and flexibility of visceral organs or gross specimens to some extent. The injecting of organs and bodies with colored epoxy provides a new alternative for enhancing the color of plastinated specimens for added aesthetic value to plastinates specifically for demonstrations and exhibitions (Iliff et al. 2019). The use of plastinated organs decreased the need to euthanize dogs and sheep which are usually required for teaching every year. This reduces concerns about biohazard exposure and takes into account ethical considerations associated with cadaver use in veterinary education (Mohamed and John 2018). Whole plastinated cadavers help students understand the complex anatomical structures and their topographic relations with other organs, which is helpful for self-directed, hands-on learning, and anatomical exploration. The visceral organs of different species like ox, pig, horse, dog, and chicken have been plastinated by impregnating Biodur S10/S3 mixture to facilitate teaching, research, and learning (Mohamed and John 2018). Initially, the size of plastinates used to be comparatively smaller, however, presently large specimens can also be plastinated by using large vacuum chambers which can accommodate even whole bodies of domestic or exotic animals like blue whales, making the study of exotic animal anatomy exclusive and interesting (Nader et al. 2019). Transparent plastinated slices (E12/P40 technique) are also an invaluable illustrative learning aid for students and allow a better, more accurate understanding of the complex cross-sectional study of anatomical structures (Sora et al. 2019). Their transparency in slices allows for the identification of anatomical structures in great detail, and with the inclusion of vascular injections it is possible to identify not only the main vessels but even small arterioles and capillaries. In fresh specimens, i.e., without fixation or vascular injection, capillaries containing blood can still be identified (Latorre et al. 2019).

2.5.2 Glycerin Method

Organ preservation by glycerin technique in veterinary anatomy is also known as Elnady technique (Elnady 2016). The process of preservation by plastination, however, requires specialized laboratory facilities and expensive chemicals such as polymer, epoxy, or resin, which are sometimes less accessible to academic or research institutions due to being costly or availability constraints. The Elnady glycerin technique is a preferred method used in specimen preparation for long-term study and display as this can be achieved at room temperatures with inexpensive chemicals. This removes the need for expensive laboratory and chemical costs. It creates realistic, durable, soft, odorless, and flexible preserved specimens with negligible shrinkage (Elnady 2016). After formalin fixation of tissues, glycerin preservation is achieved in three basic steps i) acetone-based dehydration (for 4–5 days each in three changes of acetone), ii) glycerin impregnation (2–3 weeks), and iii) curing with corn starch (for 4–5 days) which facilitates the absorption of excessive glycerin. Therefore, the whole process is slow and lengthy as it takes 16–41 weeks to be completed. The modified Elnady glycerin technique is used by some scientists to overcome the drawback of slight tissue discoloration in other preservation techniques. By this method, the canine and equine heart specimens are preserved with fine anatomical details, such as coronary arteries. The thoracic and abdominal viscera of dogs are pigmented by this modified method which creates brilliant contrast between the visceral organs such as esophagus, trachea, lungs, thoracic aorta, diaphragm, stomach, kidneys, small intestine, and spleen (Bernal et al. 2022). Similarly, the musculoskeletal specimens including the forelimb of goats are also preserved by researchers with this improved method of tissue preservation which can be utilized for easy handling and manipulation for learning anatomical structures like muscles, nerves, and blood vessels in limbs (Islam and Sultana 2021).

Whole-body midsagittal section of cats and dogs can be prepared by this method to demonstrate intact, well-preserved in situ topographic anatomy of all the thoracic, abdominal, and pelvic visceral organs. The visceral neurovascular supply is also visible by this method if arteries are simultaneously injected with colored latex (El- Shafey et al. 2022).

Recently, Pichardo-Molinero et al. (2024) have successfully devised another modified technique, i.e., alcohol propylene glycol technique to retain dimensions, color, and texture without losing anatomical details in sheep hearts while allowing smooth dissections after applying the technique. This glycerin method preserves the structural integrity of organs, muscles, and complex anatomical features, such as vascular or nervous tissues and other soft tissues that are prone to decay. It is a preferred non-toxic alternative keeping the flexibility and durability of anatomical structures intact. If stored appropriately, specimens preserved by this technique can maintain their structural composition intact for many years in laboratory settings, thereby providing hands-on expertise on dissections and comparative anatomy studies for longer periods.

2.5.3 Clay Modeling

Clay modeling is a less explored pedagogy in professional education. It can be efficiently used as a supplement in teaching veterinary anatomy in conjunction with practical sessions. The use of colored play-dough clay for preparing anatomical models by students is an active, tactile learning tool utilized to improve student engagement and enhance their understanding of anatomical relationships in 3D. It allows students to learn while creating anatomical structures or organs as the mind cannot forget what the hands have learned. It stimulates a novel "haptico-visual observation" (HVO) as reported by Reid et al. (2018) to improve their cognitive understanding and memorization of the 3D form of the anatomical part. Similarly, working with clay provides hands-on learning to students, i.e., learning by doing that engages tactile senses, which can enhance retention of complex anatomical knowledge through kinesthetic learning. 3D HVO enables the learner to improve their 3D spatial awareness of the anatomical parts they are studying (Shapiro et al. 2023).

Clay modeling is being used as a popular alternative method by medical students to review their personal knowledge regarding organ structure by referring to pictures and figures in a textbook or atlas while constructing the clay models of various organs (Oh et al. 2009; Curlewis et al. 2021). When the surfaces of prepared clay models are cut, it clearly depicts the cross-sectional arrangement of internal parts or structures which resemble CT and MR images taken at similar levels (Oh et al. 2009). In veterinary anatomy, clay modeling has recently gained the attention of some anatomists. A study involving clay modeling among students studying bones of ruminant forelimbs, joints, muscles, and nerves depicted a twofold increase in student achievement and learning as compared to the routine method (Onuk et al. 2019). In practical muscle examinations, it was revealed that students who clay-modeled muscles of cats performed better than those who studied these muscles by dissection (Motoike and Kane 2009). Another study was conducted on second- and third-year medical students who made clay models of the heart, liver, carpal, and tarsal bones (hand and foot bones, respectively), the brain, stomach, duodenum, pancreas, spleen, and gallbladder with students choosing colored clay to distinguish different segments of the organs. The students with the clay model group activity scored overall better than the group that had not used the clay modeling activity (Oh et al. 2009). Undergraduate and postgraduate students in our department's osteology laboratory learn comparative osteology by creating clay models of dog mandibles similar to actual dog mandibles while studying from the real bone (Fig. 2.11).

Similarly, the clay models of various organs or organ systems can be developed by educators as well as learners from real specimens or from pictures available in atlas or textbook models, especially for rarely available specimens. This also

Fig. 2.11 Clay modeling activity by postgraduate student for studying comparative osteology using (**a**) Fevicryl Mouldit clay as modeling material (**b**, **c**) sculpting the clay model by visualizing the real mandible, and (**d**) finally created clay model of mandible of a dog vs real mandible (*created in our department by students)

promotes interactive teaching where academicians can demonstrate anatomical concepts that are difficult to visualize by sculpting the associated structures in real time. This in turn encourages student participation and discussion. Specifically, while learning embryonic development, students struggle to mentally reconstruct and visualize the process involving the 3D structure of developing embryos when only observing prepared slides of cross-sectional slices. Students who create clay models of embryos and slice them into cross-sections, have a much easier interpretation of slices of real embryos and gain a deeper understanding of their development in 3D, as sectioning clay models makes developmental anatomy tangible (Howell and Howell 2010; Chandrasekhar et al. 2011). A similar study indicates that learning veterinary embryology by developing clay models of organogenesis as a student and teacher team-based learning method, followed by a test, verifies that student creative skills are refined along with their gaining of knowledge through clay modeling (Chandrasekhar et al. 2011).

Development of these models stimulates and educates the creators and once created provides a long-lasting teaching resource for future use in didactic teaching. Clay modeling enhances veterinary anatomy education as it boosts critical thinking and problem-solving skills in students by allowing them to compare their creations with actual anatomical structures which allows them to determine how accurately their creations represent actual anatomical structures in 3D. This fosters a deeper understanding and retention of knowledge (Oh et al. 2009).

2.6 Adopting Online Platforms to Teach Veterinary Anatomy

The traditional classroom way of learning anatomy has been upgraded to virtual classroom-based learning for worldwide outreach by students, especially during COVID-19 pandemic. This was achieved via online platforms like Zoom, Google Classroom, Teams, and E-museum portals. Virtual classrooms are an amalgamation of the traditional classroom with practical learning where the teacher can impart basic anatomical knowledge by using 3D pictures or scans with PowerPoint animations while explaining theoretical aspects of gross anatomy and histology. The virtual classroom has several advantages where learners can have live interactive sessions with teachers on their subject by observing the specimens demonstrated to them by the teacher. Additionally, they can record these live sessions delivered on virtual platforms and save them for later reference (Kapoor and Singh 2022). Previously shown videos by teachers on prosected specimens can either be shown to students in live virtual classrooms or uploaded on YouTube, Moodle, Blackboard, and Canvas as videos for their future reference and self-study. Scholars can utilize several VR aids through their routine devices including desktops and mobile devices. Some studies indicate that students who used mobile VR had significantly higher test scores than those who used text, 2D pictures, and graphs (Küçük et al. 2016). Moreover, veterinary anatomy is a subject which necessitates continuous exposure to specimens and organs. In order to facilitate this, a one-stop zone has been provided on various online platforms of veterinary universities designated as E-Museums or Online Veterinary Anatomy Museum (Kapoor and Singh 2022). For that purpose, several veterinary institutes from around the world are contributing anatomy resources that have been aggregated and methodically arranged into a readily searchable environment making them easily discoverable by learners globally. The user interface design of E-museums provides a spontaneous, engaging, and dynamic environment, which represents an in-house museum experience. Such a museum contains a compilation of over 1000 different resources including images, PowerPoint presentations, flash animations, and videos covering an array of topics related to veterinary anatomy including histology, dissection, osteology, and embryology (for animated embryology—https://www.uco.es/~an1-gamoj/MyWeb/animated_embryology.html) (Phillips et al. 2012). These abundant learning resources are accessed in the form of interactive "one click" online activities including multiple-choice questions, visual compilations of organ images, prosected specimen images, and cross-sectional CT or MRI images. Once the student selects the name of a particular organ or structure they wish to visualize, it will navigate them to the related images and animations. The knowledge attained can be used further to amalgamate clinical correlations with CT scans and test their retention by a questionnaire provided on the same site. Such sites can be easily accessed at https://www.rvc.ac.uk/e-anatomy or https://www.rvc.ac.uk/static/review/equine-distal-limb/index.html. The material is structured systematically on various e-platforms by species, body systems, and body regions accommodating different learning styles and preferences. It serves to provide a useful tool for learners to search content by using the search function for "keywords" if some instructional material is lacking at their own institutions. However, the sharing of such platforms of an E-museum is under common licensing of each university which has developed it, to permit the use of materials to supplement lectures and enable scholars across the globe to benefit from access to a variety of materials. Specifically, gross anatomy E-museums will continue to play an important role as "One-stop Zone" in anatomy education.

2.7 Use of Digital Interface Like Artificial Intelligence (AI) in Combination with Anatomical Teaching

The integration of AI in veterinary anatomy teaching by anatomy educators offers numerous possibilities to enhance the student learning experience. The most commonly used AI-powered tool in anatomy teaching is the virtual dissection tool which provides an immersive real-life experience where students can virtually dissect anatomical structures in a simulated environment. Broadly, AI-driven tools are currently integrated into online anatomy learning systems for student monitoring and personalized learning platforms (Inuwa et al. 2011), deployed through 3D-anatomical model development (Li et al. 2018), virtual dissections (Afsharpour et al. 2018), and anatomy gamification (Ang et al. 2018). These simulated environments are employed to enhance students' experiences with VR and AR (Ferrer-Torregrosa et al. 2015; Preim and Saalfeld, 2018; Erolin et al. 2019). The 3D reconstruction and visualization via CT and MRI images discussed above have been developed utilizing AI systems, for deeper learning and computer-aided vision which can be further used to reconstruct virtual 3D models of anatomical structures and finally physical 3D printed models. These models can then be visualized in immersive VR environments or AR applications to explore anatomy in an artificial environment in virtual laboratories or via mobile phones. Our institute has developed such a virtual laboratory where students can learn veterinary anatomy by exploring the comparative anatomy of bones and organs via these applications in simulated environments (Fig. 2.5). These virtual environments have proven to be highly interactive and engaging for students. AI is being used to develop digital anatomical atlases and E-museums with global access to detailed anatomical knowledge. Although, by exploiting the capabilities of AI, veterinary anatomy teaching has been greatly transformed as a more dynamic and immersive learning experience, it presents some perils. AI presents a set of different assumptions relating to the nature of knowledge, autonomy, and computability which serves to effectively minimize the uncertainties, inaccuracies, and limitations of AI-generated images due to an often misplaced trust in the "accuracy" of AI output (Lazarus et al. 2024).

2.8 Conclusion

The process of gaining anatomical knowledge is crucial in becoming a sound clinician for animal studies as a veterinarian. The teaching of veterinary anatomy has greatly evolved due to the inclusion of numerous technological interventions like CT and MRI visualizations, 3D reconstructions, 3D animations, and the development of virtual and augmented reality. 3D animations are constructed by converting intricate anatomical knowledge using computer software to provide actual representation of anatomy closer to life-like experience of actual 3D structures. This has transformed traditional anatomy teaching into more engaging, interactive, and hands-on learning experiences for students. Additionally, the transition to 3D visualizations has addressed ethical considerations which prevent the euthanizing of animals. The amalgamation of 3D printing technology with 3D visualizations or imaging modalities like CT or MRI is proving to be a pioneering step in teaching and learning veterinary anatomy. These non-invasive imaging techniques play an integral role in understanding the clinical orientation of animal anatomy and increase the effectiveness in positive clinical development. 3D printing can create models even from living animals as well as of less accessible organs. Other proliferating methods of preservation such as plastination and the glycerin method of preservation also pave the way for learning anatomy that can achieve long-lasting learning. These new interventions have greatly transformed the outlook of veterinary anatomy education from monotonous, traditional teaching pedagogy to an entirely new immersive experience in this hybrid era of education.

Acknowledgments Authors are thankful to Guru Angad Dev Veterinary and Animal Sciences University and Rashtriya Krishi Vikas Yojana (RKVY) for providing the facilities and funds for this project to procure 3D printers to develop models and for the development of an anatomy AR application for educating veterinary students.

References

Abumandour MMA, Bassuoni NF, El-Gendy S, Karkoura A, El-Bakary R (2019) Comparative morphological studies of the stifle menisci in donkeys, goats and dogs. J Morphol Sci 36:72–84

Afsharpour S, Gonsalves A, Hosek R, Partin E (2018) Analysis of immediate student outcomes following a change in gross anatomy laboratory teaching methodology. J Chiropr Educ 32:98–106

Alcântara Leite Dos Reis D, Gouveia BLR, Júnior JCR, Neto ACA (2019) Comparative assessment of anatomical details of thoracic limb bones of a horse to that of models produced via scanning and 3D printing. 3D Print Med BMC 5:13

Ang ET, Chan JM, Gopal V, Li Shia N (2018) Gamifying anatomy education. Clin Anat 31:997–1005

Bakıcı C, Akgün RO, Oto C (2019) The applicability and efficiency of 3-dimensional printing models of hyoid bone in comparative veterinary anatomy education. Veteriner Hekimler Dernegi Dergisi 90(2):71–75

Bejdić P, Turkić N, Škaljić E (2021) Cost-effectiveness of 3d printing of anatomical models. Veterinaria 70:9–14

Bernal V, Aburto P, Pérez B, Gómez M, Gutierrez JC (2022) A technical note of improvement of the Elnady technique for tissue preservation in veterinary anatomy. Animals 12:1111

Blázquez-Llorca L, Morales de Paz L, Martín-Orti R, Santos-Álvarez I, Fernández-Valle ME, Castejón D, García-Real MI, Salgüero-Fernández R, Pérez-Lloret P, Moreno N et al (2023) The application of 3D anatomy for teaching veterinary clinical neurology. Animals 13(10):1601

Borroni C, Pimentel-Ávila A, Stoore C et al (2021) A unique approach to project-based learning (PjBL) in a veterinary anatomy course. Med Sci Educ 31: 511–517

Bozkurt Y, Karayel E (2021) 3D printing technology; methods, biomedical applications, future opportunities and trends. J Mater Res Technol 14:1430–1450

Chandrasekhar L, Maya S, Rajani CV, Raja TV, Ravindran R (2011) Clay modeling—a method of teaching embryology to veterinary undergraduates. Indian J Anim Sci 81(11):1121–1123

Collins JP (2008) Modern approaches to teaching and learning anatomy. British Med J 337:a1310

Curlewis K, Leung B, Perera M, Bazira PJ, Sanders KA (2021) Clay-based modeling in the Anatomist's toolkit: a systematic review. Anat Sci Educ 14:252–262

Czeibert K, Baksa G, Grimm A, Nagy SA, Kubinyi E, Petneházy Ö (2019) MRI, CT and high resolution macro-anatomical images with cryosectioning of a Beagle brain: creating the base of a multimodal imaging atlas. PLoS One 14(3):e0213458

da Silveira EE, da Silva Lisboa Neto AF, Carlos Sabino Pereira H, Ferreira JS, Dos Santos AC, Siviero F, da Fonseca R, de Assis Neto AC (2021) Canine skull digitalization and three-dimensional printing as an educational tool for anatomical study. J Vet Med Educ 48(6):649–655

Darras KE, Spouge R, Hatala R, Nicolaou S, Hu J, Worthington A et al (2019) Integrated virtual and cadaveric dissection laboratories enhance first year medical students' anatomy experience: a pilot study. BMC Med Educ 19(1):366

De Bose K (2020) Virtual anatomy: expanding veterinary student learning. J Med Libr Assoc 108:4

Edelmers E, Kazoka D, Bolocko K, Pilmane M (2022) Different techniques of creating bone digital 3D models from natural specimens. Appl Syst Innov 5:85

El- Shafey A, Magdy Y, Hamad A, Ahmed O (2022) Modified form of the Elnady technique for tissues preservation. Benha Vet Med J 41:56–60

El-Gendy SAA, Alsafy MAM, Rutland CS et al (2022) Ossa cordis and os aorta in the one-humped camel: computed tomography, light microscopy and morphometric analysis. Microsc Res Tech 86(1):1–10

Elias' NDS, Pereira HCS, Lisboa Neto ADS, da Silveira EE, dos Santos AC, de Assis Neto AC (2021) Bioprinting of 3D anatomical models of flat and long thoracic limb bones of domestic cats (Felis catus linnaeus, 1758). Brazilian J Vet Med 43(1):e106020

Elnady FA (2016) The Elnady technique: an innovative, new method for tissue preservation. ALTEX 33:237–242

Erolin C, Reid L, McDougall S (2019) Using virtual reality to complement and enhance anatomy education. J Vis Commun Med 42:93–110

Evans DJ, Bay BH, Wilson TD, Smith CF, Lachman N, Pawlina W (2020) Going virtual to support anatomy education: a stopgap in the midst of the COVID-19 pandemic. Anat Sci Educ 13:279–283

Ferrer-Torregrosa J, Torralba J, Jimenez MA, García S, Barcia JM (2015) ARBOOK: development and assessment of a tool based on augmented reality for anatomy. J Sci Educ Technol 24:119–124

Fitria TN (2023) Augmented reality (AR) and virtual reality (VR) technology in education: media of teaching and learning: a review. Int J Comp Info Syst 4(1):14–25

Gielen I, Caelenberg A, Bree H (2012) Clinical applications of computed tomography (CT) and magnetic resonance imaging (MRI) in small animals. Eur J Companion Anim Pract 22(4):84–103

Gomez O, García-Manzanares M, Chicharro D et al (2023) Application of blended learning to veterinary gross anatomy practical sessions: students' perceptions of their learning experience and academic outcomes. Animals 13(10):1666

Gross BC, Erkal JL, Lockwood SY, Chen C, Spence DM (2014) Evaluation of 3D printing and its potential impact on biotechnology and the chemical sciences. Anal Chem 86(7):3240–3253

Haacke EM, Brown RW, Thompson MR, Vankatesan R (1999) Chap 15: signal, contrast, and noise, chap 23: motion artifacts and flow compensation. In: Magnetic resonance imaging. Wiley, New York, pp 331–380. 669–702

Haroglu D, Iscan B, Duzler A (2022) Use of three dimensional (3D) printed models of sheep brain in online veterinary anatomy education. Int J 3D Printing Technol Digi Ind 6(3):370–381

Hespel AM, Wilhite R, Hudson J (2014) Invited review-applications for 3D printers in veterinary medicine. Vet Radiol Ultrasound 55:347–358

Howell CE, Howell JE (2010) Sectioning clay models makes anatomy & development tangible. Am Biol Teach 72(5):313–314

Iliff S, Concha I, Chereminskiy V, Henry RW (2019) Coloring plastinated specimens. Anat Histol Embryol 48:552–556

Inuwa IM, Taranikanti V, Al-Rawahy M, Habbal O (2011) Perceptions and Attitudes of Medical Students towards Two Methods of Assessing Practical Anatomy Knowledge. Sultan Qaboos Univ Med J 11(3):383–390

Islam R, Sultana N (2021) Preservation of musculoskeletal specimens of goat by the Elnady technique: an innovative low-cost alternative to plastination. J Plast 33(2):1–13

Ito T, Tanaka T, Kiyatake I, Izawa T, Furuya M, Sasai K (2023) Contrast-enhanced computed tomography and cross-sectional anatomy of the trunk in the brown-banded bamboo shark (*Chiloscyllium punctatum*). Anat Histol Embryol:1–11

Jiang N, Jiang Z, Huang Y, Sun M, Sun X, Huan Y et al (2024) Application of augmented reality models of canine skull in veterinary anatomical education. Anat Sci Educ 17:546–557

Kapoor K, Singh A (2022) Veterinary anatomy teaching from real to virtual reality: An unprecedented shift during COVID-19 in socially distant era. Anat Histol Embryol 51:163–169. https://doi.org/10.1111/ahe.12783

Kapoor K (2024) 3D visualization and printing: an "Anatomical Engineering" trend revealing underlying morphology via innovation and reconstruction towards future of veterinary anatomy. Anat Sci Int 99:159–182

Karkera S (2021) Impact of COVID-19 on the future of cadaveric dissection and anatomy teaching. Int J Res Med Sci 9(3):908–913

Kathirvel SS, Paramasivan S, Narmatha N, Selvaraju M (2022) 3D printing—dawn of a new era in learning veterinary anatomy. Indian J Vet Anat 34(1):50–54

Küçük S, Kapakin S, Gökta SY (2016) Learning anatomy via mobile augmented reality: effects on achievement and cognitive load. Anat Sci Educ 9(5):411–421

Latorre R, de Jong K, Sora M-C, López-Albors O, Baptista C (2019) E12 technique: conventional epoxy resin sheet plastination. Anat Histol Embryol 00:1–7

Lauridsen H, Hansen K, Wang T, Agger P, Andersen JL et al (2011) Inside out: modern imaging techniques to reveal animal anatomy. PLoS One 6(3):e17879. https://doi.org/10.1371/journal.pone.0017879

Lazarus MD, Truong M, Douglas P, Selwyn N (2024) Artificial intelligence and clinical anatomical education: promises and perils. Anat Sci Educ 17:249–262

Lessner EJ, Holliday CM (2022) A 3D ontogenetic atlas of *Alligator mississippiensis* cranial nerves and their significance for comparative neurology of reptiles. Anat Rec 305:2854–2882

Li F, Liu C, Song X, Huan Y, Gao S, Jiang Z (2018) Production of accurate skeletal models of domestic animals using three-dimensional scanning and printing technology. Am Assoc Anat 11(1):73–80

Little WB, Dezdrobitu C, Conan A et al (2021) Is augmented reality the new way for teaching and learning veterinary cardiac anatomy? Med Sci Educ 31:723–732

Lombardero M, Yllera MM (2023) Advances in animal anatomy. Animals 13:1110

Martin JF, Arnold OR, Linton A et al (2023) How virtual animal anatomy facilitated a successful transition to online instruction and supported student learning during the coronavirus pandemic. Anat Histol Embryol 52:36–49

Massari C, Ferreira-Silva A, Riceti-Magalhães H, Souza-Silva D, Miglino M (2021) Computed tomography examination of the os cordis in a lamb (Ovis aries Linnaeus, 1758). MVZ Cordoba 27(1):e2153

Memarian P, Pishavar E, Zanotti F et al (2022) Active materials for 3D printing in small animals: current modalities and future directions for orthopedic applications. Int J Mol Sci 23:1045

Mizer LA, Farnum CE, Schenck AD (2002) The modular resource center: integrated units forthe study of the anatomical sciences in a problem-based curriculum. Anat Rec 269:249–256

Mohamed R, John R (2018) Production and use of plastinated anatomical specimens as teaching and learning tools in veterinary gross anatomy in the Caribbean. J Adv Vet Anim Res 5(1):44–52

Motoike HK, O Kane RL, Lenchner E, Haspel C (2009) Clay modeling as a method to learn human muscles: a community college study. Anat Sci Educ 2(1): 19–23

Motta T (2023) College of veterinary medicine using 3D printers to mass produce PPE, available at: https://vet.osu.edu/aboutus/news/college-veterinary-medicine-using3d-printers-mass-produce-ppe

Nader P, Henry RW, Sui H (2019) Plastination of larger and massive specimens—with silicone. Anat Histol Embryol 00:1–5

Niedermair JF, Antipova V, Manhal S, Siwetz M, Wimmer-Röll M, Hammer N, Fellner FA (2022) On the added benefit of virtual anatomy for dissection-based skills. Anat Sci Educ 00:1–13

Oh CS, Kim JY, Choe YH (2009) Learning of cross-sectional anatomy using clay models. Anat Sci Educ 2:156–159

Onar V, Chroszcz A, Poradowski D, Nazan Gezer I, Mehmet Ali P, Rahmi A (2022) Computed tomography and osteological examination of the nuchal perforations in byzantine cattle skulls. Turkish J Vet Anim Sci 46(5):764–778

Onuk B, Çolak A, Arslan S, Sizer SS, Kabak M (2019) The effects of clay modeling and plastic model dressing techniques on veterinary anatomy training. Kafkas Univ Vet Fak Derg 25(4):545–551

Phillips GG, Short N, Stanikov B (2012) Taking veterinary anatomy online. Altern Lab Anim 40(6):24–25

Pichardo-Molinero MR, Jardon-Xicotencatl S, Oliver-González MR, García-Tovar CG (2024) Implementation of a new solution for the preserva-

tion of anatomical specimens made of non-toxic substances. Open J Vet Med 14:56–67
Preece D, Williams SB, Lam R, Weller R (2013) Let's get physical: advantages of a physical model over 3D computer models and textbooks in learning imaging anatomy. Am Assoc Anat 6:216–224
Preim B, Saalfeld P (2018) A survey of virtual human anatomy education systems. Comput Graph 71:132–153
Rachmatullah R, Sukihananto (2020) Efektivitas Penggunaan virtual reality (VR) Dalam Pendidikan Keperawatan: literature review. J Kesehatan 9(2)
Reid S, Shapiro L, Louw G (2018) How haptics and drawing enhance the learning of anatomy. Anat Sci Educ 00:1–9
Sadeghinezhad J (2023a) Online veterinary anatomy education during Covid-19 pandemic in Iran: challenges and opportunities. Vet Med Sci 9:1869–1880
Sadeghinezhad J (2023b) Dissection videos as a virtual veterinary anatomy peer learning tool: Trialled at the University of Tehran during the Covid-19 pandemic. Anat Histol Embryol 52(1):55–61
Senos R, Ramos Leite CA, dos Santos TF, Roberto-Rodrigues M, Pérez W (2022) Using videos in active learning: an experience in veterinary anatomy. Anat Histol Embryol 52(1):50–54
Shapiro L, Hobbs E, Keenan ID (2023) Transforming musculoskeletal anatomy learning with haptic surface painting. Anat Sci Educ 16:677–693
Singh N, Banga HS, Singh J, Sharma R (2024) 3D printing: a novel technology for livestock sector knowledge dissemination. Libr Hi Tech News 41(2):18–24
Smith CM, Curthoys IS, Mukherjee P, Wong C, Laitman JT (2022) Three-dimensional visualization of the human membranous labyrinth: the membrane limitans and its role in vestibular form. Anat Rec 305(5):1037–1050
Sora MC, Latorre R, Baptista C, López-Albors O (2019) Plastination—a scientific method for teaching and research. Anat Histol Embryol 00:1–6
Stirling AC, Moro C (2020) The use of virtual, augmented and mixed reality in anatomy education. In: Chan LK, Pawlina W (eds) Teaching anatomy. Springer, Cham
Turgut N, Bahar S, Kılınçer A (2022) CT and cross-sectional anatomy of the paranasal sinuses in the Holstein cow. Vet Radiol Ultrasound:1–13
Urata S, Okabe S (2023) Three-dimensional mouse cochlea imaging based on the modified Sca/eS using confocal microscopy. Anat Sci Int 98(3):309–317
von Horst C, von Hagens R, Sora C-M, Henry RW (2019) History and development of plastination techniques. Anat Histol Embryol 48:512–517
Wilhite R, Wölfel I (2019) 3D printing for veterinary anatomy: an overview. Anat Histol Embryol 48:609–620
Wulster K (2022) Use of 3D printing in veterinary medicine VMD, Dip. ACVR assistant professor of large animal diagnostic imaging New Bolton Center, University of Pennsylvania
Xiberta P, Boada I (2019) IVET, an interactive veterinary education tool. J Anim Sci 97(2):932–944
Xu X, Kilroy D, Kumar A et al (2023) Mixed reality applications in tertiary veterinary education: a systematic review. In: Cai Y, Mangina E, Goei SL (eds) Mixed reality for education, Gaming media and social effects. Springer, pp 241–264
Yitbarek D, Dagnaw GG (2022) Application of advanced imaging modalities in veterinary medicine: a review. vet med (Auckland, N.Z.), 13:117–130. https://doi.org/10.2147/VMRR.S367040
Youssef S (2021) Different approaches for teaching and learning anatomy and future directions. Indian J Clin Anat Physiol 8(1):1–6
Zhang X, Yang J, Chen N, Zhang S, Xu Y, Tan L (2019) Modeling and simulation of an anatomy teaching system. Vis Comput Ind Biomed Art 2:1–8

Bakeng se Afrika: Digital Skeletal Repository: Advancing Biological Anthropology and Medical Research in South Africa

3

Ericka N. L'Abbé, Amanda Alblas, Jan H. P. Ackermann, Lunga Bam, Amélie Beaudet, Pearl N. Bothma, Angelique Carmichael, Marine Cazenave, Eugenia Cunha, Frikkie de Beer, Edwin de Jager, Meg-Kyla Erasmus, Alieske C. Hagg, Miksha Harripershad, Patria C. Hartman, Yann Heuzé, Jakobus Hoffman, Nhlanhla Japhta, Reabetswe Khonyane, Ruth Kobedi, Gabriele C. Krüger, Maritza Liebenberg, Leandi Liebenberg, Marius Loots, Chantelle Marais, Thandolwethu Mbonani, Michaela Middleton, Samantha Muller, Robert Nshimirimana, Anna C. Oettlé, Rachel Pieterse, Alison F. Ridel, Okuhle Sapo, Franci Swanepoel, Charlotte E. G. Theye, Clarisa van der Merwe, Dirk Vandermeulen, and Rudolph Venter

E. N. L'Abbé · P. N. Bothma · A. Carmichael · E. de Jager · M.-K. Erasmus · A. C. Hagg · M. Harripershad · R. Khonyane · G. C. Krüger · M. Liebenberg · L. Liebenberg · M. Loots · T. Mbonani · S. Muller · A. F. Ridel (✉) · O. Sapo · C. E. G. Theye · C. van der Merwe
Forensic Anthropology Research Centre, University of Pretoria, Pretoria, South Africa
e-mail: alison.ridel@up.ac.za

A. Alblas · C. Marais · M. Middleton · R. Pieterse · R. Venter
Stellenbosch University, Cape Town, South Africa

J. H. P. Ackermann · P. C. Hartman · N. Japhta · R. Kobedi · A. C. Oettlé · F. Swanepoel
Sefako Makgatho Health Sciences University, Pretoria, South Africa

L. Bam · F. de Beer · J. Hoffman · R. Nshimirimana
The South African Nuclear Energy Corporation, Elias Motsoaledi Street Extension (Church Street West), Brits, South Africa

M. Cazenave
Paris, France

A. Beaudet
Paris, France

E. Cunha
Faculty of Medicine, Universidade de Coimbra, Polo I – Edifício Central Rua Larga, Coimbra, Portugal

Y. Heuzé
Paris, France

Univ. Bordeaux, CNRS, Ministère de la Culture, PACEA, UMR5199, Pessac, Cedex, France

D. Vandermeulen
ESAT, Center for Processing Speech and Images, KU Leuven, Leuven, Belgium

L. Shapiro (ed.), *How to use 3D Printing Innovations and Digital Storage to Democratize Anatomy Education*, Biomedical Visualization 6, https://doi.org/10.1007/978-3-031-68501-9_3

Abstract

The *Bakeng se Afrika* (meaning "for Africa" in Sotho) digital skeletal repository is an Erasmus+ cofunded project amongst four national (South African) and four international institutions. The repository is based on twentieth-century skeletal material curated in three South African osteological collections: the Pretoria Bone Collection, the Human Osteological Research Collection, and the Kirsten Skeletal Collection. The skeletal material was scanned at the Micro-Focus X-ray Radiography and Tomography facility (MIXRAD) at the South African Nuclear Energy Cooperation, and X-Sight X-Ray Services, using a Nikon XTH 225 ST industrial computed tomography system (Nikon Metrology, Belgium). For each skeletal element, the best scanning parameters were selected according to the size and density of the bone. The original and reconstructed data is stored on a physical server at the University of Pretoria. The repository contains 3400 micro-XCT scans of both complete and partial skeletal elements, including crania, mandibles, maxillary dentition, humeri, radii, ulnae, hands, os coxae, femora, patellae, calcanei, vertebrae, and ribs. Researchers can enquire about accessing the data at the Forensic Anthropology Research Centre (forensicanthro@up.ac.za). Digital data can be applied across numerous disciplines, including biological and forensic anthropology, health sciences, and medicine.

Keywords

Digital data · Skeletons · micro-XCT · Africa Ethics · Erasmus+

3.1 Introduction

Human skeletal collections are invaluable to research and education, as they offer insights into skeletal anatomy, human diversity, diseases, and trauma (L'Abbé et al. 2021). Additionally, modern technology provides researchers the opportunity to establish digital skeletal repositories as a means to enhance inter- and trans-disciplinary partnerships and to create a global access to research (L'Abbé et al. 2021).

From 2018 to 2022, the European Union (EU) funded an Erasmus+ capacity-building project in higher education (CBHE) called *Bakeng se Afrika* (BsA), broadly meaning "For Africa" in Sotho which created the first digital repository of human skeletal remains in southern Africa. The collaborative project consisted of eight partners: four South African institutions including the University of Pretoria (UP), Sefako Makgatho Health Sciences University (SMU), Stellenbosch University (SUN), South African Nuclear Energy Corporation (Necsa), and four European institutions, namely, the Université de Bordeaux (UB, France), the Universidade de Coimbra (UC, Portugal), the Katholieke Universiteit Leuven (KUL, Belgium), and the Centre National de la Recherche Scientifique (CNRS, France).

The 4-year project aimed to establish a digital repository of micro-XCT scans composed of commonly utilized skeletal elements in research, specifically the cranium, maxilla, mandible, femur, and radius. Additionally, skeletal elements, such as ulnae, humeri, vertebrae, hands, and ribs were integrated into the repository, as these bones had been either scanned for specific research projects or collected from previous projects. The repository currently contains 3400 micro-XCT scans, which include 548 crania, 460 mandibles, 310 focus on maxillary dentition, 135 humeri, 260 radii, 128 ulnae, 259 femora, 59 os coxae, 35 patellae, hands (102 sets of distal phalanges, 45 sets of carpals, 50 sets of metacarpals), 15 calcanei, 98 vertebral columns, 57 sets of ribs, as well as 32 cases of bone trauma and 99 cases of bone pathology (Table 3.1).

Skeletal elements were selected from modern human skeletal collections of known demography at three South African medical schools, namely the Pretoria Bone Collection (PBC) in the Department of Anatomy of UP; the Kirsten Skeletal Collection (KSC) at the Division of Clinical Anatomy at SU; and the Human Osteological Research Collection (HORC) in the Anat-

Table 3.1 Skeletal elements from the pretoria bone collection, human osteological research collection, and Kirsten skeletal collection that were micro-XCT scanned for inclusion in the Bakeng se Afrika skeletal repository

Pretoria bone collection			Human osteological research collection			Kirsten skeletal collection		
	Total			Total			Total	
	n	*n*		*n*	*n*		*n*	*n*
Elements	Bones	Individuals	Elements	Bones	Individuals	Elements	Bones	Individuals
Skull and dentition			**Skull and dentition**			**Skull and dentition**		
Cranium	275	275	Cranium	68	68	Cranium	283	283
Mandible	408	408	Mandible	34	34	Mandible	30	30
Maxilla	140	140	Maxilla	37	37	Maxilla	133	133
Focus on dentition (Mandible/Maxilla)	41	41	Focus on dentition (Mandible/Maxilla)	0	0	Focus on dentition (Mandible/Maxilla)	0	0
Postcranial skeleton			**Postcranial skeleton**			**Postcranial skeleton**		
Vertebral column[a]	-	10	Vertebral column	0	0	Vertebral column☼	-	111
Ribs	0	0	Ribs	0	0	Rib[b]	-	107
Humerus[c]	100	-	Humerus	35	35	Humerus	0	0
Focus on humerus[c,d]	-	21	Focus on humerus	0	0	Focus on humerus	0	0
Radius[c]	225	-	Radius	35	35	Radius	0	0
Ulna	100	-	Ulna	28	28	Ulna	0	0
Hands[e]	102 sets	-	Hands[c,f]	95 sets	54	Hands	0	0
Os coxae[c]	51	51	Os coxae	8	4	Os coxae	0	0
Femur	129	129	Femur[g]	32	32	Femur	0	0
Focus on femur[c,d]	80	67	Focus on femur[d]	18	18	Focus on femur	0	0
Patella[c]	27	16	Patella	8	4	Patella	0	0
Feet[h]	15	15	Feet	0	0	Feet	0	0

[a] Cervical, thoracic and lumbar vertebrae were scanned. ☼cervical, thoracic and lumbar vertebrae were scanned: each vertebral column was scanned with 3 to 10 scans (391 scans in total)

[b] Entire ribcages were scanned with 1 to 6 scans per individual 53 L and 54 R(128 scans in total)

[c] For some individuals, left AND right bones scanned

[d] Only proximal and/or distal parts were scanned

[e] 51 individuals with distal phalanges

[f] 45 sets of carpals from 26 individuals, 50 sets of metacarpals from 28 individuals

[g] For some individuals, only proximal part and diaphysis scanned

[h] 15 right calcanei only

omy and Histology Department at SMU. Whilst the majority of the skeletal elements scanned, even those from previous projects, were micro-XCT scanned at South African Nuclear Energy Corporation of South Africa (Necsa), a subset of the KSC was micro-XCT scanned at X-Sight X-Ray Services.

The purpose of this paper is to discuss the history of these skeletal collections, the collaborative workflow involved in scanning and reconstructing the skeletal elements, and the ethical considerations involved in using three-dimensional (3D) images and models of skeletal remains from South Africa both nationally and internationally.

3.2 History of Twentieth-Century Skeletal Collections in South Africa

The inception of modern skeletal collections in South Africa dates back to the twentieth century when the University of Cape Town, formerly known as the South African College, established the first medical school in 1912 (Louw 1979). Since this time, modern skeletal collections in South Africa have served the educational needs of medical, dental, and healthcare students in human osteology. The medical school at the University of Witwatersrand, (Johannesburg, South Africa) was the first in 1919 to compile a human skeletal collection for the purposes of medical education and research, which is now historically known as the Raymond A. Dart Skeletal Collection (Dayal et al. 2009). These skeletal collections have played a vital role in the education of medical, dental, and healthcare students in human osteology and human skeletal variation.

In 1942, UP (Pretoria, South Africa) established a medical school and an Anatomy Department on the Prinshof campus in Prinshof, Pretoria (L'Abbé et al. 2005, 2021). Dissection room cadavers were integral in training medical and healthcare students in structural and morphological gross anatomy and osteology. In the 1990s, skeletal remains of recent cadavers were consolidated into a research collection, namely the PBC. This collection has been used to study human variation, anatomy, dentistry, bone pathology, and various research areas in biological and forensic anthropology.

At SU in the Western Cape, the intention to establish a medical school was expressed in 1946, with the Faculty of Medicine and the Department of Anatomy at Stellenbosch University only being officially founded in 1955. The University of Witwatersrand donated the first cadavers to be used for teaching medical students at SU (Alblas et al., 2018). In later years, the Department of Anatomy at SU moved to the Tygerberg campus that was established next to the newly built Tygerberg Hospital in Parow, Cape Town. Human body donations to the SU medical school are mainly from the South African Coloured communities in the Northern suburbs of Cape Town. The skeletal collection, deriving from these cadavers, was by then well established and was later named the Kirsten Skeletal Collection (KSC) in honour of the first Professor of Anatomy, Prof. JF van E Kirsten (Alblas et al., 2018). The collection primarily represents an adult population from the mid-to-late twentieth century, distinct in population origin from other collections such as the Raymond A. Dart Collection in Johannesburg and the Pretoria Bone Collection in Pretoria (Alblas et al. 2018).

The Human Osteological Research Collection (HORC) at SMU (Ga-Rankuwa) was initiated in 1978 and was named after Mr. HR Human, who had been the previous head of the Anatomy and Histology Department at SMU. Similar to the other three medical schools, the skeletal material was from dissection room cadavers and was originally used for teaching until 2018, when the material was repurposed for research. The remains in the HORC largely originate from the affiliated George Mukhari hospital in Ga-Rankuwa serving the neighbouring regions of the North West and Limpopo provinces in South Africa.

The cadavers at UP, SU, SMU, and all cadavers at all medical schools in South Africa mentioned above and the skeletal remains within the *Bakeng se Afrika* repository came from either 1) self-donations or family donations and 2) third-party donations, which is a donation from the Health Officer (aligned with the Director General), of an individual who died from natural causes in hospitals but was not collected by their family within 48 h (NHA 2003). As the person was deceased, a third party provided consent for donation of the body to the nearest medical school (NHA 2003). Human bodies that are a third-party donation are stored for up to one year prior to entering the dissection hall. Family members of the deceased can claim the body at any point in time that the body, or the associated skeletal remains, are stored at the medical school.

However, the collection of known persons by their family or the identification of unknown persons is difficult in South Africa, where approximately 1300 people are incinerated each year in Gauteng due to a lack of identification (Bloom 2015). Variable socio-economic circumstances amongst modern South Africans, migrant labour, including undocumented foreigners, and a lack of finance and infrastructure are problems for the identification of missing persons. These issues are reflected in the historic demographic profile of dissection room cadavers in medical schools and in the pressures placed upon the Victim Identification Centre (VIC) of the South African Police Services (SAPS) to identify unknown human remains.

The social classification system created within South Africa during the middle of the twentieth century largely influenced the demographic composition of these skeletal collections. Social classification had a complex and torrid history in apartheid South Africa, but for the purposes of redress and personal identification, the democratic government retained this system after 1994. Today, South Africa is home to around 62.02 million individuals, who are categorized into four socially defined groups: Black South Africans (81.4%), Coloured South Africans (8.2%), White South Africans (7.3%), and Asian/Indian South Africans (2.7%) (Statistics South Africa 2022).

Both population affinity and age noticeably affected the composition of the PBC, and other South African skeletal collections, and are a consequence of demographics (mentioned above), socio-economic circumstances of the family and regulations of the National Health Act 61 (2003) regarding self-donations, family donations, and third party donations to medical schools. Workflow for Selecting, Scanning and Reconstructing Skeletal Elements

A team of instrument scientists, who are experts on the operation of scientific instruments and data produced by the instruments, from Necsa, biological anthropologists, and medical scientists established a workflow for the micro-CT scanning of the skeletal elements (Fig. 3.1). Standard Operating Procedures (SOPs) https://www.up.ac.za/bakeng-se-afrika/article/2739025/standard-operating-procedures or sent an email (forensicanthro@up.ac.za), were written collectively amongst the South African institutions involved in the project. The SOPs were divided into four main themes, namely SOP 1 (pre-scanning procedures), SOP 2 (scanning procedures), SOP 3 (post-scanning procedures), and SOP 4 (metadata—photography).

SOP 1 Pre-scanning Procedures The first SOP describes the processes of selection, transportation, and preparation of skeletal elements for scanning. Section 1.1 of the SOP describes how the selection of skeletal elements from the three collections was done, with specific exclusion and inclusion criteria to ensure that the most complete and undamaged skeletal elements were selected for the research repository. A total of 8 teeth had to be present (4 in the maxilla and 4 in the mandible) for the maxilla and mandible to be scanned. If the crania had less than 8 teeth, or was edentulous, the cranium was scanned without the dentition. Metal dental fillings in the dentition were excluded. The long bones were selected from the left side. The right side was used only if the left side was missing. Specific skeletal elements were also selected for bone pathology and trauma, but these skeletal elements are not part of the research collection.

Section 1.2 outlines how the skeletal material was transported to the scanning facilities Necsa (for UP, SMU, and SUN), and X-Sight (for SUN).

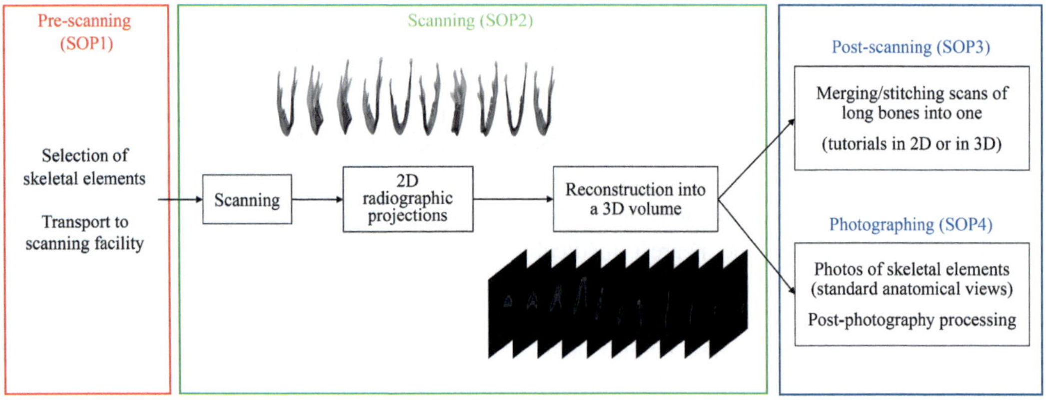

Fig. 3.1 Workflow for XCT scanning of the Human Skeletal Elements for the Bakeng se Afrika repository

Before transport, all skeletal material was checked for proper labelling, and packed tightly to avoid any potential damage. Section 1.3 explains how skeletal elements are reconstructed prior to scanning. The instrument scientists stabilized bone with Prestik, polystyrene, foam, or cardboard prior to placing them into the XCT scan. A skull was scanned alone, but the long bones were scanned two bones at a time and at various parts of the bone. For example, two proximal radii were scanned and then two distal radii—the two parts of the two bones were then stitched together using VGSTUDIO MAX.

SOP 2 Scanning Procedures The second SOP describes the different scanning modalities used, the individual scanning and naming conventions, and reconstruction procedures after scanning. Section 2.1 outlines various scanning modalities available for 3D imaging of skeletal remains. More details about the equipment can be found in Hoffman and De Beer (2012) and De Beer (2018). For each skeletal element, the best scanning parameters and the optimal number of scans per element were selected according to the size and density of the bone as per section 2.2. During acquisition, a stack of two-dimensional (2D) radiographic projections was created and reconstructed into a 3D volume using the in-house Nikon CT-Pro 3D software (Nikon Metrology, Belgium). The reconstruction was done following optimized transformation parameters adjusted for each skeletal type (e.g. cranium and femur). Guidelines for reconstruction were written for each type of bone and are available in section 2.3 of the SOP. The original stack of radiographic projections as well as the reconstructed volume of the bones were saved and stored on a physical server, which is housed and maintained at UP (Fig. 3.2). Backups of the data are also stored in a separate space within the UP system.

SOP 3 Post-scanning Procedures To ensure optimal resolution, long bones were scanned in several parts (e.g. three scans per femur, two scans per humerus, and radius). For the third SOP, the team developed step-by-step tutorials detailing how to merge and stitch several scans into one—in either 2D or 3D using ImageJ and VG Studio Max.

SOP 4 Metadata: Photography To give researchers an overview of each skeletal element, crania, and postcrania were photographed in standard anatomical views. The fourth SOP details the equipment and setup for photographing cranial and postcranial elements, as well as the post-processing of these photographs. Upon requesting use of the data, researchers are allowed access to these photographs in the form of a shared Google Drive in order to establish which individuals will be suitable for their research purposes.

To record all the information on the scans and their associated metadata, a database was created. For each skeletal element, the database includes information about the skeletal collection and

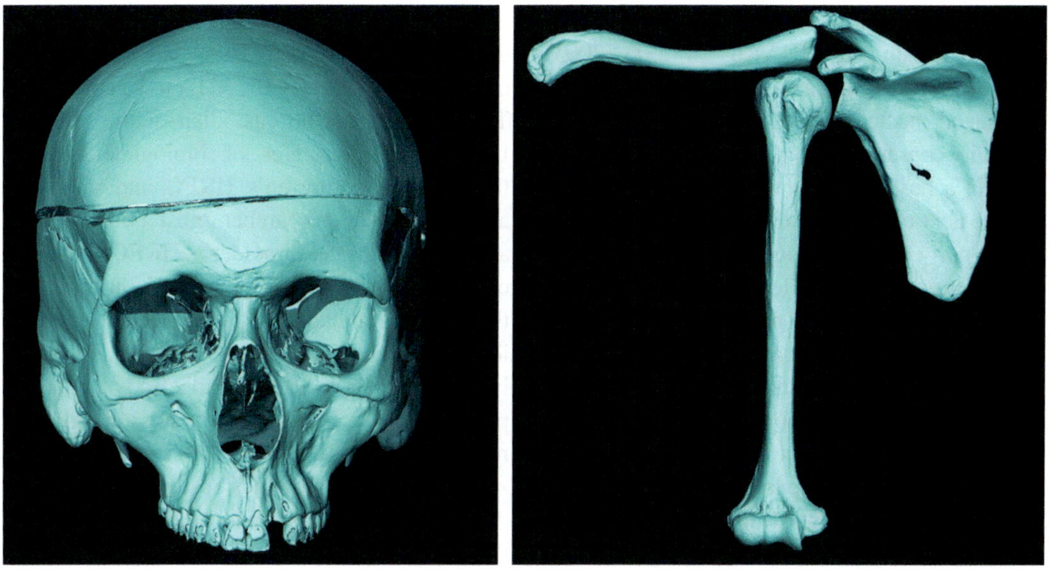

Fig. 3.2 3D models of a cranium and right shoulder joint from the *Bakeng se Afrika* repository

institution of origin, the individuals' demographics as recorded in the cadaver books of the institutions (sex, population affinity, age-at-death, cause of death, height, weight [not always available]) completed by other information recorded on the skeleton (i.e. dentition status, side, and presence of pathology/trauma). Additionally, scan information was also recorded in the database (date of scanning, name, scanning parameters, institution, collection name, cadaver number, sex, age, population origin, cause of death, date to NECSA, date of return back to the institution collection, format of the scans, recon information and date, and their storage).

3.3 Research Ethics and the Use of Human Skeletal Remains from South Africa

In the late nineteenth and early twentieth centuries, skeletons and body parts from South Africa were highly sought by researchers and collectors. During this time, indigenous South Africans, both Khoi and San, were exhumed and transported to several European museums reflecting a history of unethical treatment (Morris 1987). The historical use of human remains, body parts, and skeletons from Africa greatly contributed to South Africa's current legislation for national and international research on deceased and living humans. However, the question about the transfer and sharing of digital data is not mentioned in South African legislation. We provide an overview of the current legislation in the country that can affect the sharing of digital copies of skeletal remains of living and deceased persons, which includes the National Health Act 61 of 2003, the Protection of Personal Information (POPI) Act of 2013, and the Electronic Communication and Transaction Act 25 of 2002. We also address how a researcher can access our Human Research Ethics Committee (HREC) at UP to submit a research proposal for access to the digital skeletal collection.

3.3.1 National Health Act 61 of 2003

The National Health Act 61 of 2003 (2003), which was amended in 2013, governs healthcare structures and services in South Africa. Pertinent to the *Bakeng se Afrika* repository is the acquisition of the cadavers at three medical schools: UP, SMU, and SU.

Under Chap. 8, Sections 62 and 64 of the National Health Act (NHA) 61 of 2003 (2003), individuals can voluntarily donate their bodies to designated medical schools through their will or by family consent. Alternatively, if a person passes away in a hospital under destitute circumstances and the next of kin cannot be located within 48 h, a third-party donation from the Director General, with approval from the Health Officer, may allocate the remains to the nearest medical school (L'Abbé et al. 2021; National Health Act of 2003). Following section 64, a donation made under S64 is allowable for: (a) the purposes of the training of students in health sciences; (b) the purposes of health research; (c) the purposes of the advancement of health sciences; (d) therapeutic purposes, including the use of tissue in any living person; or (e) the production of a therapeutic, diagnostic, or prophylactic substances.

Deceased persons, whether self-donated, family donated, and/or a third-party donation become part of the "Whole Body Donation Programme" at one of the medical schools across the country. Whole-body donations operate under broad consent, which permits the use of bodies and associated data for tissue transplantation, research, and education in the future. "Into the future" can refer to, inter alia, advances in medical technology, particularly 3D modalities and imaging and the digital storage of data. Researchers must seek approval from the Human Research Ethics Committee (HREC) at the relevant institution responsible for storing and curating the data to use information from these donations (L'Abbé et al. 2021). Notably, consent from the donated person is not required in this process of research and education because the person is part of the "Whole Body Donation Programme" (NHA 2003).

3.3.2 Material Transfer Agreements of Human Biological Materials in the NHA

The Material Transfer Agreement (MTA) serves as an addendum to the NHA 2003/2013 and regulates the sharing of human material (gametes, tissue, and bone) between institutions either within South Africa or globally. Both the provider of the data and the recipient of the data have several obligations to fulfil for the MTA to be completed. The provider, as the custodian of the material, facilitates the transfer of biological material through the institution's HREC and maintains anonymity. The researcher must obtain ethical approval from the provider's HREC, maintain confidentiality and anonymity, refrain from further sharing, and notify the provider if the donor's identity is revealed. With deceased persons under broad consent within the "Whole Body Donation" programme, the individual donors' consent is not required to use the data.

According to bioethics researchers in South Africa, the MTAs are inadequate for regulating the transfer of digital data such as micro-XCT scans, CT scans, or other digital modalities (Thaldar et al. 2021). The HREC lacks the authority to enforce compliance with regulations, and the agreements offer no guidance on the intellectual property rights of the recipient or the provider. Furthermore, MTA's does not align with the Protection of Private Information (POPI) Act of 2013. Whilst digital transfer agreements (DTAs) may be proposed by the HREC, these mandates do not directly address human remains. Lastly, the digital data, such as a micro-XCT scan, are not considered a biological material under the NHA2003/2013, rendering MTAs inadequate for the *Bakeng se Afrika*'s repository on deceased and living persons (Thaldar et al. 2021). As a 3D model or a 3D print are not biological materials, the ethics of its use in research and teaching has been largely unexplored but could have implications for cultural sensitivity, ethical practice, and legal obligations (Cornwall 2016; Cornwall et al. 2016). Currently, no legislation exists in South Africa, or anywhere in the world, regarding the use of non-biological materials that represent a copy of an anonymous person in research or teaching. As one can expect, the use of AI-generated images of a human face, body parts, organs, and/or skeletal material needs to also be considered in the future.

The Electronic Communication and Transaction Act 25 of 2002 (2002) may apply to the transmission of digital copies of human remains, treating them as digital data. This entails protecting the personal information of the deceased and encrypting the data during the transfer process to ensure secure delivery and reception. Compliance with the act will also ensure that integrity of the data is preserved. A Data Utilisation Agreement (DUA) has been formulated through the Technology Transfer Office (TTO) of the UP. The DUA facilitates the sharing of digital data between providers and recipients. A copy of this DUA is supplied in Appendix 1. Each institution UP and the receiving party have to agree on how the data will be transferred, handled, and what it can be used for in the research/teaching environment. The student has to also supply their research protocol and a list of specimens which they would like to use from the repository. The DUA is part of the Ethics Constitution of the *Bakeng se Afrika* project that received ethical approval from the HREC at UP in 2021 (619/2021) and is renewed annually.

3.3.3 Protection of Personal Information (POPI) Act of 2013

The Protection of Personal Information Act 4 of 2013 (2013), commonly referred to as the POPI Act, governs, and mandates the consent, privacy, and confidentiality of scanned data (CT, CBCT, MRI etc.) associated with living individuals. However, it is important to note that this act does not apply to deceased individuals who have donated, or been donated, via broad consent, as discussed in the NHA. When personal data of living persons is de-identified and securely stored within repositories like *Bakeng se Afrika*, research collaboration becomes less restrictive, requiring ethical approval from the relevant HREC, in this case, the University of Pretoria's HREC. The POPI Act makes provision for and allows researchers to apply to South African HREC's for access to digital data of living individuals housed within repositories at the institution in which that repository is housed. Transfer of personal data outside the borders of South Africa is permissible, provided that the receiving country offers equal or better levels of protection for processing the personal information, as stipulated in Chap. 9, of the POPI Act (2024) regarding transborder information flows. Similar to The Electronic Communication and Transaction Act 25 of 2002 (2002) on digital data, data transfer occurs under the condition that the recipient agrees not to further transfer the data and to maintain data anonymity.

3.3.4 National Heritage Act 25 of 1999

The National Heritage Act 25 of 1999 (1999) safeguards heritage artefacts, including archaeological material and archaeological skeletal material. This Act also covers all human remains not previously covered in the National Health Act of 2003 (2003). However, the Act does not specifically address the use of 3D digital modalities for collecting and preserving material, particularly skeletal remains. Archaeologists require a permit from the South African Heritage and Resource Agency (SAHRA) to excavate sites older than 60 years, which may include human skeletal remains. In order to use excavated archaeological skeletal material in research and/or education or for its preservation, the researcher must also submit a research protocol to SAHRA for approval. This research protocol should detail the type of research being conducted (invasive or non-invasive) and specify any 3D surface scans, micro-XCT scans, or other types of imaging modalities that will be used. The use of digital data is a way to conserve and preserve historic skeletal material, which is in alignment with the National Heritage Act 25 of 1999 (1999).

3.3.5 Access to the Bakeng se Afrika Data Repository

A data repository is considered "*a central secure place, or digital warehouse, where data together with their metadata are stored and maintained. A repository can be a place where multiple databases or files are located with access through the Internet*" (Data Management Policy,

University of Pretoria). All data collected, created, or generated by students and employees of UP through the *Bakeng se Afrika* project belongs to UP and must be stored and maintained in perpetuity at the institution. The data is managed and curated effectively on its own server at UP. Access and use to the repository are aligned with the research data management policy at UP, the NHA 2003/2013, and the POPI Act of 2013 (2024).

3.3.6 Human Research Ethics Committee (HREC) in the Faculty of Health Sciences at the University of Pretoria

In accordance with section 72(3) of the NHA, any researcher seeking access to data on humans, living or deceased, must obtain ethical approval through an HREC at the institution housing those remains. For access and use of the *Bakeng se Afrika* repository, a researcher must submit an online application to the HREC of the Faculty of Health Sciences at the University of Pretoria (https://www.up.ac.za/healthethics). The rules of the HREC are aligned with the Department of Health Guidelines (DOH) for Ethics in Research, the NHA, and the POPI Act of 2013. A research proposal is required from the applicant and the following stipulations need to be met in the applications:

- Verification of applicant / supervisors/principle / institution.
- All researchers need to acknowledge the *Bakeng se Afrika* repository in their published outputs and social media.
- All researchers and their institutions need to enter a Digital Utilisation Agreement (DUA) with the UP undertaking not to: share the data with other parties, create commercial use of the data or derivatives of the data (see Appendix 1).
- All researchers need to maintain anonymity of the data whilst it is in their possession.
- Data cannot easily be shared outside of UP due to firewall restrictions. Researchers can contact the Forensic Anthropology Research Centre (FARC). (forensicanthro@up.ac.za) and decide for data to be shared via hard drives.
- All projects prior to submission to the HREC at the UP will be peer reviewed within the Forensic Anthropology Research Centre for adherence to good practice. To obtain an approval letter from the BsA repository for the purposes of your HREC application, please send your research proposal to the Forensic Anthropology Research Centre, forensicanthro@up.ac.za

3.4 Expected Roles of the Researcher

- Supply publications and outputs of research from the *Bakeng se Afrika* Repository.
- Acknowledge Bakeng se Afrika repository within publications and dissemination.
- Add the following to the acknowledgements of their manuscript:
- *Any opinions, findings, and conclusions or recommendations expressed in the material are those of the authors, and therefore, Bakeng se Afrika repository does not accept any liability in regard thereto.*

3.4.1 Code of Conduct in Ethics

The *Bakeng se Afrika* repository housed within the Faculty of Health Sciences at the University of Pretoria is compliant with the UP Code of Ethics in Research. Researchers who use this repository need to comply with the code of ethics for researchers in South Africa. This ensures the responsible conduct of research and upholds ethical standards. The latest version of the UP Policy and Procedures for Responsible Research (S4083/00) and UP Code of Ethics in Research (Rt429/99) can be obtained from: forensicanthro@up.ac.za.

3.5 Discussion and Conclusion

The *Bakeng se Afrika* repository contains more than 550 individuals, which comprises 3400 micro-XCT scans, and is housed at the University of Pretoria in Gauteng, South Africa.

Through a collaborative effort involving multiple South African and European institutions, and with funding from the EU Erasmus+ Capacity Building project (597924-EPP-1-2018-1-ZA-EPPKA2-CBHE-JP), we developed the first digital repository of modern human skeletal remains in southern Africa. Six national researchers and one international researcher utilized the data in the Bakeng se Afrika repository during its development (2019–2022). Since the launch of the repository in 2023, an additional four international research teams have received approval to use the data.

We consolidated data from three valuable skeletal collections in South Africa. In completing this project, we had to address the historical context of skeletal collections in South Africa, the collaborative workflow involved in scanning and reconstructing digital skeletal elements, and the legal and ethical considerations required to use 3D images and models in research and education.

South Africa's medical school history outlines the evolution of medical education and research practices within the country's legislation, both past and current. The three medical school skeletal collections have collectively contributed to advancing knowledge in human osteology, and understanding of human variation for establishing biological profiles, medicine, bone pathology, and secular trends in the region, to name but a few.

All skeletal elements scanned and collected during the *Bakeng se Afrika* project are the IP of the UP, where the data is stored and maintained into perpetuity. Access and use of the collection are aligned with UP's research data management policy, NHA, and POPI Act of 2013. Researchers who want to access the data must obtain ethical approval from a HREC in the Faculty of Health Sciences at UP, as per section 72(3) of NHA. A research proposal is required, with various stipulations aligned with good data management practice. This digital archive not only facilitates research within the participating institutions but also enables global access to this invaluable resource.

Micro-focus X-ray computed tomography (XCT) technology provides high-quality micron-level data for non-invasive and non-destructive research, making this the gold standard modality for external and internal structure assessments (Hoffman and De Beer 2012). The instrument scientists at Necsa designed a workflow for scanning the selected skeletal elements (e.g. crania, radi, and femora) and skeletal elements for project-specific data (e.g. vertebra, patellae, and humeri). The Nikon XTH 225 ST XCT system provides a resolution between 5 and 127 microns, which also varies for bone type. For example, a skull is between 90 and 120 microns, whereas a femur is between 40 and 90 microns (Bouxsein ML et al. 2010). Furthermore, in order to avoid rescanning specimens for other research projects, scanning protocols were optimized and standardized for each skeletal element, following detailed SOPs. The scanning parameters for each specimen are available to researchers in the form of an XTEKCT file which includes the voxel size (x,y,z).

Ethical considerations surrounding the use of human skeletal remains in research and education are paramount in South Africa, especially given the historical context of exploitation and illegal practices with human bodies and human skeletons. Whilst South African legislation does not dictate how a 3D image or model can or should be used in research and education, our approach offers an important foundation for the development of legislation around the appropriate use of 3D models and images of human skeletal remains into the future in South Africa and abroad. All skeletal elements in the repository were legally obtained in accordance with the NHA 2003/2013 in South Africa. The repository is representative of three socially identified South African groups, namely Black, White, and Coloured South Africans (L'Abbé et al. 2021; Statistics South Africa 2022). In the coming years, new skeletal material is to be added from the PBC, from living persons (anonymized CBCT data), and from skeletal material with bone trauma and pathology.

Worldwide, substantial variation exists in opinions concerning the acquisition, display, and utilization of human skeletal remains, and the considerations regarding public and private use of human bones are ongoing debates within diverse socio-cultural and socio-political contexts and agendas. In establishing a well-defined ethical framework through the *Bakeng se Afrika* project, digital copies of human remains are preserved and used within teaching and research environments; require ethical approval from the host institution (UP); and require a DUA that addresses how the digital remains can be appropriately shared and used in research. Additionally, a digital copy of a skeletal element, whether a 3D image or a 3D print, is not a biological material and is a copy of the original biological material. A copy without a biological source cannot be considered the same as a biological skeletal element, as shown by bioethics discussions on the transfer of biological materials with MTAs and digital materials with DUAs.

Ethical considerations remain paramount in the use and sharing of digital skeletal data, necessitating adherence to legislation and responsible conduct in research. Moving forward, continued efforts to address the use of skeletal remains in South Africa will advance scientific knowledge and will promote ways in which we can define the use of digital data of human skeletal remains in research and education into the future.

The *Bakeng se Afrika* digital skeletal repository is a milestone in human skeletal research and education in southern Africa. The digital database is a valuable resource for national and international researchers and educators. The *Bakeng se Afrika* repository has been used by researchers with applications in advancements in medical technology such as dentistry for the conception of dental implants as well as forensic anthropology and biological anthropology for the development of advanced methodologies for biological profile estimation and evaluating human variation. Modern digital human remains have also been used in paleoanthropology as a comparative sample for fossil anatomy. Additionally, over 60 postgraduate southern African researchers obtained advanced skills in 3D modelling, 3D printing, and statistical analysis during this EU Erasmus+ project, which enabled students to work with large datasets and to apply their knowledge to biological anthropology, medicine, and healthcare sciences.

Acknowledgement The development of the Bakeng se Afrika digital skeletal repository and all supplementary materials was co-funded by the Erasmus+ programme of the European Union [grant number 597924-EPP-1-2018-1-ZA-EPPKA2-CBHE-JP]. Its contents are the sole responsibility of Bakeng se Afrika consortium and do not necessarily reflect the views of the European Union.

References

Alblas A, Greyling LM, Geldenhuys E (2018) Composition of the Kirsten skeletal collection at Stellenbosch University. S Afr J Sci 114(1):1–6. https://doi.org/10.17159/sajs

Bakeng se Afrika Standard Operating Procedures Manual. Available at: [https://www.up.ac.za/bakeng-se-afrika/article/2739025/standard-operating-procedures]. Accessed 9 March 2024

Bloom J (2015) 1272 unidentified bodies in Gauteng Mortuaries in 2014/2015. [WWW Document]. Politicsweb. URL:http://www.politicsweb.co.za/politics/1272-unidentified-bodies-in-gauteng-mortuaries

Bouxsein ML, Boyd SK, Christiansen BA, Guldberg RE, Jepsen KJ, Müller R (2010) Guidelines for assessment of bone microstructure in rodents using micro–computed tomography. J Bone Miner Res 25(7):1468–1486

Cornwall J (2016) The ethics of 3D printing copies of bodies donated for medical education and research: what is there to worry about? Australas Med J 9(1):8

Cornwall J, Callahan D, Wee R (2016) Ethical issues surrounding the use of images from donated cadavers in the anatomical sciences. Clin Anat 29(1):30–36

Dayal MR, Kegley ADT, Strkalj G, Bidmos MA, Kuykendall KL (2009) The history and composition of the Raymond A. Dart collection of human skeletons at the University of the Witwatersrand, Johannesburg, South Africa. Am J Phys Anthropol 140:324–335. https://doi.org/10.1002/ajpa.21072

De Beer FC (2018) Neutron and X-ray tomography as research tools for applied research in South Africa (Doctoral dissertation, North-West University)

Electronic Communication and Transaction Act 2002 (South Africa). Available at: [https://www.gov.za/documents/electronic-communications-and-transactions-act]. Accessed 08 March 2024

Hoffman JW, De Beer FC (2012) Characteristics of the micro-focus X-ray tomography facility (MIXRAD) at Necsa in South Africa. In: 18th world conference on non-destructive testing, 16 to 20 April, Durban, South Africa, pp 16–20

Human Tissue Act of 1983 (South Africa) Available at: [https://pubmed.ncbi.nlm.nih.gov/12289616/] Accessed 8 Mar 2024

L'Abbé EN, Loots M, Meiring JH (2005) The Pretoria bone collection: a modern south African skeletal sample. Homo 56:197–205. https://doi.org/10.1016/j.jchb.2004.10.004

L'Abbé EN, Krüger GC, Theye CE, Hagg AC, Sapo O (2021) The Pretoria bone collection: a 21st-century skeletal collection in South Africa. Forensic Sci 1(3):220–227. https://doi.org/10.1016/j.jchb.2004.10.004

Louw JH (1979) A brief history of the medical faculty, University of Cape Town. S Afr Med J 56(22):864–870

Morris AG (1987) The reflection of the collector: San and Khoi skeletons in museum collections. S Afr Archaeol Bull:12–22

National Health Act (2003) Chapter 8 (South Africa). Available at: [https://www.gov.za/documents/acts/national-health-act-61-2003-23-jul-2004] Accessed 8 Mar 2024

National Heritage Act 25 of 1999 (South Africa). Available at: [https://www.gov.za/documents/acts/national-heritage-resources-act-25-1999-28-apr-1999]. Accessed 8 Mar 2024

Protection of Personal Information Act 2013 (South Africa) Available at: [https://www.gov.za/documents/protection-personal-information-act] Accessed 8 Mar 2024

Research Data Management Policy (2018) University of Pretoria. [https://www.up.ac.za/media/shared/12/ZP_Files/research-data-management-policy_august-2018.zp161094.pdf] Accessed 1 May 2024

Statistics South Africa (2022) [Census 2022 Results]. Statistics South Africa. Available at: [https://www.statssa.gov.za/?m=2021] Accessed 8 Mar 2024

Thaldar D, Townsend B, Botes M, Shozi B, Pillay S (2021) A virtual deliberative public engagement study on heritable genome editing among south Africans: study protocol. PLoS One 16(8):e0256097

Bone Flute: An Art-Science Research Project

4

Ralph Borland

Abstract

This chapter explores an art-research project that uses biomedical imaging to produce an artwork and exhibition while investigating a research topic. The author is the artist-researcher who designed and produced the project in collaboration with an orthopedic surgeon and a musician, while a research fellow investigating the use of emerging technologies in healthcare in Africa at the University of Cape Town. The author became a medical patient due to a diagnosis of illness during the course of the project, and brought this material into their work. The project centers on the production of the artwork *Bone Flute*, a replica of the author's femur made into a flute, that was made in a public hospital using the surgeon's tools and three-dimensional printing (3Dp) processes. This object formed the centerpiece of a public exhibition in conclusion of the research fellowship, in an artist-run gallery space. The chapter takes the form of a personal essay which explores each of the author's roles as researcher, artist, and patient. It focuses on a number of collaborative relationships that enabled the realization of the artwork, and frames the project, and the chapter itself, as a form of storytelling.

Keywords

Art-science · Interdisciplinarity · Music · Sculpture · Collaboration · 3D-printing

R. Borland (✉)
HUMA - Institute for Humanities in Africa,
University of Cape Town, Cape Town, South Africa

L. Shapiro (ed.), *How to use 3D Printing Innovations and Digital Storage to Democratize Anatomy Education*, Biomedical Visualization 6, https://doi.org/10.1007/978-3-031-68501-9_4

An artwork in tabloid poster style, highlighting the sensational aspect of the artwork *Bone Flute*, on the exhibition *AIAIA—Aesthetic Interventions in Artificial Intelligence in Africa* (2023)

4.1 Introduction

In a laboratory in a public hospital in Cape Town, three experts[1] are handling a replica of a human bone. They measure, cut, and drill. Tools pass from hand to hand. In the background, a 3D printer whirrs, building another replica of a bone, layer by layer in white plastic. The fourth expert in the room, a filmmaker, documents this operation from behind their video camera. Weeks later, the film they make will be on exhibition in an art gallery in the city of Cape Town, South Africa. Here in this room, hours pass. One of the experts, a musician, picks up the object they have worked on—a femur—and holds it to their lips. Blowing across a hole near the ball of the hip, their fingers move over a series of holes spaced along the shaft of the bone. Musical notes fill the room with a breathy, haunting sound. Their work complete for now, all four-the filmmaker, the musician, a surgeon and an artist-pack up their gear, and the room empties. The only sound remaining is the continued whirr of the printer, blindly completing its task (Fig. 4.1).

I'm an artist and researcher, and one of the experts in the scene described above, which took place at Tygerberg Hospital in Cape Town in early 2023, toward the end of a research fellowship taking place over the preceding 2 years. My expertise is in the investigation and communication of concepts to audiences using symbolic language—especially objects as symbols. The replica bone we worked on is 3D printed from an MRI scan of my femur, and this operation was part of the final stages of turning it into a flute. The artwork, *Bone Flute*, is a memento mori: a reminder of death. My collaborators are an orthopedic surgeon, Rudolph Venter, whose laboratory we worked in, and a musician, Alessandro Gigli—a classically trained flutist and instrument maker. The filmmaker behind the camera is my partner, Dara Kell, then 6 months pregnant with our first child, collaborating with me to produce a short film for exhibition. And, at the time of this performance, I was in my first round of chemotherapy, having been recently diagnosed with colon cancer, for which I underwent surgery and would need a few months of treatment with powerful drugs.

This book chapter is a personal essay, in which the emphasis is on my exploration of my own experience, using the knowledge I acquired while on a journey as a researcher, artist, and patient, with the object *Bone Flute*. It's about the uses of biomedical imaging for art and in healthcare, and the relationships that enabled the work; and it is both about storytelling and an act of storytelling itself.[2]

[1] I'm influenced here by the surgeon Roger Kneebone's book *Expert* (2021) which looks at multidisciplinary collaborations between experts. His "Thread Management" workshop, for example, brought together surgeons, embroiderers, puppeteers, fly-fishermen and engineers around the theme of surgical suturing, via the expert handling of thread.

[2] This sentence lists three topics: technology, social relationships, and storytelling. This triumvirate is a touchstone in my work, inspired by an account I read at a museum of humankind near Cape Town, which proposes these three characteristics as the essential attributes of being human: we are social, storytelling creatures that use technology.

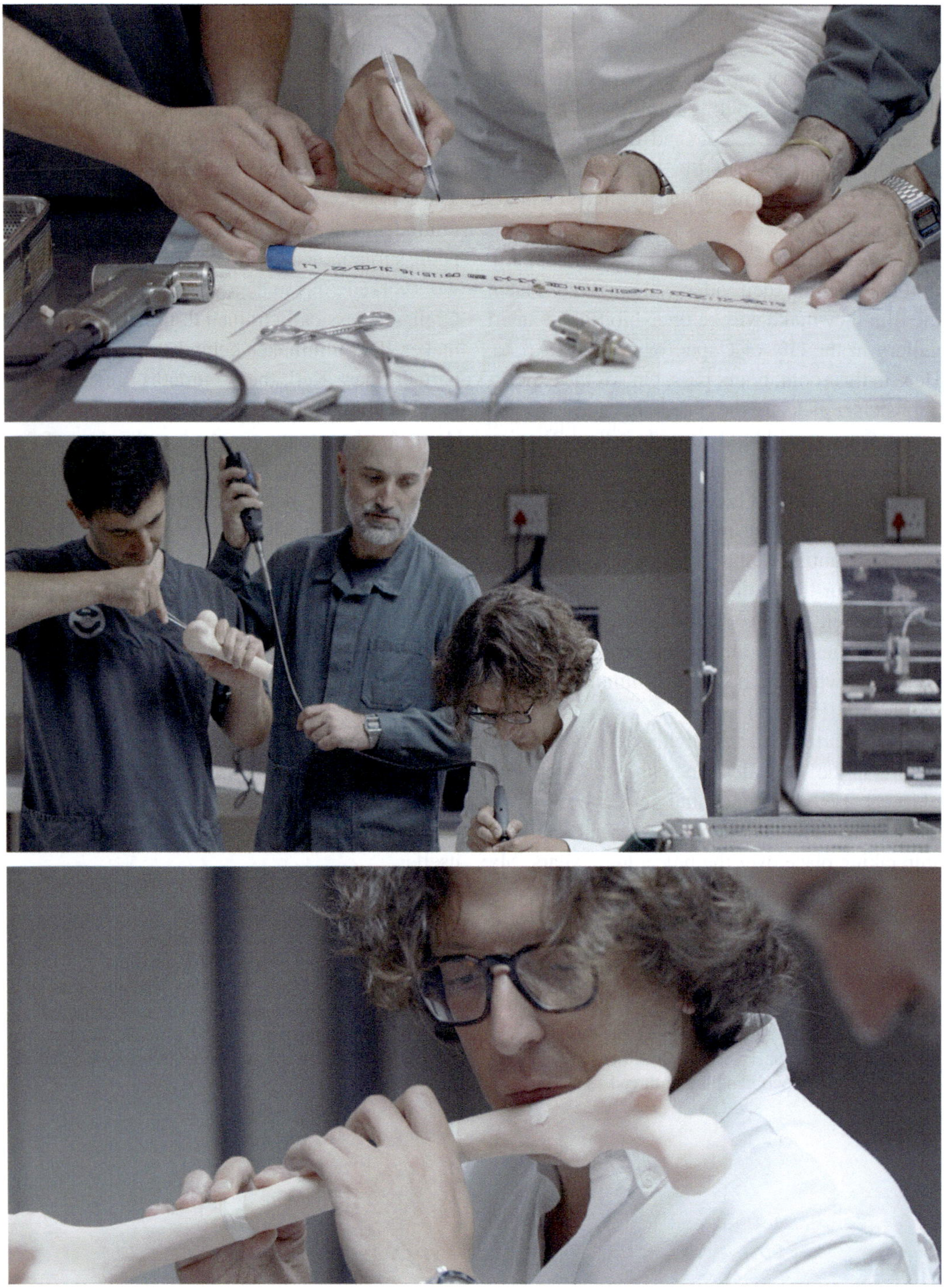

Fig. 4.1 Composite image showing stills from a short film by Dara Kell about the making of Bone Flute (2023) showing Rudolph Venter, Alessandro Gigli, and Ralph Borland

4.2 Researcher

I undertook this work as a research fellow in a humanities research institute at the University of Cape Town. HUMA, the Institute for Humanities in Africa, was in 2022 in the process of reinventing itself as a Pan-African research institute under new leadership, after having fallen dormant since its establishment some 10 years previously. I was among the first cohort of new recruits, joining a research project funded by the Carnegie Corporation of New York called "Future Hospitals - 4IR and Ethics of Care," which asks how new technologies (particularly those of the "4th Industrial Revolution") are changing the work of healthcare professionals in Africa. The "fusion between the human body, new technologies and the experience of being human," it proposes, requires critical attention (HUMA 2024).

My proposal to HUMA was to devise creative and collaborative ways of working with hospital healthcare workers to elicit their ideas about the future of healthcare. Soon after I joined the institute, I posted a call on my Instagram account asking for leads to collaborators. An art school colleague introduced me to their friend Rudolph Venter, an orthopedic surgeon and lecturer in the Faculty of Medicine at Stellenbosch University, who was making low-cost 3D-printed replicas of his patient's bones in his laboratory at Tygerberg Hospital in Cape Town, to use in rehearsals of complex surgical operations (Branson et al. 2021). I already worked as an artist with computer-controlled fabrication technologies, including 3D printing, and anticipated the potential for collaboration this introduction might offer.

My first conversations with Rudolph were informed by some of my reading at the start of my fellowship. One of my key texts in learning more about the impact of emerging technologies on work is Nicholas Carr's book *The Glass Cage* (Carr 2015). Carr's concern is not just with the replacement of human workers by machines, but with the impact on our experience of work when even some functions are taken over by machines. In *The Glass Cage* he explores automation, as a particular type of technological process that takes over functions that might previously have been executed by a person, focusing particularly on computer-controlled automation.

Carr's entry point is in the field of aviation—the "glass cage" of the title is a play on the glass box that contains the pilot in an airplane, surrounded by instrumentation and feedback displays. In early airplanes, the pilot had direct control of their flying machine. Through systems of wires and rods, they could receive direct haptic feedback from distant parts of the airplane, and they could control the movement of the plane directly too, using their bodies. When sensor devices and actuators were developed that intervened in this direct feedback and control, pilots lost part of their felt relationship to their machines. The onward trajectory for airplanes was increasing automation, leading to the role of today's airline pilot, who is largely a custodian of a mostly computer-controlled machine (Carr 2015).

One of the results of this process is the deskilling of the human operator of technology, who ends up alienated from their work, and Carr goes through other examples in which people's skills are diminished by having tasks taken over by machines: from architectural software which tends toward uniformity in building design, to the ubiquitous map applications on our mobile devices, which tend to reduce our way-finding abilities with potentially wide-ranging impacts on our cognitive abilities (2015). One of our desired "experiences of being human," to incorporate the Future Hospitals rubric, is to experience satisfaction in work, including enjoying states of "flow" whether in work or in other experiences (such as playing a musical instrument, running, or rock climbing). One of Carr's concerns with computer-controlled technologies is that by automating aspects of our experience, they diminish the potential for achieving flow.[3]

Working with his surgical team, Rudolph would practice surgical operations several times

[3]The principal researcher and writer on flow, who Carr references, is psychology professor Mihaly Csikszentmihalyi, who published *Flow: The Psychology of Optimal Experience* in 1990.

on 3D prints of patients' bones, so that they would have already solved potential problems that might come up in the surgery itself. This assisted surgeons with their work, especially with patients with atypical anatomy or complex cases, by reducing theater time, reducing blood loss, and reducing X-ray usage in theater (Morgan et al. 2020; Venter et al. 2022). Rudolph hadn't invented this technique but was focused on making it more affordable and so accessible to patients in public hospitals in South Africa. In describing his work, Rudolph foregrounded "embodiment" as one of the benefits of this approach. Surgeons develop skills in and with their bodies that do not reside just in the mind: their ability to manipulate tools and assess material, with trained hands, eyes, ears as well as cognition. These skills are built through experiences that utilize the whole multisensory person of the surgeon—and so surgical simulations should aim to reproduce, or stimulate, as many of these sensory cues as possible.

Simulations are useful for allowing the surgeon to work out approaches to the operation without risk to the human patient, but the more "haptic" (felt) cues that a simulation contains, the better for engaging these accumulated, embodied skills. This distinguishes a screen-based reference for pre-operative planning from approaches using physical models, which allow the surgical team to work together using the instruments they are accustomed to, on the physical copy of the patient's anatomical part or feature. Rudolph's description of the surgeon's state of total absorption in their work during operations (R. Venter, personal communication, September 29, 2021) reminded me of Carr's description of flow, and the use of 3D prints as a way of employing computer-aided technology in ways that did not de-skill the user—retaining the hand skills of the surgeon rather than replacing them—served as a counterpoint to the negative impacts of automation identified by Carr. My reflection on this to Rudolph resonated with how he understood his work.

Rudolph had been introduced to me as a doctor interested in the work of artists. In what our mutual friend who introduced us described as "a lovely balance of technology," Rudolph and his students took haptic drawing classes with an artist I know[4] in which they explored the 3D form and textures of bones using their sense of touch and drawing with graphite. Our first conversations were wide ranging, from sharing work experiences, to discussing the social impacts of technology, and talking about art, music, and fiction. An issue I've long been interested in is technology choice: the potential for people to choose which technologies they use based on their own assessment of their social benefit, rather than simply availability or fashion. Rudolph responded by recommending that I reread the science-fiction novel *Dune* (1965) for its theme of technology choice. I did just that, encountering the book's depiction of a multiplanetary society which has a prohibition on "thinking machines," rejecting computers entirely—something I'd missed in my first reading of the book as a teenager.

In our first conversations, I didn't know precisely how I would work with Rudolph on my research project—just that so far he represented an *entre* to the hospital, and a firsthand informant about the use of emerging technologies in healthcare. I hoped that we could work on something creative together, where I could bring my skills as an artist into a relationship with him as a surgeon working with technology. But I knew that there was a special "something" to our relationship: I enjoyed the multiple levels at which we were able to engage, and I appreciated his warmth and openness to discussing ideas. He appeared to me, from my experience of interdisciplinary collaboration, as one of those valuable figures found across disciplines who are interested in reaching out to people from other fields: and he spoke about collaboration across disciplines as one of his practices, saying "chatting opens doors."

I related too to his identification with human-centered applications of emerging technologies, as this resonated with my own background and ongoing practice as an artist critically engaged with the application of new technologies for cre-

[4] Leonard Shapiro, who teaches the Haptico-visual observation and drawing (HVOD) method, is the editor of this volume.

ative purposes; with art as a medium for collaboration and research; and an interest in the body as material for art. Within a few months of meeting Rudolph, I had revived a past idea for an artwork that involved scanning and reproducing my femur, and determined that we could realize it using his processes: I would have my leg scanned in the hospital, and printed under his supervision. This seemed like a useful confluence of objectives, in representing a concrete project we could work on together that would require me to spend time in his workplaces, and find out more about his ways of working, while making an artwork that could direct the attention of wider audiences both to his work and that of my research project.

4.3 Artist

Bone Flute started out as an idea for another artwork, *Club*, some 10 years previously. I had sketched it out and mocked it up back in 2011 as an independent artwork soon after finishing my PhD (a critical design study focused on the South—see Borland 2011). *Club* was intended to present a replica of my femur as the most basic of weapons, the club. I felt that this artwork might perhaps carry implications about our innate potential for violence. When I shared the idea, several people pointed out to me the opening scenes of Stanley Kubrick's iconic science-fiction movie *2001:A Space Odyssey* (1968) in which an early hominid is introduced to the use of violence, represented by their wielding of a femur as a weapon (I'd seen the movie a long time before, so perhaps the image was embedded in my mind already). The hominid tosses the bone up into the air after smashing a pile of bones with it, and as it spins in mid-air, the film cuts to a space station orbiting the earth, suggesting the links between violence, weaponry and advanced technology.

I have a long relationship with technology in my artwork as an artist and researcher, and particularly with the relationship between technology and the body. As an art student, in completion of my undergraduate degree in sculpture, I used salvaged electronics and synthetic material waste to build bone-like sculptural sound installations, through which I deejayed live, on an industrial site in Cape Town (*Transwerk*, 1997). As a Master's student in Interactive Telecommunications (New York University, 2002), I produced a series of projects in which I turned myself into a human turntable stylus, crafting attachments to my arm that enabled me to "read" the color and textures of my environment and turn them into sound, via a microphone, a miniature video camera, electronics, and computer software (*IRSampler*, 1999–2001). My Master's thesis work was a protective suit for protestors, which amplified the wearer's heartbeat via a loudspeaker and made it audible outside their body (*Suited for Subversion*, 2002).

This last project (*Suited for Subversion*, 2002) led me to focus in my PhD work on objects designed to have functions for the user, and also to communicate to audiences: including types of functional and interventionist art, and some forms of design, such as "discursive design," which focus explicitly on communication to audiences, as well as other forms of design that are ostensibly user-centered but rely on their storytelling capabilities to audiences for attracting funding and support. *Suited for Subversion* was particularly successful at communicating with audiences. Selected for the exhibition *Safe* at the New York Museum of Modern Art in 2005 and later bought for their permanent collection, it attracted a great deal of media attention, and was continuously on exhibition for many years afterward. Through being on exhibition, it reached much wider audiences than in its immediate context of use (street protest) and conveyed information about tactics, situations, and subcultures via a designed object that functions in some ways like a text, using form and symbol (and also, accompanied by written text). The object (the protective suit) has a relationship both to its user and to wider audiences.

Suited for Subversion had led directly to my concept for *Club*, and later *Bone Flute*: they were the result of a conscious intention to extend a theme in my art production that it represented: the repurposing of body parts or bodily functions. *Suited for Subversion* captured my heartbeat and made it audible in real time outside my body; my heartbeat became audible material for the artwork. What other body parts or functions could I use in this way? I thought about bone as material—this hard, persistent material that outlives us, and can be made into functional objects (I was later to learn from my medical collaborators that orthopedic surgeons are sometimes called "the carpenters of the medical world" because of this quality of bone). The femur is the largest bone in our bodies and so seemed to offer the most potential utility.

I revived this artwork and these concepts for my research project for *Future Hospitals*, using it as a vehicle for collaboration with Rudolph. In the intervening years since the idea first came to me, I had refined the concept of *Club* further, telling a different story with it—a more sophisticated use for the bone replica—a project to make it into a flute. I knew that I'd seen the image of a flute made from a human femur somewhere; that it was an object that was out there in the world. The first visual images that formed in my mind's eye were of a skeleton playing a flute made from a human bone, but I couldn't recall where I had seen it.

Through my research, I recognized that I was tapping into a range of influences that I'd been exposed to: such as Hans Holbein's woodcuts, from his sixteenth-century book *Dance of Death*, which includes many music-making skeletons—an example of the medieval *danse macabre*. Holbein's later painting *The Ambassadors* (1523) is an iconic example of the memento mori with its distorted image of a skull intruding into a portrait of wealthy and powerful men, reminding us that they are as mortal as the rest of us. I'd been introduced to this painting at art school. The *danse macabre* appears in more recent popular culture, such as early black and white Disney cartoons in which skeletons in a graveyard dance and play a range of instruments made from human bone (*The Skeleton Dance*, 1929). Fritz Lang's early science-fiction movie *Metropolis* (1927) of which I had seen a screening with live instrumentation, includes an incidental scene in which a human skeleton plays a femur flute. Rudolph later sent me a scan of a page from a Tintin comic (*Prisoners of the Sun*, 1948) that his child was reading, in which an Incan whistle made from a human tibia appears—and I realized that I had read the same comic book when I was a child, and that it was still on my bookshelf.

I came across archaeological and anthropological material documenting flutes made from human bone used in a range of locations and cultures, some still in practice today such as the Tibetan *Kangling*, a ceremonial flute made from a human femur, and I found confirmation of human bone flutes found in many different parts of the world, from North and South America to Europe, Asia, and Oceania. *Kangling* is made from the bones of holy men or of criminals, and this dichotomy is surmised in other examples of the use of human bones too: they might be the bones of loved ones to keep them close, or the bones of enemies to hold power over them after death. And apart from human bones, flutes made from the bones of animals are the earliest examples of musical instruments yet identified by archaeologists.

All of this is to say that I found confirmation that this idea for an artwork was tapping into a rich vein of symbolic material, spanning wide histories and geographies, and across cultures. These are signs to me of good art-making material: resonant images and symbols that will effectively attract the attention of wide audiences, by reminding them of "something," even if they are not consciously aware of what that is (as I wasn't at first) appealing to us on a subconscious level with images that are proven to be persistently memorable, and that can draw audiences into the more specific information I might want to expose them to. These types of symbols are useful for arresting attention, and for evoking emotions, through sets of associations. This is archetypal and provocative material: bones normally become visible only after death, and human bones are objects charged with symbolism and significance—making them into functional objects produces a frisson, a boundary-breaking that evokes both sacrilege and ceremony.

Making my own bone (albeit a 3D-printed replica) into an instrument carries its only particular meanings. Making someone else's bone into an instrument is fraught terrain: it depends very much on context and relationship to determine the ethics of doing so. I think that making use of one's own body for material is disarming. It's an act of self-possession rather than of exploitation. Such an act can still be provocative, and evoke feelings of shock, horror, or revulsion in audiences, as in body-based works by artists such as Stelarc, who uses technological apparatus to pierce, manipulate, and permanently alter his flesh. However, I intended to use technology to work on my body in a different way: to use digital technology's capacity for reproduction and simulation to painlessly extract and reproduce a hidden part of my anatomy. While a copy of my bone could now be seen and manipulated outside my body, the original would remain in place.

Normally, a person would have to be dead to have a tune played on their bone. I intended to create a situation in which a living person could do so. While Rudolph's use of technology allows a preview into a living body before opening it up through surgery, in order to heal, my use of the technology implies a preview of the body after death, when the flesh has disintegrated from around the bone. A phrase that came into my mind in devising the work is that I would be making myself "the Pied Piper of my own demise," playing on another association that the flute evokes—as an instrument for charming and enticing. The Pied Piper of Hamelin in the European fairy tale I heard as a child (recorded in many sources, e.g., Grimm 1816/1818) stole the villagers' children, vanishing them into the earth. Was I telling a story through my own artwork that I was leading myself toward my own death? I imagined myself naked, performing in a spotlight on a dark stage, playing a haunting tune on a flute made from my own bone. This was certainly an artwork about mortality in general and my own in particular: the bones beneath the skin, and the inevitably of death—a memento mori. But this work would also be about music and emotion, about the poetry that comes from our human awareness of death, and the desire to celebrate life in the face of it—one of the messages contained in the *danse macabre* or the Mexican *Day of The Dead.* The "healing" in my artwork, in comparison to Rudolph's healing of the physical body through surgical intervention, would not be healing for the patient in a hospital, but perhaps a psychic healing for individuals in audiences as a result of the story it tells. It would also, it turned out, become a source of healing for me (Fig. 4.2).

Most of these thoughts, ideas, and interpretations of what this artwork would mean, appeared to me through the process of research and design over many years, before the object viz. the flute, was made. In order to move the making of the flute from proposal to object would require more work, drawing in further material, and producing unforeseen outcomes. This required opportunity, collaboration, relationship building, processes and technologies, objects, and interactions. By the time I was introduced to Rudolph, I was primed for the opportunity to make this artwork. I also needed a musician to work with, and this introduction came via an unexpected route: because the medical scan of my femur needed to be made suitable for 3D printing, Rudolph referred me to the medical technology company CranioTech (https://www.craniotech.com), who had the skills and software needed to perfect the digital model for 3D printing. Their director Bernard Swart was a musician in his spare time, and his main accompanist and friend was a flutist and instrument maker, Alessandro Gigli—which meant that when I acquired the means to print the femur, I was also introduced to a musician who I could collaborate with on turning it into a flute.

One of the elements of the work that I hadn't anticipated until entering into a relationship with Rudolph in order to realize it, was the chance to make a performance of the process of fabricating the flute. My concept for turning the 3D-printed bone into a flute—drilling holes into it in the right places—was to operate on it in Rudolph's laboratory, in the same way as he would operate on his 3D prints in rehearsal of surgical operations. Though in this case, the simulation was not a rehearsal for the real thing, but the thing itself. I wasn't sure what it would yield, but I felt intuitively that echoing Rudolph's processes, on the same site and using some of the same technologies, would open up opportunities for bringing further ideas into the project. One of the outputs

Fig. 4.2 Ralph Borland playing Bone Flute in the exhibition *AIAIA—Aesthetic Interventions in Artificial Intelligence in Africa* (2023)

of this performance, apart from the bone flute itself, would be a film of the operation that could accompany the artwork on exhibition. My partner Dara Kell is a filmmaker, and she agreed to film and edit a video for the project.

Dara and I worked before the operation to plan out aspects of how we would film it. We looked over photographs of the laboratory that I'd taken on my first visit there, reviewing the colors and forms in the space. We asked Rudolph if he'd be wearing his green surgical scrubs, and Alessandro if he'd be wearing a plain white shirt as he had been so far in our meetings to prototype the flute—and I selected a worker's jacket which I wear regularly, as a complementary color. During the operation, part of my role was to keep thinking about how what we were doing would be captured, and checking in with Dara that she was getting what she needed for her part of the project. So it was that the four of us met in the orthopedics 3D-printing laboratory of Tygerberg Hospital, my partner behind the camera, and Rudolph, Alessandro and I at the operating table, to produce both *Bone Flute* and a film about its making.

The three of us worked together harmoniously on the femur; measuring and drilling the holes, determining what needed to be done at what stage during the process, consulting with one another, picking up on our own self-determined actions and assigning tasks to each other. There were some lovely moments. We needed to drill a hole in the "knee-bone end" of the femur that had to be in line with the internal cavity of the bone, for the outlet of the flute, and Rudolph showed us how he would locate this on a patient, drawing a line between two anatomical protuberances: a distinct overlap between the concerns of the project both medical and musical. Rudolph told me how impressed he was with Alessandro's ability with the tools to carve into the bone—he'd be an asset in the operating theater, he said.

Rudolph's enthusiasm and excitement at how we were applying some of his materials and processes for an unfamiliar, creative purpose was palpable. I was impressed by how well we all worked together for the first time—in my experience it's not certain that this will happen the first time collaborators work together. Something that occurred to me looking back on the experience was the generosity displayed by all participants; we were reaching out, each from a different domain—surgery, music, art, and filmmaking—and finding common ground. Medicine, as Roger Kneebone points out in his book *Expert* (2021) "is not a science, a craft or an art, but a practice which combines all three"; and there is a commonality between experts from different fields: "being expert is about how you think and see things… it's not simply defined by what you create" (p7).

4.4 Patient

On my first visit to Rudolph's laboratory at Tygerberg, I was there as an artist and researcher, to discuss his work and the creative project I wanted to use his processes to make, but I was also there as a patient. A few months previously, I had injured my hip while running. I'd run a difficult course on a mountainside, in the rain, and later that night in bed my right hip and thigh were hot and stiff. I had several sessions with a physiotherapist and was waiting to see if the injury would heal by itself, but it didn't. I suspected it may be a labral injury, but a scan would help to determine this. I asked Rudolph if he could help me book a scan, and so before I saw him, my first stop was at the patient registration counter to get myself registered on Tygerberg Hospital's administrative system. After I'd met with Rudolph at his laboratory, he directed me to where I needed to drop off my forms to book an MRI scan. As a public health patient, it would be a wait of several months to receive it.

I chose MRI over CT scan because of my concerns about exposing myself to radiation—an example of the kind of agency in relation to technology choice that I'm interested in seeing people practicing more widely. The day before my scan, I started to think: I know what I'm aiming to avoid with a CT scan—X-ray radiation—but how does MRI scanning actually work? I set up a call with Rudolph to ask him more about it. "Well," Rudolph explained, "basically it reorganizes all of your atoms. It moves your most basic particles around" (R.Venter, personal communication, December 10, 2021). That wasn't very reassuring—I wondered whether I'd made the right decision after all. I felt an existential fear around what it meant to have all my atoms reorganized—would I still be the same person afterward, and how would I tell?

My other concern was about money; I wasn't sure how much the scan was going to cost. The hospital staff weren't able to tell me before I had it. As a public health patient without medical insurance, it would be based on my declared income (my fellowship salary). Waiting for the scan in a hospital corridor with my partner on a Saturday morning, we were concerned about what bill I might be landed with. I needed the scan both to find out more about my hip injury and in order to progress my art-research project. It had taken three months from booking it to getting the scan, and I couldn't afford to delay it further on either count. I decided to risk it, and to go ahead with the scan without knowing what it would cost. Fortunately, when I received the bill a month or so later, it was affordable: about 10% of my monthly salary.

My partner left, and I waited my turn for the scan, which was likely to be only in a few hours. My turn came, and I was taken into the scanning room. I lay down on the stretcher that is inserted into the scanner, and the attendant asked if I'd like to listen to music through headphones. She placed the headphones on my head, and I heard elevator muzak. I was nervous, and the music didn't help—the contrast between my mood and the saccharine sounds of panpipes, as if I was on call waiting, was dissociating. I needed grounding. I asked if they could turn the music off, but I would keep the headphones on. With the music off, there was still a hum through the headphones which they couldn't do anything about, but that was preferable to the music. They pushed me headfirst into the machine. The metal interior wall of the cylinder I was in was inches from my face, and I had to breathe mindfully to contain the claustrophobic panic I felt rising in me. Then the sounds started.

Rudolph had told me about the noises I was likely to hear inside the machine. "It has a beat" he told me, "it's like being inside this big drum'n'bass machine that's set on random." Like a Berlin techno club at 4 in the morning, "a big synthesizer that pumps out beats." He told me I might quite enjoy it, knowing that I was into sound art and electronic music, and he was right. The sounds of the machine were quite extraordinary: pings and grunts like machinic whale sounds, juddering beats like a jackhammer, and deep bass rumbles. In the machine, you must keep completely still, so for an hour and 45 minutes I lay still, breathing as I've learnt to in yoga and meditation, and listening to the sounds of the machine. I thought, I would love to sample these noises and produce a soundtrack for patients to listen to on the headphones that is sympathetic to what they hear from the scanner, rather than completely disconnected.

Weeks after my scan, I went back to the hospital to get a copy of the scanned data, to take to a specialist for my hip. I was interested in the process of acquiring my own data as a patient. As a public hospital, the system was quite antiquated. I made a request by email to an official, who let me know when I could come by to pick up a CD of my data. I came to the office in Tygerberg where the CD was ready for me, and they gave me an invoice to go pay a fee in another part of the building. I went to the cashier, paid the fee, and brought a slip back to the office, to receive my CD. This process was in contrast to my later experience of having scans in a private hospital, when I received a login to an online system, where my data was uploaded and I could access and download it from home.

I'd injured my hip within the first few months of starting my fellowship, and had my MRI scan, which fed into my plans for my work, several months after that. A few months before the end of my fellowship, I was brought back into the hospital for further scans as a patient—a different hospital this time, and for a different purpose. This time, it was for CT scans, and I wasn't in any position to choose otherwise.

My GP had booked me for a colonoscopy, based in part on a family history of cancer—my grandfather had died long before I was born, before my own father was an adult, of colon cancer, at 49 years, 1 year older than I was now. My parents had never spoken about us children having an inherited cancer risk, but as I started to reach the same age as my grandfather, and as my partner fell pregnant and I was set to become a father myself for the first time, I started to think about it—and so I came to be sitting with my partner, soon after coming out of sedation following the examination, in front of a surgeon's desk as he explained that they had found a large tumor almost completely blocking my colon. I was sent for a CT scan the next day to see whether the cancer extended beyond the parts the doctor could see.

Fortunately, the CT scan showed the tumor contained to my colon, and it was operable. A week later I was on the surgeon's operating table, where they removed 20 cm of my colon and connected the ends back together again. It was done via laparoscopy, cameras, and instruments inserted through several small incisions in my abdomen, in addition to one larger cut above my pubic bone at the main site of the operation. This larger cut is almost identical to the bikini-line scar left by a Caesarian. Six months later, I found myself in the same operating theater at the same hospital, though this time it was my partner on the table. I was holding her hand, and we were singing together as our son Ira was born through an emergency C-section, lifted out through a matching incision in her abdomen.

My own surgery took place on the same day that the simulated operation on my *Bone Flute* was first due to take place in Rudolph's laboratory. I had to cancel this initial appointment with Rudolph and Alessandro after I received the results of my colonoscopy. It had taken a long time to find a date that worked for all three of us (and my partner as the fourth) and I was loath to cancel it. It would be another two months before we could reconvene, just in time to complete the flute for exhibition. As it was, on the same day I was first due to drill holes into a replica of my own femur, it was instead my body that was pierced by the surgeon's instruments, on an operating table in another hospital in the same city.

By the time of my diagnosis, I had taken up private health insurance for the first time in my life, encouraged by my partner. This meant that there was a very short time between my diagnosis and my operation. If I had still been in the public healthcare system, it might have been much longer—and if I had wanted it to take place sooner, it would have cost a great deal of money. Instead, I could get top-quality care, in a comfortable hospital with good facilities, almost immediately. And this may have saved my life.

Something else that supported my recovery was the resonance between my art-research project and this sudden internal accident that had brought me as a patient into the same field I was creating and researching in. My diagnosis did not bring me suddenly into the hospital and into contact with medical technologies from a place outside of this realm: I was already immersed in it. It felt meaningful—this was a story, there was something in it that made sense. And that feeling of meaningfulness can, I think, be a powerful support through a difficult time, as the next few months proved to be.

My engagement as a researcher and artist, working from a professional position, in collaborative, equitable relationships with medical workers, and with a public exhibition to come as a platform for sharing my work, I think gave me power as a patient which might otherwise have been harder for me to feel. I had already exercised choice as a patient in selecting the type of scan for my hip. When I first learnt that a tumor had been identified in my bowels, I wanted to hear from the surgeon about his experience, and his approaches to my surgery. He was willing and able to share this with me. In retrospect he was still exercising his legitimate power as a doctor in how he led me, the patient, into unfamiliar terrain. Reading Kneebone on what he learnt as a doctor, he recounts taking a patient gradually into the world they are entering through sudden diagnosis, as my surgeon did: it was one step at a time, first surgery, and chemotherapy may not be needed. After the surgery, the biopsy—and now chemotherapy is advised.

Beyond the basic questions to establish what the surgeon planned to do, I at first consciously ceded power to the doctors. This was serious business. They reassured me of their experience (around 500 abdominal cancer surgeries for my surgeon) and we'd talked about what they intended to do. Chemotherapy—it sounds daunting, and no one ever says it's easy, but I was not going to find out too much about it beforehand—I'm just going to do it. It was only once I'd started and experienced how unpleasant it felt, that I started to find out more. First was talking to other people who had been through it (surprising once you have cancer how many people you know have had it too, and you didn't know about it). They told me about the permanent side-effects chemotherapy had left them with: neuropathy of their hands and feet. I'd been prescribed four months of chemotherapy, which is less than the standard 6 months on my particular treatment, but halfway in I was concerned enough about the risks I was facing of permanent side-effects versus the risk of a recurrence of the cancer, to put some research into it.

The relationship between duration and efficacy of chemotherapy is notably under-researched—I think historically the idea has been to see it as a life or death struggle in which rather do more than less, and damn the side-effects. But gradually there has been push-back against this approach, and one significant study conducted by International Duration Evaluation of Adjuvant Chemotherapy (IDEA) and published in 2018, which investigated multiple sites and over ten thousand patients to look at the difference in outcomes between shorter and longer durations of chemotherapy treatment (Grothey et al. 2018). This study concluded that for particular risk categories of colon cancer, and particularly drug regimes—both of which I fell into—3 months duration of treatment showed no significantly worse outcomes for survival than 6 months of treatment, but significantly decreased risk of neuropathic side-effects. I took this paper to my oncologist and explained what I'd been thinking: that weighing up the risk of a permanent disabling neurological condition vs no discernibly increased chances of survival with longer treatment, I was considering shortening my course. "Ah, I was at the conference where they announced these results" he told me, "and it's a good study." He told me that he supported my line of thinking, and he would not object to me deciding to stop at 3 months of treatment instead of the four he had prescribed. I completed my course of chemotherapy at the end of March 2023, and a week later my son was born.

4.5 Story

One of the ways in which I see my work as an artist and a researcher, and how I frame it in this paper, is as an act of storytelling. Stories can be told in many ways: through spoken word, in person, through text and image, and—my personal favorite—through objects. As well as making my own objects, and having my art and design work selected for exhibition by others, I've staged my own exhibitions of my work, and I have worked as a curator selecting work by other artists, designers, scientists, technologists, and the public for exhibitions along a theme.

As an artist and curator, I love to tell stories through collections of objects, texts, images, and other media and experiences. We can define "stories" here broadly, to include factual explorations of real-life phenomena. Most of my curatorial work has been for the institution Science Gallery, which was founded at Trinity College in Dublin in 2008, and now has other university-associated galleries all over the world. Their remit is to explore science through a combination of art and design objects, and scientific experiments and demonstrations. At the end of my PhD, I co-curated *Surface Tension: The Future of Water* (2011) drawing on my thesis, which had as a central case study a novel method for pumping water in developing world contexts. Several years later, I co-curated *Design and Violence* (2016) as a co-production between Science Gallery and NY MoMA.

I see exhibitions, whether of my own work or the work of others, as functioning like object-based essays, which use a constellation of exhibits to explore a theme, supported by texts and events. Part of the work of research and art is in making public—of sharing the results of your work (and sometimes the process of it) with audiences. I knew that with this project, my main output would be an exhibition that would allow me to work with a range of media, with *Bone Flute* at the center; and that in this exhibition my experiences as researcher, artist, and patient would combine (Fig. 4.3).

The exhibition, titled *AIAIA—Aesthetic Interventions in Artificial Intelligence in Africa*,[5] was staged on the threshold of my personal and professional life—in a street-level artist-run gallery called *Brutal* in the complex in Cape Town where I've had my studio for the last 10 years or so. I exhibited *Bone Flute* as an object on a pedestal, near a wall-based installation of images showing some of the references for the project (from memento mori to automata), with the short film we'd made of the operation on a large TV screen (Fig. 4.4). Another smaller TV screen in the space showed a loop of images from my MRI and CAT scans, and a large window served as a light box for a several-meter-square collage of my medical scans. In the center of the space was a table and chairs, for meeting people during the exhibition. I intended to use the exhibition space as a live laboratory for the few weeks it was open, working on material to add to the display, conducting interviews, and talking about my work. During the exhibition, I wrote into a large timeline chart covering the 2 years of my research fellowship. It has columns for the 24 months, and two rows: one row for my personal experiences and one for my art-research work, so that connections along the same timeline can be seen between the two. The flute was performed at the opening and closing events.

At the opening of the exhibition, Rudolph gave a short speech about our work together (Fig. 4.5). "The thing that connected us was the way we think about technology" he said—and he credited me for bringing a perspective to his work that he hadn't yet realized: where his focus had been on the benefit to the patients of his work, and his awareness of how his processes enabled the embodied experience of the surgeon within

[5] While Artificial Intelligence (AI) was intended as the main focus of the *Future Hospitals* research project, few in our team, working at a range of sites across the continent, found much direct evidence for the use of AI in healthcare. My exhibition contained some of my creative and theoretical exploration of AI alongside *Bone Flute*.

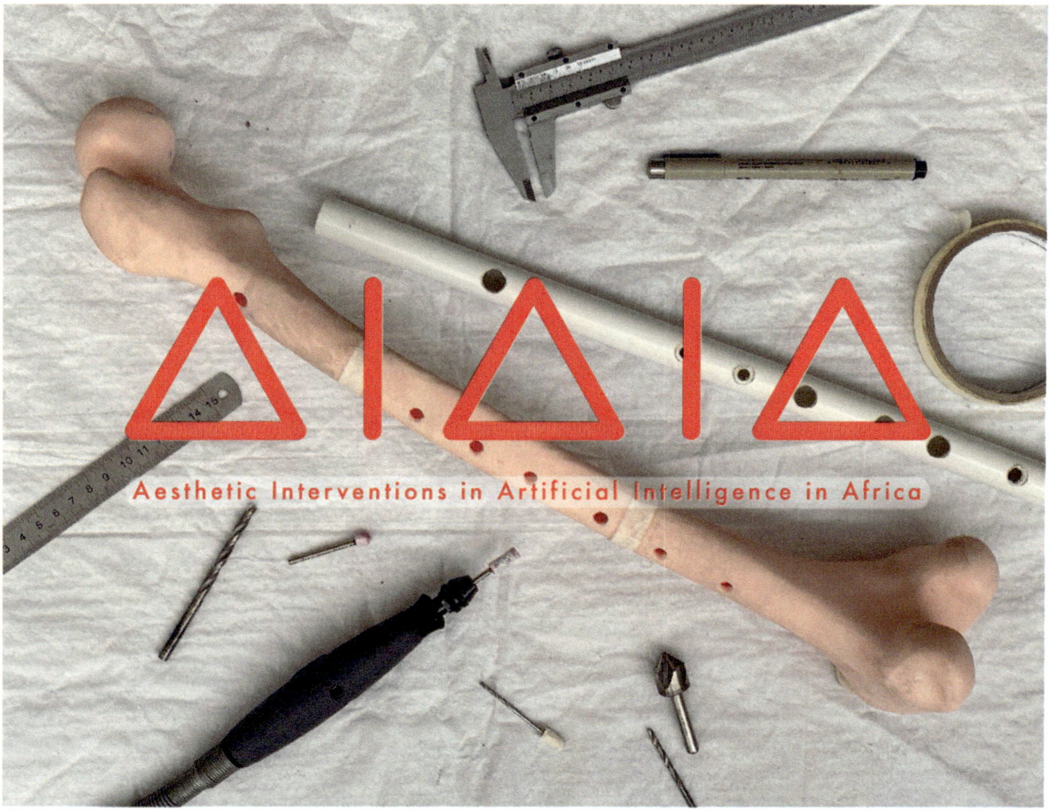

Fig. 4.3 The flier promoting the exhibition *AIAIA—Aesthetic Interventions in Artificial Intelligence in Africa* (2023)

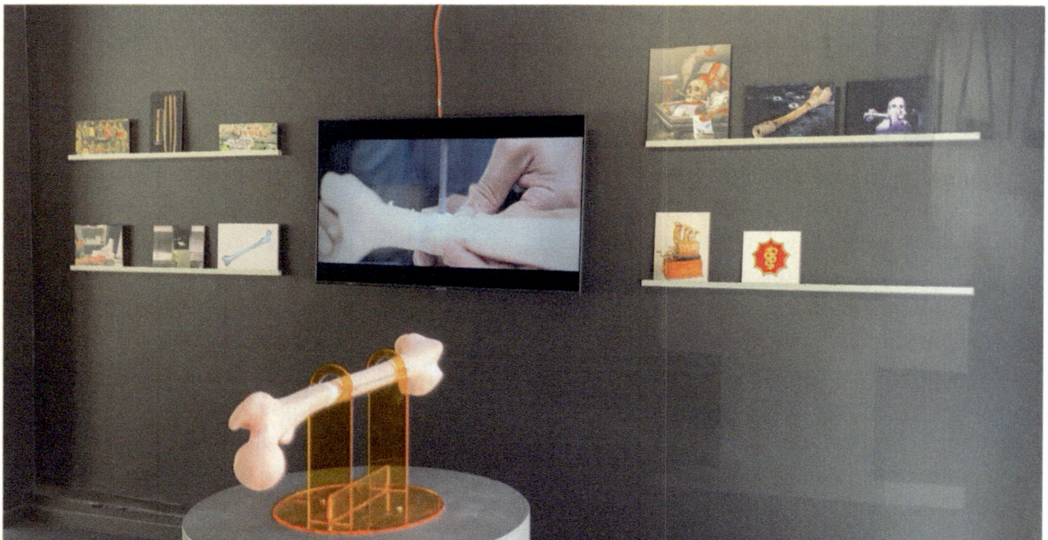

Fig. 4.4 Installation view of *Bone Flute* on the exhibition *AIAIA—Aesthetic Interventions in Artificial Intelligence in Africa* (2023)

Fig. 4.5 Rudolph Venter talking about our collaboration at the opening of the exhibition *AIAIA—Aesthetic Interventions in Artificial Intelligence in Africa* (2023)

pre-operative rehearsal, he hadn't thought about it in terms of the upskilling or de-skilling tendencies of technology in relation to work. That was the frame I had brought to his work, by conveying what I had learnt from reading *The Glass Cage* at the start of my fellowship.

Something interesting for me was that I had not realized before that this was a new perspective for Rudolph; in my memory, I was reflecting back to him something he had already said, perhaps in a different way—yet this had given him something he hadn't seen before. I think there's something here about the significance of framing information in particular ways, and in the nature of equitable collaboration: through conversation and reflecting one another, we both learnt to see the same material differently. I Rudolph found the perspective I had shared with him enriching. I was touched by his credit and noticed that this was the first time he had articulated it to me. As I interpret it, the opportunity for a ceremony and an audience elicited more from him than might have appeared just between us. In telling the story of our collaboration to an audience (which included me) he revealed something new.

I transcribed some of Rudolph's speech at the opening, in which he refers to storytelling:

> The most precious thing that I've got out of this collaboration so far, is learning a lot from you about how to communicate about my work. Because doctors would go: "Wow this is really cool. I segment DICOM images into 3D-printed models to rehearse my procedures" and audiences go "what's he talking about?" Ralph tells a story, and I've literally been taken along for the ride. I've learnt so much from you about how to communicate about my work, how to think about it, how to draw people in. I really appreciate that.

How *to* draw people in, as Rudolph refers to it? How I think about what I'm doing in a project like this is that I'm constructing layers of information (such as object, image, and story) that have both surface sensational appeal and that point to wider fields of knowledge. The story closest to Rudolph, and the premise for my research project, is a documentary of his work: the fact that he is developing low-cost methods appropriate to public hospitals in South Africa for scanning and 3D printing patients' bones in order to conduct pre-operative rehearsals. This is a story I recount when talking about *Bone Flute*—it's the starting point for our collaboration.

Then there are the things we both have to say about what his work does: the enhancement of the surgeon's capabilities; the sympathetic relationship of the haptic mode of imaging to surgeon's embodied skills; the counterpoint his example offers to automation's tendency to de-skill workers and detract from their embodied engagement with it. This information is mainly conveyed orally (to audiences at the exhibition, and in my other presentations to audiences-conferences, workshops, and so on) and in text (such as on placards in the exhibition, or on my website or other platforms). I designed the exhibition to have rearrangeable elements, which I added to through the course of the 2 weeks it was open, and inhabited the space as much as I could, to be available to give tours of the show. I used the exhibition as a real-life memory palace, with signs to guide my storytelling about the material.

Bone Flute is itself a story, told in object form, supported by image, text, and video. It is a story about the capabilities of technology to perform a kind of magic—to extract hidden parts of ourselves and bring them outside the body; about the goods achievable through collaboration—a surgeon, an artist, and a musician working together on the same project; about art as both a reminder of death, and our celebration of life in the face of it—the memento mori and the *danse macabre*. It also, as a friend pointed out to me at the closing event of the exhibition, points to the potential for objects to be animated, and to speak.[6] What struck them at seeing me play the flute was the significance of the breath, rather than the bone; the breath that animates the bone, and is shaped by it to make sound.

When they told me this, it brought back something I'd read in Meghan O'Gieblyn's excellent book about the mapping of religious feelings onto technology, *God, Animal, Human, Machine* (2021) in which she refers to one of the chapters of Ezekiel in the Bible: the story of the valley of dry bones. It contains the redolent image of bones refleshed and brought to life by divine breath. My friend's observation set me on a path to identify animism using bodily remains in other places: such as the Mongolian practice of turning a favored horse after death into a musical instrument, using their hair, skin, and bone to make a type of lyre, through which horse-like sounds are made. So, the stories told by *Bone Flute* were multiplied by its public exhibition and performance, as audiences reflected back to me their reception of the work.

One of the elements of the story told in the exhibition is the coincidence between medical technology as the object of my art and research, and myself as the object of medical technologies once I became a patient in the hospital through illness. Not only that coincidence but the ones that preceded it: at the start of my fellowship my introduction to a surgeon who uses technologies that would help me to realize an idea for an artwork that I'd had some 10 years previously; the injury to my hip soon after meeting him, making a medical scan necessary that would help both to diagnose my condition and obtain the data for the artwork; and finding the other collaborator I needed, an expert musician and flute-maker, thanks to his being friends with the medical technologist printing my femur. And further, perhaps the most resonant overall coincidence: I was working on an artwork about mortality—with my own mortality the immediate subject, given it was my own femur I was reproducing—when I was diagnosed with a life-threatening illness.

One more anecdote: when I was waiting for that first, fateful colonoscopy in the surgeon's hospital rooms, I was reading a book I'd picked up recently on the bargain table outside a bookshop in a shopping mall. Titled *Red Thread: On*

[6]My friend here is the sculptor Gerhard Marx, who I thank for his observation.

Mazes and Labyrinths (Verso Books, Higgins (2021)) it's an account of labyrinths and mazes by Catherine Higgins which becomes increasingly… labyrinthine as it follows a thread from historical and mythical descriptions of labyrinths to the author's own experiences of them. I'm not sure, or I can't remember, exactly what drew my attention to the book. It's a subject that interests me, and I was already intrigued in my research by the figure of the minotaur, who inhabited the labyrinth on Crete, in relation to some of my reading around AI and technology: in James Bridle's book about technology *New Dark Age: Technology and the End of the Future* (2018) he refers to "minotaur chess," in which a computer and a human collaborate to play chess against another human–machine duo (as now practiced by Gary Kasparov, famously the first world chess champion to lose to a computer, IBM's Big Blue, in 1997).

On the cover of *Red Thread* is an iconic image of a labyrinth, as on an antique coin, circular in outline, made up of quadrants of coils turning back on themselves. I was looking at it and thinking about the resonance with my present situation, in which the doctor's instruments would soon be exploring my own internal labyrinth, my intestines, when I reopened the book and turned to the next page. This is what I read:

> To feel trapped within the labyrinth's intestinal coils ushers in thoughts of entrails, of the strange unremembered red tunnels out of which we all, once, emerged … The labyrinth is, then, both a symbol of the body and its fragile mysteries, and a gesture of optimism that a corner of the universe can be mastered and given pattern and order by the human mind. (p.23)

A few hours later, sitting in the doctor's surgery and hearing what he had found inside me with his snake-like scope, I entered the labyrinth of surgery, treatment and recovery, and the image of the labyrinth became, like a *mise en abyme*, a thread which guided me through it. A text on the wall of my exhibition describes this motif, along with the snake, a symbol that entered my work through my plans for a snake-like automaton that will play the flute, wound around it like the Rod of Asclepius, and that came to resonate with my experience of medicine as I grappled with the voluntary poisoning of chemotherapy. I wrote:

> Looking back over the last two years, I feel as though I am following threads through the labyrinth, with the strands of my own personal fate woven into my art and research. The labyrinth became one of the motifs to my experience, along with the bone flute, the most ancient musical instrument we've found, and the snake: the next symbol I intend to combine with the flute. The snake is a symbol of medicine, combining toxicity with the potential for renewal, as it sheds its skin and emerges anew.

Coincidences lend themselves to stories. "The reason we respond positively to accounts of coincidence," write the authors Martin Plimmer and Brian King in their book *Beyond Coincidence* (2019) "ever eager to give them the benefit of the doubt, is because they make such good stories. They have the resonance of myth and fairy tale, with their dramatic shifts of fortune, their spectacular life and death events and their many enchanted objects and preoccupations… a good coincidence story has the gravitas of Greek drama, the difference being that it is true" (p.7).

They describe here the pleasure for audiences in factual stories of coincidence. But before the story reaches an audience, it has meaning for the person experiencing the coincidence. In the real-life stories collected by Plimmer and King, "when coincidences resonate personally and intimately it becomes hard to dismiss them as mere chance, events that are slung together by the wind." Whether or not the coincidences they experience "have a deeper meaning in any truly objective sense," they carry meaning for the person perceiving and experiencing them, for whom they may produce "emotional catharses and transformations." So it was for me. It felt meaningful that I should find myself brought deeper as a patient into the world I was researching and making art in, and I felt supported through a destabilizing and frightening experience by the feeling that my experiences were contributing to a story, and to my work.

4.6 Conclusion

This is a personal essay: a story I have told about my experiences and my ways of working. Biomedical imaging is at the core of the enquiry: used as a means for innovative, low-cost methods for rehearsing complex surgeries in a public healthcare setting, which are sympathetic to the embodied skills of the surgeon—a positive example of human–machine collaboration; as a way of producing a sensational artwork that makes available a hidden part of a living body; as a means of detection and diagnosis of illness; and as a sensory experience for a patient. The work I conducted as a researcher and artist was made possible by collaboration with others from different disciplines: a surgeon, a musician, and a filmmaker. When the work was exhibited, more relationships were formed as audiences engaged with it. As a patient, I worked with others for my treatment and recovery, negotiating the choices available to me.

Storytelling runs through my professional and personal experiences: using my skills to direct attention to the work of another expert, and to create compelling experiences for audiences that have value in their own right. My narrativizing and making sense of my own experience within the context of my research and art, and my acquisition of symbols to support me, helped me on a difficult personal journey—one that I was accompanied on by my partner, and met by my son.

I'm almost a year and a half post diagnosis and surgery, and no cancer has been detected in the regular checks that I've had since—blood tests every three months, a CAT scan every six, and colonoscopy once a year. I'll continue with these tests for the next three to 5 years, after which they consider my risk to have reduced to the level of life-long remission. My son Ira is turning one in a week's time, and my partner Dara and I are getting married a month after that. Who doesn't love a happy ending? My *Bone Flute* is on a shelf at home, and I practice on it whenever I can. In the end, stories and death are intertwined, with one meant to transcend the other. Art is our medicine in the face of mortality.

Animals die,
friends die,
and thyself, too, shall die;
but one thing I know that never dies
the tales of the one who died.

Gestaþáttr number 77, *Hávamál*, in the thirteenth-century Icelandic manuscript *Codex Regius.*

4.7 Credits

Ralph Borland
ralph.borland@gmail.com

I made this work as a Research Fellow on the project *Future Hospitals: 4IR and Ethics of Care In Africa*, at HUMA, the Institute for Humanities in Africa at the University of Cape Town. The project was funded by the Carnegie Corporation of New York.

Project website: https://boneflute.org

References

Borland R (2011) Radical plumbers and PlayPumps - objects in development. Doctoral thesis, Trinity College, Dublin, Ireland. http://www.tara.tcd.ie/handle/2262/89797

Branson TM, Shapiro L, Venter RG (2021) Observation of patients' 3D printed anatomical features and 3D visualisation technologies improve spatial awareness for surgical planning and in-theatre performance. In: Rea PM (ed) Biomedical visualisation, Advances in experimental medicine and biology, vol 1334. Springer, Cham. https://doi.org/10.1007/978-3-030-76951-2_2

Bridle J (2018) New dark age: technology and the end of the future. Verso Books

Carr N (2015) The glass cage: where automation is taking us. The Bodley Head

Codex Regius (c.1270) Gestaþáttr number 77, *Hávamál*

CranioTech, Patient-specific solutions https://www.craniotech.com/ Accessed 23 Mar 2024

Csikszentmihalyi M (1990) Flow: the psychology of optimal experience. Harper and Row, New York

Disney W (1929) The Skeleton Dance [Film]. Walt Disney Productions

Grimm J (1816/1818) Die Kinder zu Hameln (The Children of Hameln), Deutsche Sagen 1:245

Grothey A et al (2018) Duration of adjuvant chemotherapy for stage III colon cancer. N Engl J Med 378(13):1177–1188

Herbert F (1965) Dune. Chilton Books

Hergé (1948) Prisoners of the Sun. Methuen

Higgins C (2021) Red thread: on Mazes and Labyrinths. Verso Books

Holbein H (1523) Dance of death

HUMA (2024) Future hospitals: 4IR and ethics of care in Africa. https://humanities.uct.ac.za/huma/research-huma-strategic-programmes/future-hospitals-4ir-and-ethics-care-africa. Accessed 23 Mar 2024

King M, Plimmer B (2019) Beyond coincidence. Icon Books

Kneebone R (2021) Expert. Penguin Books

Kubrick S (1968) 2001: A Space Odyssey [Film]. Stanley Kubrick Productions

Lang F (1927) Metropolis [Film], UFA

Morgan C, Khatri C, Hanna SA, Ashrafian H, Sarraf KM (2020) Use of three-dimensional printing in preoperative planning in orthopaedic trauma surgery: a systematic review and meta-analysis. World J Orthop 11(1):57–67. https://doi.org/10.5312/wjo.v11.i1.57

O'Gieblyn M (2021) God, animal, human, machine, Penguin Random House

Venter RG, Kotze L, Ferreira N (2022) A clinician-run 3D-printing laboratory for orthopaedic preoperative planning: an illustrative case series. SA Orthop J 21(3):180–186. https://doi.org/10.17159/2309-8309/2022/v21n3a7

Part II

3D Bioprinting, Clinical Practice, Imaging

5 Digital Protocol for the Bioprinting of a Three-Dimensional Acellular Dermal Scaffold

Princess Ijeoma, Alison Fany Ridel, and Hafiza Parkar

Abstract

The gold standard for treating secondary intent wounds is skin grafting, but this is limited by patient conditions (i.e. diabetes, vascular insufficiencies, and immunocompromised status), the availability of viable skin, and potential immune rejection. Decellularized acellular dermal scaffolds (ADS) are an alternative to skin grafts and are developed by removing cells from donor or animal sources, reducing the potential for rejection. Traditional decellularization processes have various limitations, producing ADS with altered three-dimensional (3D) structure, damaged proteins, and decreased tensile strength. Emerging technologies, like 3D bioprinting, can overcome challenges faced by traditional methods. This research aimed to provide a digital protocol for the generation of a 3D acellular dermal scaffold for 3D bioprinting. One optimized porcine extracellular matrix (ECM)-based ADS (P3-S3), 8 mm × 2 mm in size, from a previous departmental study was captured by Micro–X-Ray Computed Tomography (Micro-XCT) at the South African Nuclear Energy Corporation (NECSA). The micro-XCT images were imported into the Amira-Avizo 2019.3 software (Thermo Fisher Scientific, Inc.) to generate a 3D ADS model using an image segmentation procedure. This 3D model was exported in stereolithography (STL) format and will be used for 3D bioprinting of the ADS. The development of bioengineered ADS constructs is integral to the advancement of wound healing technology.

Keywords

3D bioprinting · Acellular dermal scaffold
Digital protocol · Decellularization dermis
Micro X-ray computed tomography · Tissue engineering · Regenerative medicine

Abbreviations

ADS	Acellular dermal scaffold
DFUs	Diabetic foot ulcers
DNA	Deoxyribonucleic acid
RNA	Ribonucleic acid
ECM	Extracellular matrix
FTSG	Full-thickness skin graft
Micro-XCT	Micro X-ray computed tomography
MIXRAD	Micro-focus X-ray tomography facility

P. Ijeoma · H. Parkar
Department of Pharmacology, University of Pretoria, Hatfield, South Africa
e-mail: hafiza.parkar@up.ac.za

A. F. Ridel (✉)
Department of Anatomy, University of Pretoria, Hatfield, South Africa
e-mail: alison.ridel@up.ac.za

L. Shapiro (ed.), *How to use 3D Printing Innovations and Digital Storage to Democratize Anatomy Education*, Biomedical Visualization 6, https://doi.org/10.1007/978-3-031-68501-9_5

NECSA	The south African nuclear energy corporation
NPWT	Negative pressure wound therapy
PBS	Phosphate buffered saline
RMPI	Roswell Park memorial institute
ROS	Reactive oxygen species
SCO_2	Supercritical carbon dioxide
STL	Stereolithography
STSG	Split-thickness skin graft
TIFF	Tag image file format
3D	Three-dimensional

5.1 Introduction

The burden of wound healing has been termed a silent epidemic because of the substantial impact it has on healthcare systems, consuming an estimated 2–3% of total healthcare expenditure (Lindholm and Searle 2016). One of the main problems in wound healing is the lack of sustainable treatment options for patients with chronic wounds and wounds healing by secondary intention. Secondary intent wounds are those where extensive tissue loss has occurred and the wound is allowed to heal naturally without the use of surgical closure techniques (Azmat and Council 2023; Chhabra et al. 2017).

Secondary intent wounds have proven to be costly and difficult to treat as they have an extended healing time, a higher risk of infection, re-hospitalization, and the potential to develop into chronic wounds (Parkar et al. 2021). Patients suffering with secondary intent and chronic wounds have impaired quality of life (Vogt et al. 2020). Secondary intent, non-healing, and chronic wounds create many obstacles in a patient's life, from mobility issues to reduced psychological well-being manifesting itself in human cost as well as the financial burden. The purpose of this chapter is to discuss the literature on wound healing, regenerative medicine, and tissue engineering, including the description of an innovative new digital protocol for the creation of a 3D ADS for bioprinting.

5.2 Wound Healing

5.2.1 The Phases of Wound Healing

Wound healing is a physiological process that occurs in the body after tissue injury (Wallace et al. 2023). The process is marked by consequent steps that work together to create a stable, and balanced microenvironment that ensures the wound heals timeously, and without complication. Acute wounds are healed through four "cascading" processes; haemostasis, inflammation, proliferation, and remodelling. These four steps have distinct characteristics, hormones, and proteins that trigger the next step. It is important that they occur in the correct sequence and the correct time frame for the wound healing process to be deemed successful.

Haemostasis is the first phase of the process and is characterized by vascular constriction which stops the bleeding, platelet aggregation which forms the platelet plug, the activation of the coagulation cascade, and the formation of a temporary wound matrix through fibrin clot formation (Guo and Dipietro 2010; Lapelusa and Dave 2023). This process occurs immediately after tissue injury and can last for up to 15 minutes, which is when inflammation, the next phase, begins (Schultz et al. 2011b). The next phase is characterized by white blood cell infiltration (specifically neutrophils), it occurs within the first 24 hours and can last up to 2 weeks post-injury (Schultz et al. 2011b). Once the neutrophils have cleared any possible bacteria from the injury site, they release proinflammatory cytokines that recruit and assist in the activation of fibroblasts that are important for the next phase (Schultz et al. 2011b).

The third phase is the proliferative phase, which is characterized by the proliferation of fibroblasts and synthesis of granulation tissue, components including collagen, and elastin amongst others (Schultz et al. 2011b). Extracellular matrix (ECM) components are deposited into the wound area and create a scaffold for differentiation, cell adhesion, and growth during wound repair (Landén et al. 2016). Remodelling is the final phase of wound healing that is characterized by the replacement of deposited collagens type I–type III and vascular maturation (Cundell 2016). Granulation tissue matures into scar tissue, and the wound bed increases in tensile strength (Schultz et al. 2011b). This phase is characteristically the longest as it could last for longer than 1 year after wound formation (Gonzalez et al. 2016).

5.2.2 Wound Classifications and Healing by Intent

Wounds may be classified as acute or chronic. Acute wounds are those that progress through the phases of healing in an orderly manner and in a predictable time frame resulting in complete restoration of the damaged tissue (Parkar and Cromarty 2021). Chronic wounds are those that fail to advance through the natural, overlapping phases of wound healing within the anticipated time frame, often remaining unhealed after 6 weeks. These wounds frequently persist in an inflammatory state, caused by the overproduction of reactive oxygen species (ROS), excessive levels of proinflammatory cytokines, lack of migration, senescent cells, dysfunctional stem cells, and persistent infections resulting in ongoing tissue damage at the wound site or periphery (Parkar and Cromarty 2021; Frykberg and Banks 2015). Chronic wounds may either persist without healing, only partially heal before recurring, or necessitate prolonged and intensive treatments, such as debridement, to heal (Parkar and Cromarty 2021). Impaired vascularization, desiccated tissues, and fibrotic tissues are physiological factors which influence the formation of chronic wounds (Hoversten et al. 2020). Other factors include diabetes, stress, smoking, age amongst others (Guo and Dipietro 2010).

The manner in which wounds heal and the interventions employed may be further classified as healing by intention. Wounds may heal by primary, secondary, or tertiary intentions, this is dependent on the amount of tissue loss from the injury. Primary intention wounds are those where minimal tissue loss has occurred and surgical tapes or sutures are used to close the wound (Chhabra et al. 2017). Secondary intent wounds are those where extensive tissue loss has occurred following injury, these wounds cannot be closed by simple approximation of the wound edges. This means that the wound area must heal through replacement of the lost tissue, typically from the wound bed (Parkar et al. 2021). Tertiary intent wound healing is the intentional interruption of the secondary intention wound healing, and the wound is mechanically closed (Demmer et al. 2021).

In secondary intent wound healing, like chronic wounds, the inflammatory reaction is more severe, wound contracture is increased, and granulation tissue is formed at the margins of the wound (Chhabra et al. 2017). These wounds undergo healing from the base upward, typically exhibiting a slow recovery rate and possessing an increased risk of infection (Mccaughan et al. 2018). Wounds that usually require secondary intention healing interventions include full-thickness skin wounds, severe burn wounds, and chronic non-healing wounds such as diabetic foot ulcers (DFUs) (Byeon et al. 2023; Parkar and Cromarty 2020).

5.2.3 Treatment of Secondary Intent Wounds

The current gold standard for the treatment of secondary intent wounds is skin grafting. This is a procedure that aims to restore soft tissue defects by removing healthy skin from another part of the patient's body and transplanting it to the injury site (Braza and Fahrenkopf 2023; Broccoli et al. 2020).. The use of a skin graft is dependent on the thickness of the wound. A full-thickness skin graft (FTSG), or a split-thickness skin graft (STSG) is used in the treatment of wound healing by secondary intent (Broccoli et al. 2020). FTSGs are harvested from epidermal and dermal skin (Herskovitz et al. 2016). This graft aids in full contracture and improves the aesthetic view of the scar in the long run (Braza and Fahrenkopf

2023). In comparison, STSGs contain the entire epidermis and only part of the dermis. These grafts are thinner than FTSGs which assists them to promote wound coverage and aid skin function, but not aid wound contracture (Herskovitz et al. 2016).

Complications with this method include a build-up of fluid, skin pigmentation, and haematomas and may result in partial loss of function, aesthetic disparity, severe scarring, and misaligned architecture of the skin grafts to the wound (Braza and Fahrenkopf 2023). The success of skin grafting in a wound area depends heavily on specific criteria that must be met for the graft to take effectively. The limitations of grafting include the potential for graft rejection due to the presence of antigenic material in the cells of the transplanted skin, risk of disease transmission, and the availability of donors (Elseth and Nunez Lopez 2022; Wang et al. 2022). Autografts are further restricted by the patient's accessible healthy skin and may lead to the development of a secondary non-healing wound in individuals with healing impairments. Furthermore, studies in the literature indicate that graft failure rates range from 2 to 30%, particularly in individuals with higher BMI, diabetes, peripheral vascular disease, or those taking immunosuppressive drugs (Long et al. 2019; Oganesyan et al. 2013; Reddy et al. 2014; Ramanujam et al. 2013).

As a response to these shortfalls, dermal scaffolds have been introduced as an alternative to grafting (Hosseini and Shafiee 2021). They are made from different components such as proteins and small molecules and are divided into acellular and cellular. The former is created without cells in the structure and is aimed at initiating the repair of in situ host tissue (Salih 2009). The latter are also known as bioactive scaffolds and possess cells (usually stem cells) placed on the wound healing site (Hazrati et al. 2022).

5.2.4 Acellular Dermal Scaffolds

Acellular dermal scaffolds (ADS) have been proposed as an alternative to skin transplants (Urciuolo et al. 2019). To produce an ADS, skin is obtained from human donors (cadavers) or animal sources (porcine, bovine, ovine), and the cells are then removed from donor or animal skin by decellularization. The process of decellularization makes use of combinations of physical, chemical, and enzymatic methods to remove cellular content from the skin, leaving behind the scaffold of the extracellular matrix, which is called the acellular dermal matrix. Thereafter, the ADS may either be directly implanted into a wound or it may be recellularized in vitro using autologous patient cells or donor stem cells after which it is implanted into the wound (Zhang et al. 2022; García-Gareta et al. 2020). This process is illustrated in Fig. 5.1.

Acellular dermal scaffolds are highly compatible with the body, possess exceptional biochemical properties, and feature a macromolecular structure mirroring that of native tissue.

Amongst the various scaffolds based on the biomimetic extracellular matrix, ADS are distinguished by their unique dermal microenvironment. This characteristic makes them particularly valuable for applications in skin regeneration and reconstructive surgery, due to their ability to closely mimic the natural skin environment (Rosadas et al. 2023). They serve as a replacement for ECM components such as collagen, elastin, and glycosaminoglycans in the wound environment, supporting the external influx of wound-healing cells such as fibroblasts, platelets, and epithelial cells (Schultz et al. 2011a).

To date, many studies have been conducted proving the efficacy of acellular dermal scaffolds in the enhancement of wound healing. A trial conducted by Hahn et al. (2021) tested human acellular dermal matrices in DFUs. The results showed that patients in the experimental group where the ADS was added to the negative pressure wound therapy (NPWT), performed superiorly to those who had not received the ADS implant. Reyzelman et al. (2009) conducted a 12-week, prospective, randomized, controlled multi-centre study where the clinical effectiveness of an acellular dermal regenerative tissue matrix was compared to standard wound management in healing DFUs. Results showed a statistically significant difference in non-healing

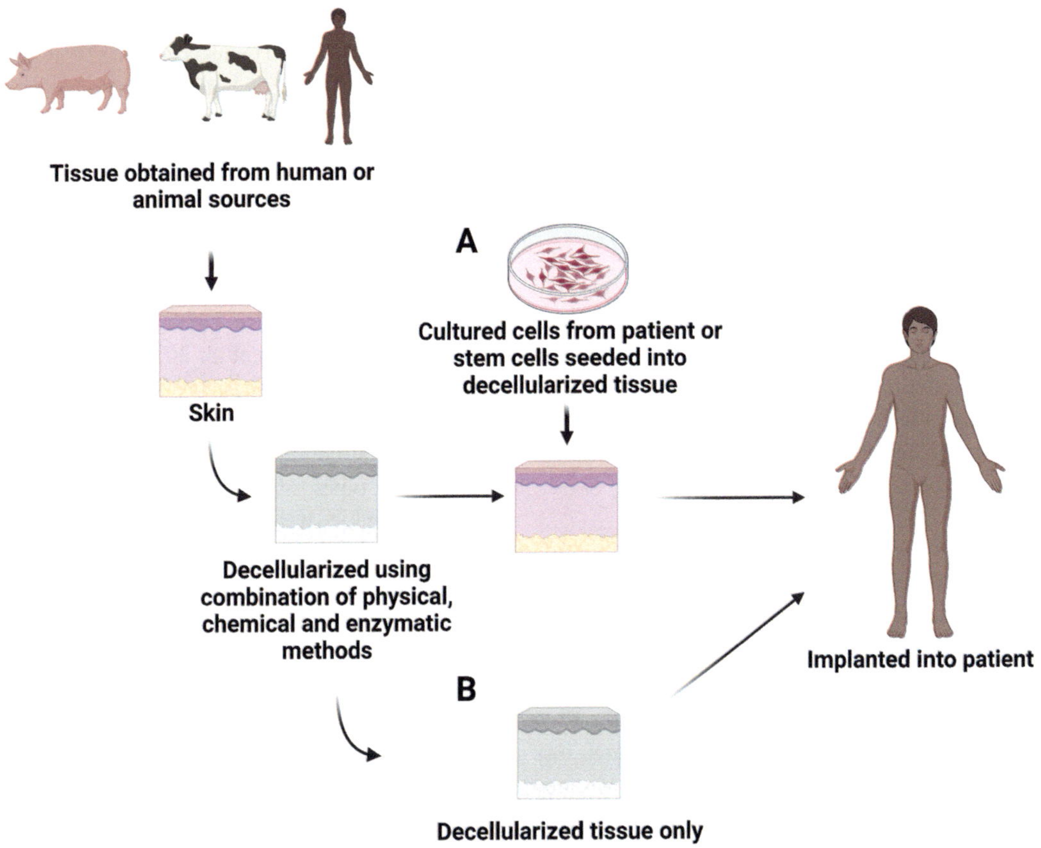

Fig. 5.1 Process of creating decellularized tissues. Tissues are obtained from xenogeneic or cadaveric sources after which it is decellularized using a combination of physical, chemical, and/or enzymatic methods. Thereafter it may A] be recellularized using cells obtained from the patient or donor stem cells or B] used on its own and implanted into the patient. Created with BioRender.com (Parkar H 2024)

rate between the treatment and groups, with odds of healing 2·0 times higher in the study versus control group.

Truong et al. (2005) conducted a study where different types of dermal substitutes including acellular dermal matrices were implanted into full-thickness skin wounds in nude mice and compared to an untreated control. The wounds were assessed 28 days after surgery, and primarily investigated for wound contraction, epithelialization, and other histological characteristics. The results of that study showed that all the wounds treated with dermal substitutes had been incorporated into the wound bed and decreased wound contracture as compared to the untreated control. The acellular dermal matrix and Alloderm produced less contraction and the thickest new "dermis" in the healed wounds when compared to the control or synthetic dermal substitutes.

Conventional methods for producing an ADS make use of a combination of physical (freeze-thaw, sonication), chemical (detergents, acids/bases), and enzymatic (nucleases, trypsin) techniques (Crapo et al. 2011; Gilpin and Yang 2017). However, each technique is not without limitations. Physical decellularization methods may disrupt the ECM micro-structures and damage the three-dimensional structure affecting the mechanical properties of the ADS (Crapo et al. 2011; Liao et al. 2020). Residual reagents from chemical decellularization methods may increase the likelihood of adverse immune reactions, disrupt collagen fibres, diminish the mechanical attributes of the ECM, and eliminate some beneficial growth factors interfering with biological functions and decreasing the tensile strength of the ADS (Liao et al. 2020; Meyer et al. 2006; Patel et al. 2008). Enzymatic methods may affect the microstructure of proteins and residual enzymes may produce negative immune responses (Gilpin and Yang 2017; García-Gareta et al. 2020).

To address the challenges introduced by the decellularization process while leveraging the potential of engineered materials, numerous researchers in the last two decades have embarked on research efforts to enhance decellularized scaffolds. One such avenue the use of is 3D bioprinting (Neishabouri et al. 2022; García-Gareta et al. 2020).

5.3 Regenerative Medicine and Tissue Engineering

Regenerative medicine is a field that merges engineering and medicine to create and promote regeneration of damaged tissue or organ loss through disease or injury (Mao and Mooney 2015). Tissue engineering is the combining of biologically active molecules that form functional tissues in vivo (Berthiaume and Yarmush 2003). Functional skin substitutes have been produced for the purpose of replacing broken cartilage, damaged in vivo cellular tissue and reducing wound healing time amongst other applications (Berthiaume and Yarmush 2003). Due to the increase in the use of tissue engineering, a model has been proposed to aid in the creation of an optimal ECM scaffold model that will bridge the gap between the current gold standard of wound care and an absolute resolve.

5.3.1 Bioprinting

Bioprinting is a type of tissue engineering and refers to printing of biomaterials, biological molecules, and viable cells into physical 3D objects using bioinks (hydrogels) that display physiological properties (Gu et al. 2020; Kačarević et al. 2018). Bioprinting is a term used to describe the synthetic production of tissues, and possibly internal organs through the layer-by-layer deposition of biomaterials (Fang et al. 2022). It has proven to be invaluable in both the healthcare and pharmaceutical manufacturing industry having uses such as the production of printlets (3D-printed tablets) that are suited to the specificity of the patient's needs, and wound dressings such as vascular graft dressings (Basit and Trenfield 2022; Jamróz et al. 2018). Bioprinting was first demonstrated in 1988 through micropositioning of cells (Gu et al. 2020). Since then, bioprinting has given rise to developments which include the bioprinting of a blood vessel, and the first cardioid structure in 2019 (Jose et al. 2016).

Different types of bioprinters are available such as the inkjet-based printer, extrusion-based bioprinter, laser-assisted bioprinter, stereolithography bioprinter, acoustic bioprinter, microvalve bioprinter, and scaffold-free bioprinter (Yu et al. 2020). The inkjet and direct ink writing bioprinting styles have been chosen more frequently to print 3D live cells because it offers a non-contact strategy of printing which reduces cross contamination with cells, is more cost effective, has a smaller printing footprint, higher printing resolution, and uses less amounts of raw materials (Gopinathan and Noh 2018; Kumar et al. 2021).

Bioprinting of the ADS allows for the combination of key structural and functional proteins as a part of the scaffold to create the optimal ECM scaffold to increase in vivo cell production and adherence. 3D bioprinting of wound healing ADS promises to be a more reproducible, feasible, and accessible option for patients suffering from chronic wounds and secondary intention wounds. It uses little to no wastage of materials and it can be customized to the size, depth, and type of wound that the patient has. As such, this is a more directed and personal approach.

The creation of autologous acellular dermal scaffolds is a more targeted personalized regenerative medicine approach. The use of autologous cells during the production of the scaffold will theoretically reduce the possibility of scaffold rejection, reduce the possibility of an undesirable inflammatory reaction, as well as a negative immunological effect (Lu et al. 2011). This has been researched by Lu et al. and showed promising results. This approach may not only improve the wound healing time, but it will also aid the psychology of the patient during post recovery which will in turn increase their quality of life and reduce the likelihood of a wound infection or slowed healing. With all these techniques in place, the intention is to simplify and improve the ways wounds are healed (Lu et al. 2011).

5.3.2 Bioinks

To print a biologically active structure successfully, bioinks are used. A bioink is a hydrogel made up of biomaterials that are selected and put together to encapsulate cells and create physiologically active and functioning 3D objects (Kačarević et al. 2018). Cell-laden bioinks are hydrogel based because they have high water contents that are necessary for cellular survival (Yang et al. 2022). Bioinks need to encapsulate some physiological and psychochemical properties for them to be considered suitable for 3D printing.

Bioinks should produce structures with good mechanical strength whilst maintaining the original properties of the tissue, such as fluidity and elasticity (Gungor-Ozkerim et al. 2018). Two main bioprinting approaches currently exist, namely: the cell-scaffold approach and the scaffold-free approach. The former uses a bioink that contains biomaterial and live cells to develop 3D structures, however, the scaffold that the bioink is printed on is biodegradable (Gopinathan and Noh 2018). Once the scaffold has been placed in vivo, the encapsulated live cells will undergo cellular growth and then fill the space of the predetermined structure. For this reason, using a bioink that has all the right properties for the expected outcome is important, as it will affect the adherence, growth, and functionality of the cells in vivo. The scaffold-free approach uses live cells to form neo-tissues that are deposited directly in vivo and end up fusing and forming their predetermined structure over time (Gopinathan and Noh 2018). Numerous biomaterials are used as bioinks for building different tissue structures. Examples are collagen and gelatin (natural polymers), and polyethylene glycol (PEG) (synthetic polymers) (Gopinathan and Noh 2018).

5.3.3 X-ray Computed Micro-Tomography for Visualizing Soft Biological Structures in Three Dimensions

Optical and electron imaging of serial sections and sectioned block faces, as well as observation of whole specimens using confocal microscopy or electron tomography, are common ways to look at the 3D structure of noncalcified tissues and biomaterials (Shearer et al. 2016). X-ray computed micro-tomography (micro-XCT) provides a fast way to capture detailed 3D structures of vast volumes of a specimen with sub-micron resolutions (Shearer et al. 2016). It also allows for the observation of real-time changes in living tissues at a microsecond scale (Shearer et al. 2016). Initially used for imaging inorganic materials and fossils, micro-XCT quickly gained popularity in the biomedical engineering field for the structural analysis of calcified tissues and scaffolds (Shearer et al. 2016). Micro-XCT generates 3D volumes that researchers can divide and analyze quantitatively, providing crucial data on volume fraction, porosity, and connectedness (Shearer et al. 2016). Case studies have shown that lab micro-XCT instruments can accurately show structures at the micron level in a variety of tissues, such as rat arteries, human skin, tendons, and ligaments. They can also visualize the distribution of cells within tissue engineering constructs (Withers 2007; Stock 2008; Bravin et al. 2012).

Micro-XCT is a valuable method for imaging cells within tissue engineered structures, as shown by Dorsey et al. The researchers used osmium tetroxide staining to identify the cells placed on porous polymer scaffolds and compared qualitative 3D micro-XCT imaging with qualitative fluorescence microscopy combined with Deoxyribonucleic acid (DNA) quantification to evaluate cell numbers (Dorsey et al. 2009). The authors suggest that micro-XCT can accurately show tissue engineered constructs with a cell density of one million cells/mL or more (Dorsey et al. 2009). This would solve the problem of obtaining full 3D images of clinically relevant constructs. Micro-XCT enables a thorough 3D observation of cell location and changes in biomaterial architecture across various sizes, in contrast to other imaging techniques limited to smaller construction areas (Shearer et al. 2016).

In the following section of this chapter, we will discuss the development of a new and innovative digital protocol for the generation of a 3D-acellular dermal scaffold using a micro-XCT scanner with a direct application for 3D bioprinting.

5.4 Digital Protocol for the Creation of 3D Scaffolds

5.4.1 Acquisition and Preparation of the Scaffold

Ethical approval for this study was obtained from the University of Pretoria Research Ethics Committee (approval number: 421/2018). Ethical approval for the use of animals was obtained from both the University of Pretoria (approval number: h010–17 (A1)) and the University of Witwatersrand (approval number: 2017/11/77/C) Animal Ethics Committees. Porcine skin was donated by the University of Witwatersrand Central Animal Services. Skin sections were harvested from three female Large White pigs following sacrifice after a trauma surgery training course for surgical students was conducted.

The harvested skin sections (4 × 10 cm) were sterilized using a 3% hydrogen peroxide solution prior to being transferred into unsupplemented Roswell Park Memorial Institute (RPMI) media for transport and storage. Sections (three from each pig) were removed from the RPMI, rinsed with phosphate buffered saline (PBS) and the subcutaneous (fatty) layer of the skin was removed using a scalpel. The skin sections were then placed in separate sterile T75 tissue culture flasks containing a hypertonic saline solution. The flasks were briefly sonicated and then placed on an orbital shaker at a medium speed and allowed to shake for 48 h. The hypertonic saline solution was discarded, and the epidermal layer of the skin was removed using a scalpel leaving behind only the dermal layer of the skin.

Dermal sections were then placed into sterile T75 tissue culture flasks and rinsed with PBS, after which they were subjected to a chemical detergent-based decellularization. A solution of Triton-X100 and cetylpyridinium chloride was added to the flask and allowed to shake on an orbital shaker at a moderate speed for approximately 72 h. Following chemical decellularization, the dermal sections were subjected to a physical decellularization process using supercritical carbon dioxide (SCO_2) following methods described by Wang et al. 2017 with modifications. Dermal sections were placed into the extraction vessel of an SCO_2 apparatus, designed and developed by Parkar and Cromarty (2024). The extraction vessel was sealed and placed into a heated water bath, after which medical-grade carbon dioxide gas was pumped into the extraction vessel at a controlled pressure using a pressurizing pump at a controlled flow rate for 24 h where it was allowed to diffuse through the dermal section, and dissolve and remove any lipophilic cellular material.

Following SCO_2 extraction, the dermal section was rinsed with PBS and subjected to final enzymatic decellularization using a benzonase nuclease hydrolysis to eradicate any remaining nuclear material. Eight-millimetre sections were cut out of the decellularized dermis using a biopsy punch and either fixed in formalin for histological analysis or lyophilized for further examination. The sections were then embedded into paraffin wax blocks, sectioned into 5 μm slices, and mounted onto glass slides. The sections were then stained using a standard haematoxylin and eosin staining procedure (Monte-Alto-Costa and Porto 2010). The resultant stained ADS sections were viewed using a light microscope (Zeiss AXIOPLAN; Oberkochen, Germany) coupled to a digital camera. Images were captured at 100× magnification (Fig. 5.2) using AxioVision Rel. 4.8.2 Software

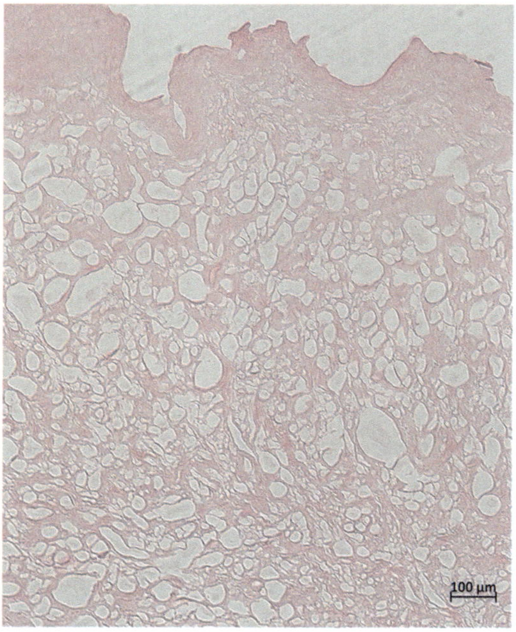

Fig. 5.2 Haematoxylin and Eosin staining of the ADS at 100× magnification (Image captured by Parkar 2019)

and saved in JPEG and Tag image File Format (Tiff) formats.

Lyophilized 8 mm biopsies of the ADS were scanned using a micro-XCT scanner. The ADS was then scanned using a micro-XCT scanner, available at NECSA, South Africa.

5.4.2 Characteristics of the Micro-Focus X-ray Tomography Facility (MIXRAD) Situated at NECSA in South Africa

The Nikon XTH 225 ST micro-focus X-ray tomography system, that is available at NECSA and was used in this research, functions within a voltage range of 30–225 kV and beam currents from 0 to 1 mA, with a maximum power capacity of 30 W. The X-ray tube has a spot size ranging from 0.001 to 0.003 mm (Fig. 5.3). The sample was secured to the sample manipulator (Fig. 5.3) within the micro-focus X-ray chamber to prevent any movement during scanning. The TIFF format of the projection images allowed the CT-Pro software to reconstruct them into a 3D volume. Then, the raw 3D volume file generated by the CT-Pro software was saved and imported into visualization software for image segmentation.

5.4.3 Image Segmentation Procedure Using Amira-Avizo 2019.3 Software (Thermo Fisher Scientific, Inc.)

Amira-Avizo 2019.3 software (Thermo Fisher Scientific, Inc.) is powerful 3D visualization software that enables users to visualize, generate 3D models, compare, and quantify 3D data efficiently. Image segmentation is an essential component in processing micro-XCT scan data, involving dividing an image into separate materials (e.g. ADS and Air) based on their, respective, grey levels to extract significant data and generate 3D mesh. The ADS used in this research was micro-XCT scanned with a resolution of 0.005 mm, and the images in TIFF format were imported into the Amira-Avizo 2019.3 software (Thermo Fisher Scientific, Inc.) for image segmentation and 3D model generation.

The following steps were followed for the image segmentation procedure:

1) The TIFF images were imported into the Amira-Avizo 2019.3 software (Thermo Fisher Scientific, Inc.) (Fig. 5.4). 2) To enhance the segmentation outputs, pre-processing steps such as denoizing, filtering, and enhancing contrast were computed. 3) The segmentation thresholds based on the intensity values for each material (e.g. ADS and air) using the "*Edit New Label Field*" module (Fig. 5.5) were manually set. To this end, the "Half Maximum Height" (HMH) quantitative iterative thresholding approach (Spoor et al. 1993) to determine the threshold levels between segmented elements was used. 4) The segmentation was manually finalized using editing tools to add or remove regions, change boundaries, or fix segments.5) Finally, the ADS 3D model was generated (Fig. 5.6) using the "*generate surface*" module, and the 3D model was exported in stereolithography (STL) format.

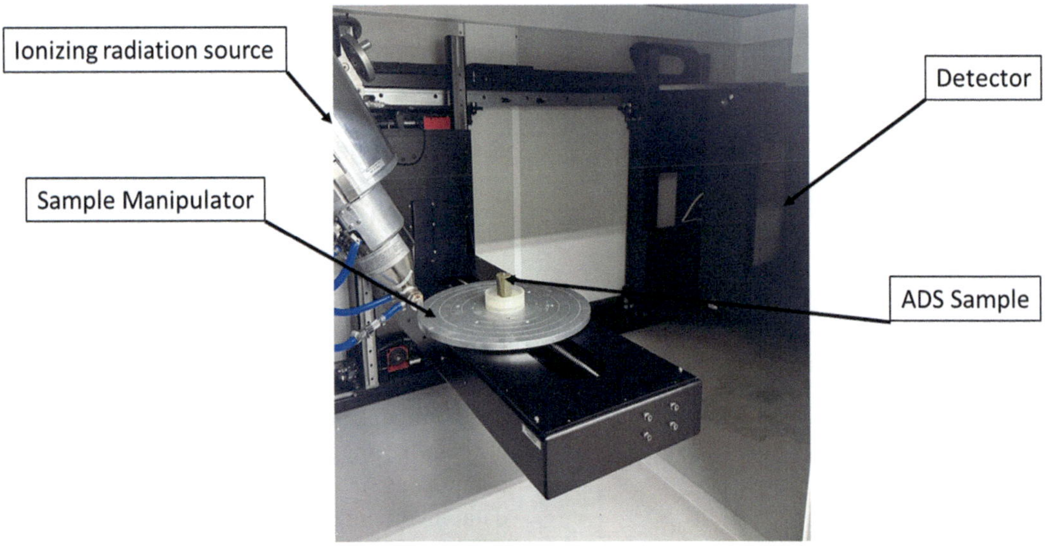

Fig. 5.3 Nikon XTH 225 ST micro-focus X-ray tomography system available at NECSA

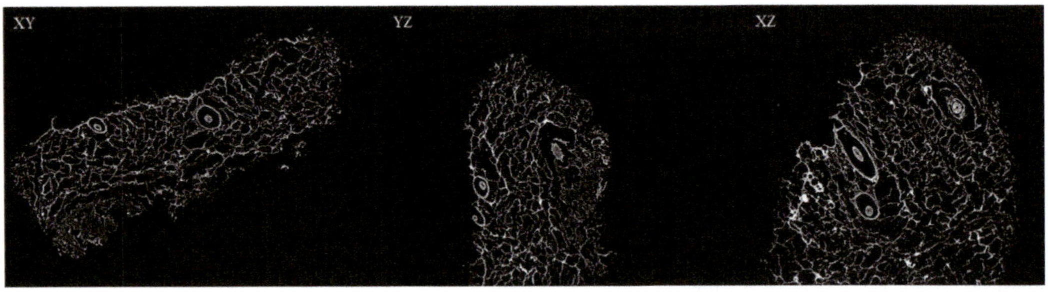

Fig. 5.4 Micro-XCT-scanned image data of the ADS in the three views (XY, YZ, XZ)

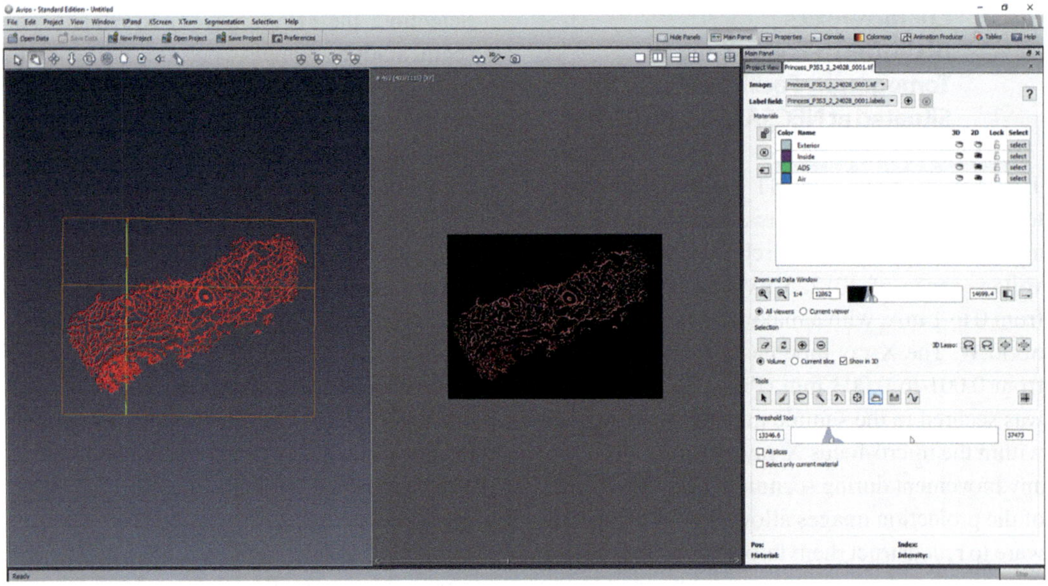

Fig. 5.5 Image segmentation procedure on Amira-Avizo 2019.3 software (Thermo Fisher Scientific, Inc.) using manual thresholding method

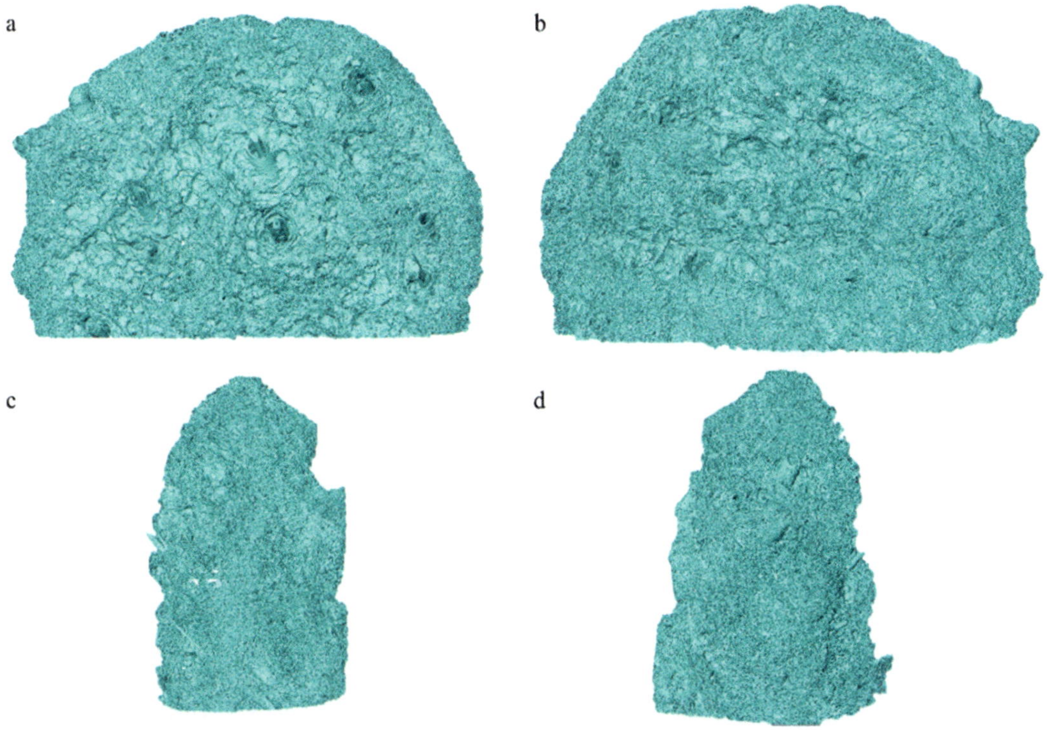

Fig. 5.6 Three-dimensional mesh of the ADS generated from image segmentation. (**a**) epidermal interface; (**b**) dermal interface; (**c**) left lateral view; (**d**) right lateral view

5.5 Discussion and Conclusion

This chapter discusses the literature on wound healing, regenerative medicine, and tissue engineering, including an innovative new digital protocol for creating a 3D-acellular dermal scaffold for bioprinting. The study involved the development of a decellularized acellular dermal scaffold from porcine dermis using a combination of physical, chemical, and enzymatic decellularization methodologies. Chemical decellularization using a combination of a non-ionic Triton X-100 and a cationic cetylpyridinium chloride detergent was followed by a physical decellularization process using SCO_2. This was followed by a final enzymatic decellularization by performing a benzonase nuclease hydrolysis.

Eight-millimetre sections were cut out of the decellularized dermis, and either fixed in formalin for histological analysis or lyophilized for further examination. The sections were embedded in paraffin wax blocks, cut into 5 μm slices, mounted on glass slides, and stained using haematoxylin and eosin. The lyophilized ADS was scanned using an X-ray micro-XCT system at NECSA, South Africa, with a resolution of 0.005 mm. TIFF images were imported into the Amira-Avizo 2019.3 software (Thermo Fisher Scientific, Inc.) for image segmentation and 3D model generation.

According to previous studies, a decellularized tissue or organ must adhere to specific criteria to be considered successful. First, it should have all cellular components, including cell membranes and nucleic acids, completely removed. Second, the process should maintain the biochemical composition, biological activity, and three-dimensional structure intact. Third, it is crucial to preserve the complex network of structural proteins that are part of the ECM. Finally, it should ensure that the native ECM-related signalling sites crucial for cell proliferation are retained (Badylak et al. 2009; Youngstrom et al. 2013; Hussein et al. 2016; Gilpin and Yang 2017). The use of a combinative method in this study produced an ADS of superior quality, retaining the intricate 3-dimensional ECM structure of the dermis whilst eradicating cells and any nuclear material which may have been present in the skin. As evidenced by Fig. 5.2, the haematoxylin and eosin staining shows the retention of only the fibrous structure of the ECM-stained pink by eosin, and the absence of cellular material, as cell nuclei would have been stained a purplish-blue colour by haematoxylin.

The use of combinative methods for decellularization of complex organs has been recommended in literature. A study conducted by Wainwright (1995) showed the combination of chemical and physical processing techniques resulted in an acellular dermal matrix with normal collagen bundling and organization and an intact basement membrane complex. The choice of decellularization agents in this study is supported by literature as being advantageous for achieving an optimally decellularized tissue. Supercritical carbon dioxide fluid has been shown to cause minimal damage to the mechanical characteristics of the extracellular matrix and has proven effective in cell removal from aortic tissue (Sawada et al. 2008). Non-ionic agents such as Triton-X100 have been extensively studied for their efficacy in decellularization by maintaining the ultrastructure of the ECM (Sullivan et al. 2012; Mendoza-Novelo et al. 2011; García-Gareta et al. 2020). Benzonase nuclease has successfully been used as a supplement to chemical and physical methods to remove any traces of residual DNA or Ribonucleic acid (RNA) adhering to the scaffolds post-decellularization (García-Gareta et al. 2020; Nag and Boyd 2023).

The advantage of producing a structurally intact ADS with no cellular content in this study allowed for the capturing of an anatomically and physiologically accurate high-resolution micro-XCT scan. As micro-XCT scanning can accurately show structures at the micron level in various tissues and can also visualize the distribution of cells within tissue engineering constructs, a full 3D image of a clinically relevant ADS construct could be produced (Shearer et al. 2016). Micro-XCT was initially used for imaging inorganic materials and fossils and gained popularity in biomedical engineering for structural analysis of calcified tissues and scaffolds. It generates 3D volumes that researchers can divide and analyze quantitatively, providing crucial data on volume fraction, porosity, and connectedness. Micro-XCT enables a thorough 3D observation of cell location and changes in biomaterial architecture across various sizes, unlike other imaging techniques limited to smaller construction areas (Shearer et al. 2016).

As illustrated by Fig. 5.4, high-resolution micro-XCT scans of the ADS could be produced, showing the ultrastructure of the ADS in intricate detail. This allowed for the image segmentation procedure to be carried out on the Amira-Avizo 2019.3 software (Thermo Fisher Scientific, Inc.) using a manual thresholding method (Fig. 5.5) which produced a detailed three-dimensional mesh of the ADS showing the epidermal interface, dermal interface, left lateral view and right lateral view as shown in Fig. 5.6. This 3D model was exported in stereolithography (STL) format which will be used for the 3D bioprinting of the ADS using the CELLINK BIO X bioprinter graciously supplied by PromoLab (Pty) Ltd. T/A Separations.

Using the generated STL file, a 3D-bioprinted ADS will be produced, this will allow for an accurate reproduction of the scanned ADS. Primary human dermal fibroblast cells will be incorporated into the print at a later stage to ensure that the bioprinted ADS can maintain cellular growth in vitro. This is with future plans to print patient-specific ADS containing autologous patient cells, which is an emerging use of this technology as reported in literature (Tran et al. 2023; Wang et al. 2022).

The scan produced in this research is a high-resolution product that can be reproduced easily and inexpensively. Whilst the workflow for this model is innovative, it is not without limitations. This method will be expensive to produce due to the high cost of bioinks. However, through regular use, ongoing research, coupled with advancements in bioprinting, and wound healing technologies, it is anticipated that this method will have reduced production costs over time, and the benefits will outweigh the initial investment.

Developing artificial dermal substitutes that replicate human skin is crucial for reconstructive engineering to facilitate the healing of secondary intent wounds (Sierra-Sánchez et al. 2021). 3D bioprinting of wound healing ADS promises a more consistent, feasible, and accessible solution for patients with chronic and secondary intention wounds. It minimizes material waste and can be tailored to fit the specific size, depth, and nature of the patient's wound, being a more specific and individualized approach (Yaneva et al. 2023). Enhancing wound healing can positively impact the patient's psychology during recovery and increase the quality of life by reducing the risk of wound infection or delayed healing. Enhancing wound healing can positively impact the patient's psychology during recovery and increase the quality of life by reducing the risk of wound infection or delayed healing (Gouin and Kiecolt-Glaser 2011).

Acknowledgements The authors would like to thank the following persons for their valuable contributions to this study:

Mr. Jakobus Hoffman and Dr. Lunga Bam, researchers at the South African Nuclear Energy Corporation (NECSA), for their assistance and support.

Professor AD Cromarty, Department of Pharmacology, University of Pretoria, South Africa, for the design and development of the supercritical carbon dioxide apparatus as well as assistance with optimization of the decellularization processes.

Mrs. Rephima Phaswane and the team at the Department of Pathology, Faculty of Veterinary Sciences, University of Pretoria, South Africa, for assistance with histology.

Competing Interests All the authors of this chapter declare no competing interests for this body of work.

Ethical Approval Ethical approval for this study has been granted by the University of Pretoria Research Ethics Committee. Ethics reference number: 470/2023.

References

Azmat CE, Council (2023) Wound closure techniques. [Updated 2023 Jun 26]. In: StatPearls [Internet]. StatPearls Publishing, Treasure Island, FL

Badylak SF, Freytes DO, Gilbert TW (2009) Extracellular matrix as a biological scaffold material: structure and function. Acta Biomater 5:1–13

Basit AW, Trenfield SJ (2022) 3D printing of pharmaceuticals and the role of pharmacy. Pharm J 308:7959

Berthiaume F, Yarmush ML (2003) Tissue engineering. In: Meyers RA (ed) Encyclopedia of physical science and technology, 3rd edn. Academic Press, New York

Bravin A, Coan P, Suortti P (2012) X-ray phase-contrast imaging: from pre-clinical applications towards clinics. Phys Med Biol 58:R1

Braza, M. E. & Fahrenkopf, M. P. 2023. Split-thickness skin grafts

Broccoli N, Rechtin M, Krishnan GD, Phero AJ (2020) Step-by-step: skin grafting. J Oral Maxillofac Surg 78:6–10

Byeon JY, Hwang YS, Choi HJ, Kim JH, Lee DW (2023) A long-term follow-up study of diabetic foot ulcer using micronized acellular dermal matrix. Int Wound J 20:1622–1637

Chhabra S, Chhabra N, Kaur A, Gupta N (2017) Wound healing concepts in clinical practice of OMFS. J Maxillofac Oral Surg 16:403–423

Crapo PM, Gilbert TW, Badylak SF (2011) An overview of tissue and whole organ decellularization processes. Biomaterials 32:3233–3243

Cundell J (2016) Diabetic foot ulcers: assessment, treatment, and management. Elsevier, Smart Bandage Technologies

Demmer W, Sorg H, Steiert A, Hauser J, Tilkorn DJ (2021) Wound healing and therapy in soft tissue defects of the hand and foot from a surgical point of view. Med Sci 9:71

Dorsey SM, Lin-Gibson S, Simon JR, C. G. (2009) X-ray microcomputed tomography for the measurement of cell adhesion and proliferation in polymer scaffolds. Biomaterials 30:2967–2974

Elseth A, Nunez Lopez O (2022) Wound grafts. [Updated 2021 Nov 3]. In: StatPearls [Internet]. StatPearls Publishing, Treasure Island, FL

Fang Y, Guo Y, Liu T, Xu R, Mao S, Mo X, Zhang T, Ouyang L, Xiong Z, Sun W (2022) Advances in 3D bioprinting. Chin J Mech Eng Addit Manufact Front 1:100011–100011

Frykberg RG, Banks J (2015) Challenges in the treatment of chronic wounds. Adv Wound Care 4:560–582

García-Gareta E, Abduldaiem Y, Sawadkar P, Kyriakidis C, Lali F, Greco KV (2020) Decellularised scaffolds: just a framework? Current knowledge and future directions. J Tissue Eng 11:2041731420942903

Gilpin A, Yang Y (2017) Decellularization strategies for regenerative medicine: from processing techniques to applications. BioMed Research International

Gonzalez ACDO, Costa TF, Andrade ZDA, Medrado ARAP (2016) Wound healing - a literature review. An Bras Dermatol 91:614–620

Gopinathan J, Noh I (2018) Recent trends in bioinks for 3D printing. Biomater Res 22:11

Gouin JP, Kiecolt-Glaser JK (2011) The impact of psychological stress on wound healing: methods and mechanisms. Immunol Allergy Clin N Am 31:81–93

Gu Z, Fu J, Lin H, He Y (2020) Development of 3D bioprinting: from printing methods to biomedical applications. Asian J Pharm Sci 15:529–557

Gungor-Ozkerim PS, Inci I, Zhang YS, Khademhosseini A, Dokmeci MR (2018) Bioinks for 3D bioprinting: an overview. Biomater Sci 6:915–946

Guo S, Dipietro LA (2010) Factors affecting wound healing. J Dent Res 89:219–229

Hahn HM, Lee DH, Lee IJ (2021) Ready-to-use micronized human acellular dermal matrix to accelerate wound healing in diabetic foot ulcers: a prospective randomized pilot study. Adv Skin Wound Care 34:1–6

Hazrati R, Davaran S, Omidi Y (2022) Bioactive functional scaffolds for stem cells delivery in wound healing and skin regeneration. React Funct Polym 174:105233–105233

Herskovitz I, Hughes OB, Macquhae F, Rakosi A, Kirsner R (2016) Epidermal skin grafting. Int Wound J 13:52–56

Hosseini M, Shafiee A (2021) Engineering bioactive scaffolds for skin regeneration. Small 17:2101384–2101384

Hoversten KP, Kiemele LJ, Stolp AM, Takahashi PY, Verdoorn BP (2020) Prevention, diagnosis, and management of chronic wounds in older adults. Mayo Clin Proc 95:2021–2034

Hussein KH, Park K-M, Kang K-S, Woo H-M (2016) Biocompatibility evaluation of tissue-engineered decellularized scaffolds for biomedical application. Mater Sci Eng C 67:766–778

Jamróz W, Szafraniec J, Kurek M, Jachowicz R (2018) 3D printing in pharmaceutical and medical applications – recent achievements and challenges. Pharm Res 35:176–176

Jose RR, Rodriguez MJ, Dixon TA, Omenetto F, Kaplan DL (2016) Evolution of bioinks and additive manufacturing technologies for 3D bioprinting. ACS Biomater Sci Eng 2:1662–1678

Kačarević ŽP, Rider PM, Alkildani S, Retnasingh S, Smeets R, Jung O, Ivanišević Z, Barbeck M (2018) An introduction to 3D bioprinting: possibilities, challenges and future aspects. Materials 11:2199

Kumar P, Ebbens S, Zhao X (2021) Inkjet printing of mammalian cells–theory and applications. Bioprinting 23:e00157

Landén N, Dongqing L, Mona S (2016) Transition from inflammation to proliferation: a critical step during wound healing. Cell Mol Life Sci 73:3861–3885

Lapelusa A, Dave HD (2023) Physiology. StatPearls Publishing, Hemostasis, Treasure Island

Liao J, Xu B, Zhang R, Fan Y, Xie H, Li X (2020) Applications of decellularized materials in tissue engineering: advantages, drawbacks and current improvements, and future perspectives. J Mater Chem B 8:10023–10049

Lindholm C, Searle R (2016) Wound management for the 21st century: combining effectiveness and efficiency. Int Wound J 13:5–15

Long Z, Tuttle V, Mckinney J, Andolina P, Ettlinger H, Kells A (2019) Rates of skin graft or flap failure in patients who have received osteopathic manipulative medicine treatment: A retrospective study. New York Med J 2154–9400:13

Lu H, Hoshiba T, Kawazoe N, Chen G (2011) Autologous extracellular matrix scaffolds for tissue engineering. Biomaterials 32:2489–2499

Mao AS, Mooney DJ (2015) Regenerative medicine: current therapies and future directions. Proc Natl Acad Sci 112:14452–14459

Mccaughan D, Sheard L, Cullum N, Dumville J, Chetter I (2018) Patients' perceptions and experiences of living with a surgical wound healing by secondary intention: a qualitative study. Int J Nurs Stud 77:29–38

Mendoza-Novelo B, Avila EE, Cauich-Rodríguez JV, Jorge-Herrero E, Rojo FJ, Guinea GV, Mata-Mata JL (2011) Decellularization of pericardial tissue and its impact on tensile viscoelasticity and glycosaminoglycan content. Acta Biomater 7:1241–1248

Meyer SR, Chiu B, Churchill TA, Zhu L, Lakey JR, Ross DB (2006) Comparison of aortic valve allograft decellularization techniques in the rat. J Biomed Mater Res A 79:254–262

Monte-Alto-Costa A, Porto LC (2010) Special stains for extracellular matrix. In: Hewitson TD, Darby IA (eds) Histology protocols. Humana Press, Totowa, NJ

Nag S, Boyd AS (2023) Decellularization of mouse kidneys to generate an extracellular matrix gel for human induced pluripotent stem cell derived renal organoids. Organoids 2:66–78

Neishabouri A, Soltani Khaboushan A, Daghigh F, Kajbafzadeh A-M, Majidi Zolbin M (2022) Decellularization in tissue engineering and regenerative medicine: evaluation, modification, and application methods. Front Bioeng Biotechnol 10:805299

Oganesyan G, Jarell AD, Srivastava M, Brian Jiang SI (2013) Efficacy and complication rates of full-thickness skin graft repair of lower extremity wounds after Mohs micrographic surgery. Dermatologic Surg 39:1334–1339

Parkar HP, Cromarty DC (2020) Wounds: an overlooked burden (Part 1)-effective wound management starts with proper wound assessment. South African General Practitioner 1:196–198

Parkar H, Cromarty AD, (2024) Proteomic characterisation of processed extracellular matrix scaffolds

for wound healing. PhD. University of Pretoria (unpublished)
Parkar H, Mlambo S, Naude L, Strydom H, Nakidien N, Cromarty A (2021) Wounds: an overlooked burden (Part 2)-wound treatment: a daunting decision. South African General Practitioner 2:24–30
Patel N, Solanki E, Picciani R, Cavett V, Caldwell-Busby JA, Bhattacharya SK (2008) Strategies to recover proteins from ocular tissues for proteomics. Proteomics 8:1055–1070
Ramanujam CL, Han D, Fowler S, Kilpadi K, Zgonis T (2013) Impact of diabetes and comorbidities on split-thickness skin grafts for foot wounds. J Am Podiatr Med Assoc 103:223–232
Reddy S, El-Haddawi F, Fancourt M, Farrant G, Gilkison W, Henderson N, Kyle S, Mosquera D (2014) The incidence and risk factors for lower limb skin graft failure. Dermatol Res Pract 2014:1
Reyzelman A, Crews RT, Moore JC, Moore L, Mukker JS, Offutt S, Tallis A, Turner WB, Vayser D, Winters C (2009) Clinical effectiveness of an acellular dermal regenerative tissue matrix compared to standard wound management in healing diabetic foot ulcers: a prospective, randomised, multicentre study. Int Wound J 6:196–208
Rosadas M, Silva IV, Costa JB, Ribeiro VP, Oliveira AL (2023) Decellularized dermal matrices: unleashing the potential in tissue engineering and regenerative medicine. Front Mater 10
Salih V (2009) Biodegradable scaffolds for tissue engineering. Cellular response to biomaterials. Elsevier
Sawada K, Terada D, Yamaoka T, Kitamura S, Fujisato T (2008) Cell removal with supercritical carbon dioxide for acellular artificial tissue. J Chem Technol Biotechnol 83:943–949
Schultz GS, Chin GA, Moldawer L, Diegelmann RF (2011a) 23 principles of wound healing. In: Fitridge R (ed) Mechanisms of vascular disease: a reference book for vascular specialists, vol 423. Springer, Cham
Schultz GS, Chin GA, Moldawer L, Diegelmann RF (2011b) Principles of wound healing. University of Adelaide Press, Adelaide
Shearer T, Bradley RS, Hidalgo-Bastida LA, Sherratt MJ, Cartmell SH (2016) Three-dimensional visualisation of soft biological structures by X-ray computed microtomography. J Cell Sci 129:2483–2492
Sierra-Sánchez Á, Kim KH, Blasco-Morente G, Arias-Santiago S (2021) Cellular human tissue-engineered skin substitutes investigated for deep and difficult to heal injuries. NPJ Regen Med 6:35
Spoor CF, Zonneveld FW, Macho GA (1993) Linear measurements of cortical bone and dental enamel by computed tomography: Applications and problems, Am. J. Phys. Anthr. 91 469–484, https://doi.org/10.1002/ajpa.1330910405.
Stock SR (2008) Recent advances in X-ray microtomography applied to materials. Int Mater Rev 53:129–181
Sullivan DC, Mirmalek-Sani S-H, Deegan DB, Baptista PM, Aboushwareb T, Atala A, Yoo JJ (2012) Decellularization methods of porcine kidneys for whole organ engineering using a high-throughput system. Biomaterials 33:7756–7764
Tran HQ, Shahriar SS, Yan Z, Xie J (2023) Recent advances in functional wound dressings. Adv Wound Care 12:399–427
Truong A-TN, Kowal-vern A, Latenser BA, Wiley DE, Walter RJ (2005) Comparison of dermal substitutes in wound healing utilizing a nude mouse model. J Burns Wounds 4:e4
Urciuolo F, Casale C, Imparato G, Netti PA (2019) Bioengineered skin substitutes: the role of extracellular matrix and vascularization in the healing of deep wounds. J Clin Med 8:2083
Vogt TN, Koller FJ, Santos PND, Lenhani BE, Guimarães PRB, Kalinke LP (2020) Quality of life assessment in chronic wound patients using the wound-QoL and FLQA-Wk instruments. Invest Educ Enferm 38:e11
Wainwright D (1995) Use of an acellular allograft dermal matrix (AlloDerm) in the management of full-thickness burns. Burns 21:243–248
Wallace HA, Basehore BM, Zito PM (2023) Wound healing phases. StatPearls Publishing, Treasure Island
Wang JK, Luo B, Guneta V, Li L, Foo SEM, Dai Y, Tan TTY, Tan NS, Choong C, Wong MTC (2017) Supercritical carbon dioxide extracted extracellular matrix material from adipose tissue. Mater Sci Eng C 75:349–358
Wang B, Qinglai T, Yang Q, Li M, Zeng S, Yang X, Xiao Z, Tong X, Lei L, Li S (2022) Functional acellular matrix for tissue repair. Materials Today Bio 18:100530
Withers PJ (2007) X-ray nanotomography. Materials Today 10:26–34
Yaneva A, Shopova D, Bakova D, Mihaylova A, Kasnakova P, Hristozova M, Semerdjieva M (2023) The Progress in bioprinting and its potential impact on health-related quality of life. Bioengineering 10:910
Yang Z, Yi P, Liu Z, Zhang W, Mei L, Feng C, Tu C, Li Z (2022) Stem cell-laden hydrogel-based 3D bioprinting for bone and cartilage tissue engineering. Front Bioeng Biotechnol 10:865770
Youngstrom DW, Barrett JG, Jose RR, Kaplan DL (2013) Functional characterization of detergent-decellularized equine tendon extracellular matrix for tissue engineering applications. PLoS One 8:e64151
Yu J, Park SA, Kim WD, Ha T, Xin Y-Z, Lee J, Lee D (2020) Current advances in 3D bioprinting technology and its applications for tissue engineering. Polymers 12:2958
Zhang C-Y, Fu C-P, Li X-Y, Lu X-C, Hu L-G, Kankala RK, Wang S-B, Chen A-Z (2022) Three-dimensional bioprinting of decellularized extracellular matrix-based bioinks for tissue engineering. Molecules 27:3442

6 Applied Anatomy for Acupuncture

Luke Welsh, Jon Hobbs, and Steven Buddle

Abstract

This chapter seeks to provide a detailed anatomical guide to assist practitioners of acupuncture techniques. The key to safe practice in any acupuncture procedure is a thorough respect and understanding of the patient's anatomy. This understanding must be accompanied by excellent palpation skills and extends to not just the specific needling point itself, but to the underlying anatomy within each region. Thus, this unique guide provides a detailed anatomical perspective of common acupuncture points (as defined and taught by a cross-section of acupuncture courses delivered within manual therapy), including how to efficiently utilize relevant surface landmarks to assist in their location. Discussion of the anatomical fundamentals of each needling point includes reference to key surrounding structures to promote a safe approach to each, and the relative safety of needling those points in clinical practice. A key feature of the chapter is the inclusion of cadaveric images for many of the acupuncture points which show needle position relevant to the underlying anatomy (e.g., in relation to neurovascular structures). This enhanced visualization provides an excellent insight into the safe practice of acupuncture. The chapter is aimed at clinicians currently practicing, or in training to practice acupuncture.

Keywords

Anatomy · Acupuncture · Cadaveric atlas
Clinical anatomy · Education

Abbreviations

BL	Bladder
CV	Conception Vessel
GB	Gall Bladder
GV	Governor Vessel
HT	Heart
KI	Kidney
LI	Large Intestine
LR	Liver
LU	Lung
PC	Pericardium
SI	Small Intestine
SP	Spleen
ST	Stomach
TE	Triple Energizer

L. Welsh (✉) · S. Buddle
Keele Anatomy & Surgical Training Centre, School of Medicine, Keele University, Keele, UK
e-mail: l.a.welsh@keele.ac.uk; s.r.buddle@keele.ac.uk

J. Hobbs
Acupuncture Association of Chartered Physiotherapists, Acupuncture Training Providers, Peterborough, UK

L. Shapiro (ed.), *How to use 3D Printing Innovations and Digital Storage to Democratize Anatomy Education*, Biomedical Visualization 6, https://doi.org/10.1007/978-3-031-68501-9_6

6.1 Introduction

One of the most fundamental components of any needling approach, whether that be traditional Chinese acupuncture, Western medical acupuncture, dry needling, or trigger point needling, is an understanding of the anatomy relevant to point location. In its most simplistic form, acupuncture point location can be considered to exist for perhaps two reasons; the first being that the location is relative to effect, the second being that once inserted, the indwelling needle misses all important structures. This is particularly evident with peripheral needling where many of the acupuncture point locations are perineural, therefore are close enough to nerves to exert an effect, while avoiding direct penetration and potential damage to neurons.

The key to safe practice of any acupuncture procedure is adherence to the triumvirate of respect; that being, respect the patient, respect the anatomy, and respect the technique. When viewing the interrelationship of these components, technique is entirely dictated by anatomy, both for safety and effectiveness. Therefore, if this triumvirate is adhered to then the potential incidence of adverse events can be significantly reduced. For this to occur, however, a practical and clinically applicable knowledge of anatomy must be in effect. Simply knowing point location as per classical text is not enough to ensure safety. This is primarily because of anatomical variance amongst individuals. The concept of N = 1 should be applied when inserting needles into a patient in the application of acupuncture, so while anatomy is shared across individuals, there remains a significant possibility of variation. Critically, understanding of anatomical variation must be supported by sound palpation skills and a profound knowledge of underlying anatomy to optimize the safety and effectiveness of any treatment.

This chapter acts as a cadaveric atlas to provide in-depth anatomical descriptions of commonly applied acupuncture points, with an aim to support practitioners in accurately locating these points using palpable and observable landmarks. Emphasis is placed on needle points requiring considerations of patient safety, typically those overlying vulnerable or intricate regions of anatomy. Annotated cadaveric images have been included to enhance visualization and understanding of the anatomy at these point locations. Each acupuncture point will be referenced by the abbreviation of their name denoting their associated meridian and the numerical value relating to the location they occupy on said meridian. For example, Large Intestine 4 will be referred to as LI4. The acupuncture point nomenclature is aligned with that of the World Health Organization's nomenclature for the naming of acupuncture points. The abbreviations list can be found below. The acupuncture point will also be named by its Mandarin name converted from the PinYin format to Wade Giles for English pronunciation. The Wade Giles system is a romanization of Mandarin Chinese developed by Thomas Francis Wade, a former professor of Chinese at Cambridge University during the nineteenth century (Chung 2016). The Mandarin name typically denotes a characteristic of the acupuncture point, such as where it is anatomically, what it does physiologically within the context of traditional Chinese medicine, or what the related anatomy looks like. For example, KI6 is known as Kunlun. It lies midway between the lateral malleolus of the ankle and the calcaneal tendon. Its name draws reference from the Kunlun Mountains of China as a comparison to the vertex of the lateral malleolus by which the point is located.

"Extra Points" are traditional acupuncture points that fall outside of the traditional 14 meridians but within traditional Chinese acupuncture they are still considered to be classical points. As these points fall outside of the traditional acupuncture meridians, they do not have a meridian name and number assigned to them. Instead, they have a specific Mandarin name which is again derived from such aspects as the local anatomy, what the point does physiologically, again within the context of traditional Chinese medicine, or what the related anatomy looks like. For example, XiYan which is the medial "eye of the knee" located at the medial arthroscopic portal of the knee, literally translates to "knee eye" in Mandarin.

All images of cadaveric specimens within this chapter display formalin-fixed tissue and were dissected by anatomy teaching staff at Keele Anatomy and Surgical Training Centre (KASTC). All cadaveric donors provided written consent to their images being used for educational and publishing purposes. Specific image permission was obtained from donors under the provision of the Human Tissue Act 2004. Needling of cadaveric specimens for photographing was performed by Jon Hobbs, Chairman of the board of Directors of the Acupuncture Association of Chartered Physiotherapists (AACP). During the dissection process of each cadaveric specimen, some acupuncture targets, particularly those consisting of connective tissue, may have been removed or cut prior to photography. In addition, some structures described as being of risk to needle points may no longer be present in full. Prosected tissue is often missing tissue structures such as adipose tissue, fascial layers, intermuscular septa, cutaneous nerves, muscular twigs of blood vessels and nerves, superficial veins, and lymphatics (Buddle 2023). Where possible, specimens have been dissected to accurately demonstrate the structures described at each needling site. All additions of color to specimens were added in post-production of images using Adobe Photoshop. Annotations of figures were then added by anatomy teaching staff at KASTC.

This chapter is divided into four main sections based on body region. These are: 1. The upper extremity, 2. The lower extremity, 3. The trunk, and 4. The head and neck. Descriptions of points within each of these regions are ordered from proximal to distal within the limbs, and superior to inferior in the head, neck, and torso. Subheadings within these sections are therefore based on anatomical sub-region and do not directly follow each meridian, however meridian lines are reflected within the nomenclature for each needle point. The following diagrammatic images detailing each meridian are for use with all chapters (Figs. 6.1, 6.2, 6.3, 6.4, 6.5, 6.6, 6.7, 6.8, 6.9, 6.10, and 6.11).

Figures 6.1–6.11 are supplied by Jon Hobbs (co-author) and used with permission.

6.1.1 A Note on Embryology

Understanding the segmental patterning of skeletal muscle, connective tissue, cartilage, and bone can provide significant insight into the interconnected nature of body morphology. While this chapter does not seek to provide significant embryological insight, it is useful to briefly review the development of key musculoskeletal tissues and consider how they rapidly entwine with the nervous system.

6.1.1.1 Somites

Somites are key structures not only in the formation of muscle, bone, and related connective tissue but also in neural tube closure and in defining the segmental pattern of vertebrates. Somites themselves are differentiated, paired masses of paraxial mesoderm which can be found immediately lateral to the midline neural tube (the precursor of the brain and spinal cord). The somites undergo further differentiation to form two distinct regions: the sclerotome and the dermomyotome.

6.1.1.2 Sclerotome

The ventrally placed sclerotome will eventually migrate around the neural tube and notochord (a midline cell population vital in guiding development) to form the bone and connective tissue of the vertebral column. The notochord will segment to become the nucleus pulposus. The sclerotome will also contribute to the formation of the proximal ribs and skull base. Vertebrae form by resegmentation of the sclerotomes with each forming from a total of four sclerotomes. This is achieved via the splitting of adjacent sclerotomes—the caudal half of the superior sclerotome subsequently unites to form a vertebra with the cranial half of the inferior sclerotome. The result of this is a reorienting of the relationship between the bony vertebrae and muscles, allowing muscles to span the joint spaces between the bones and therefore permitting movements throughout the vertebral column. Segmentation of the body during development may offer insight into the interconnectivity of certain acupuncture points, which follow similar patterns.

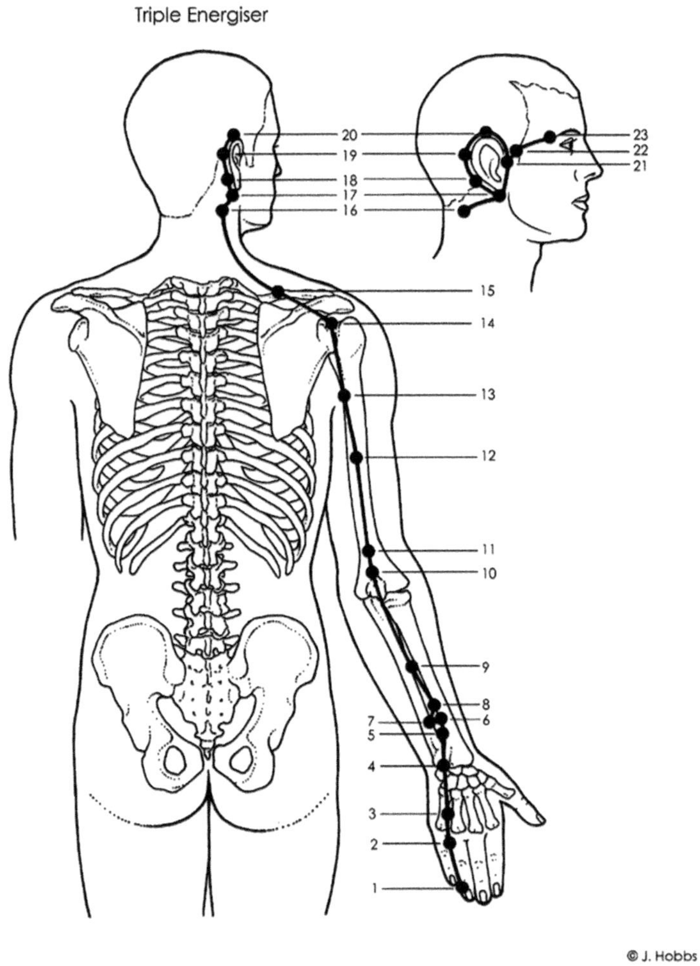

Fig. 6.1 Triple Energizer (TE) Meridian—acupuncture points

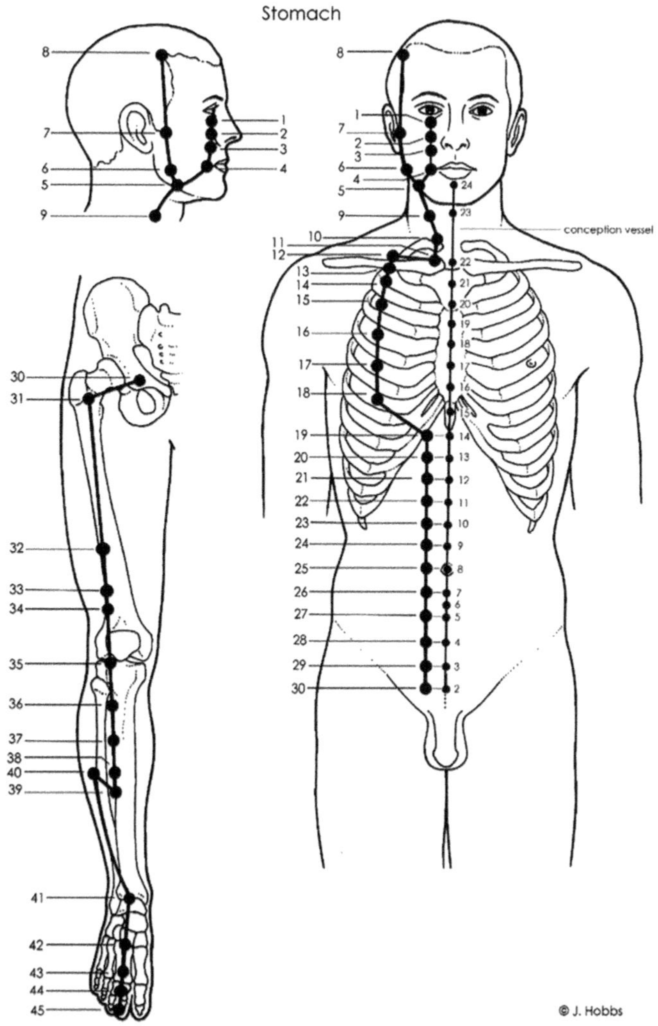

Fig. 6.2 Stomach (ST) Meridian—acupuncture points

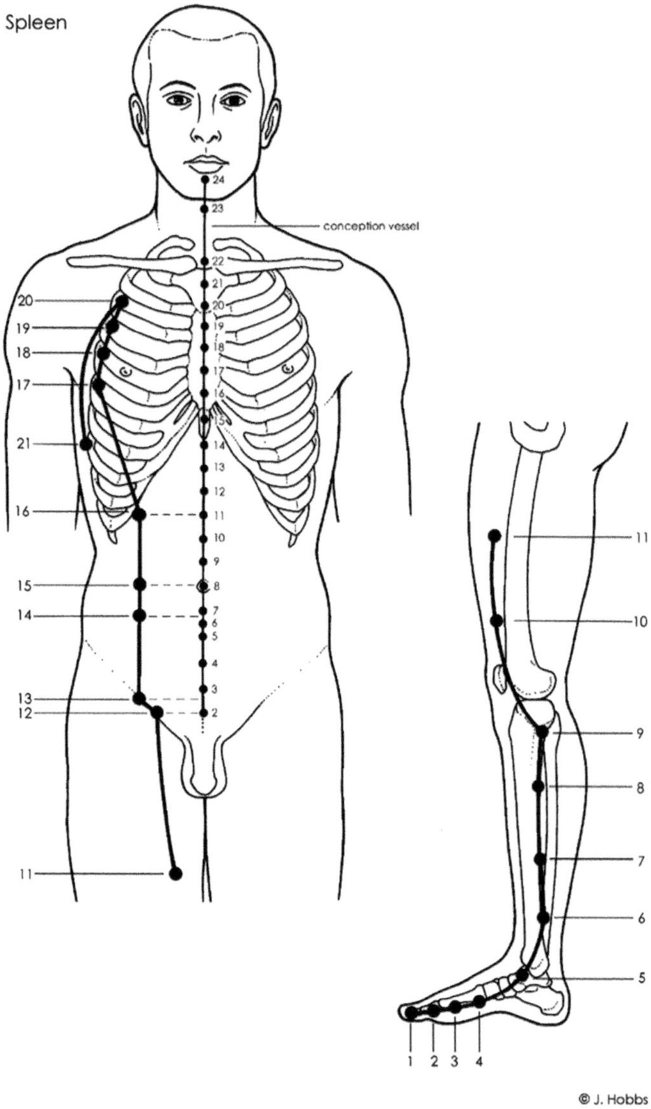

Fig. 6.3 Spleen (SP) Meridian—acupuncture points

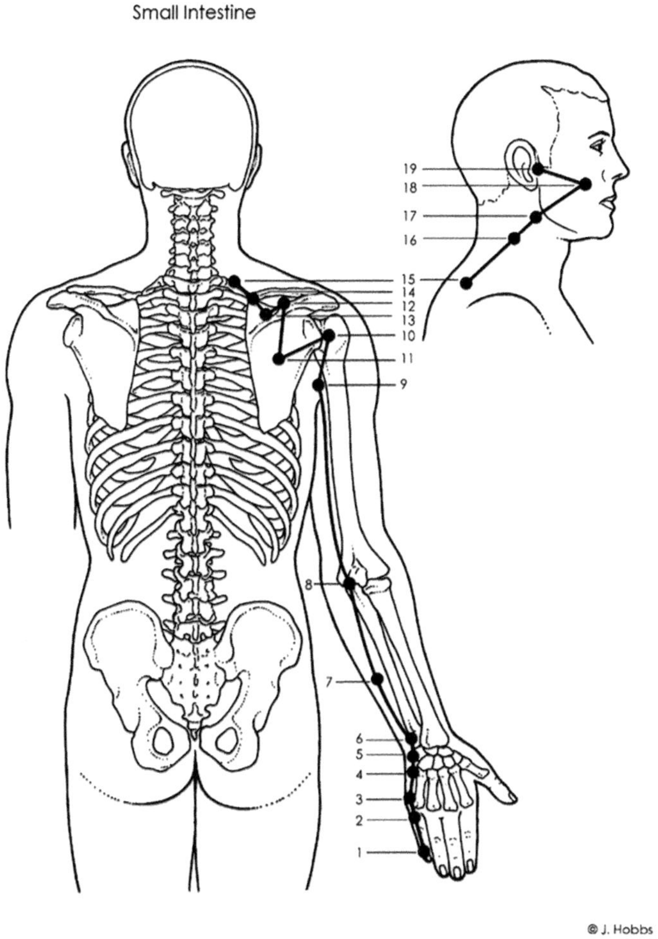

Fig. 6.4 Small Intestine (SI) Meridian—acupuncture points

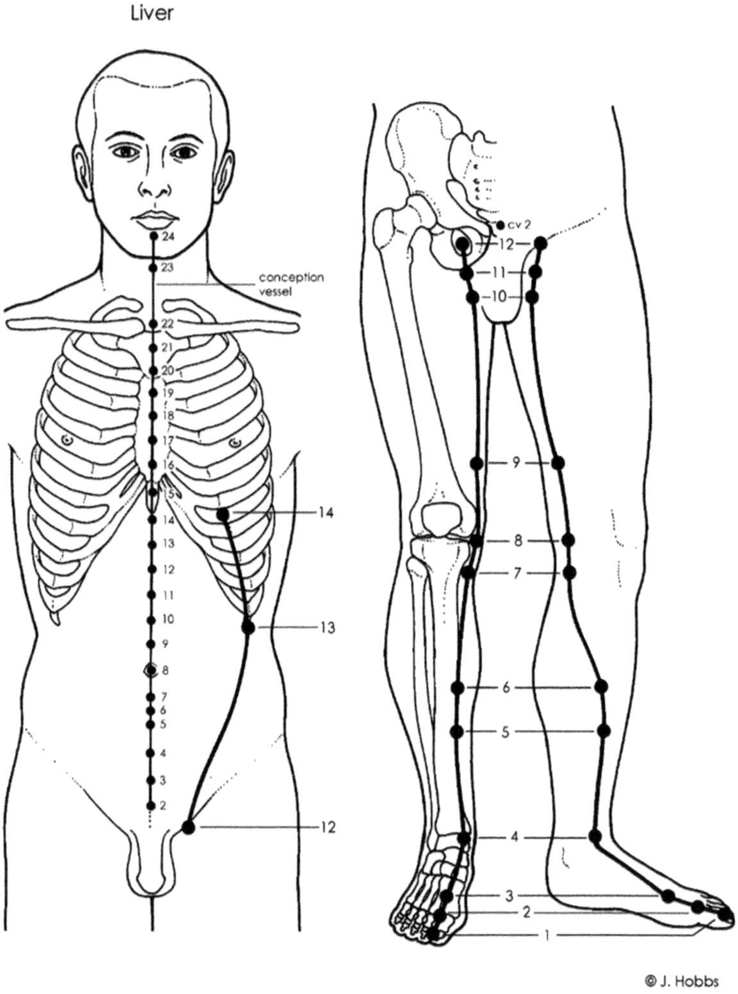

Fig. 6.5 Liver (LR) Meridian—acupuncture points

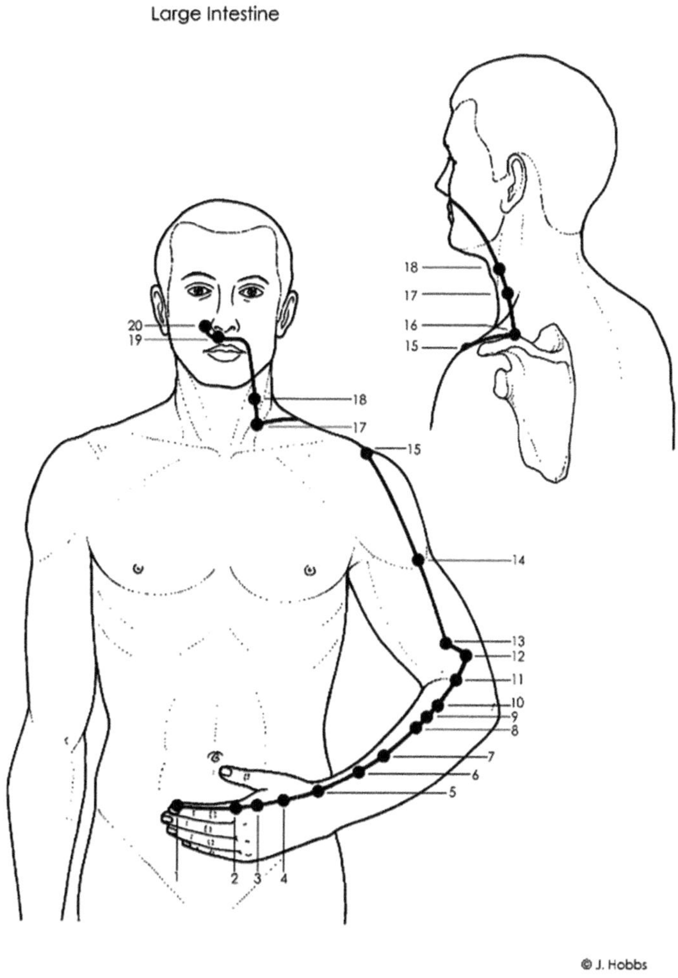

Fig. 6.6 Large Intestine (LI) Meridian—acupuncture points

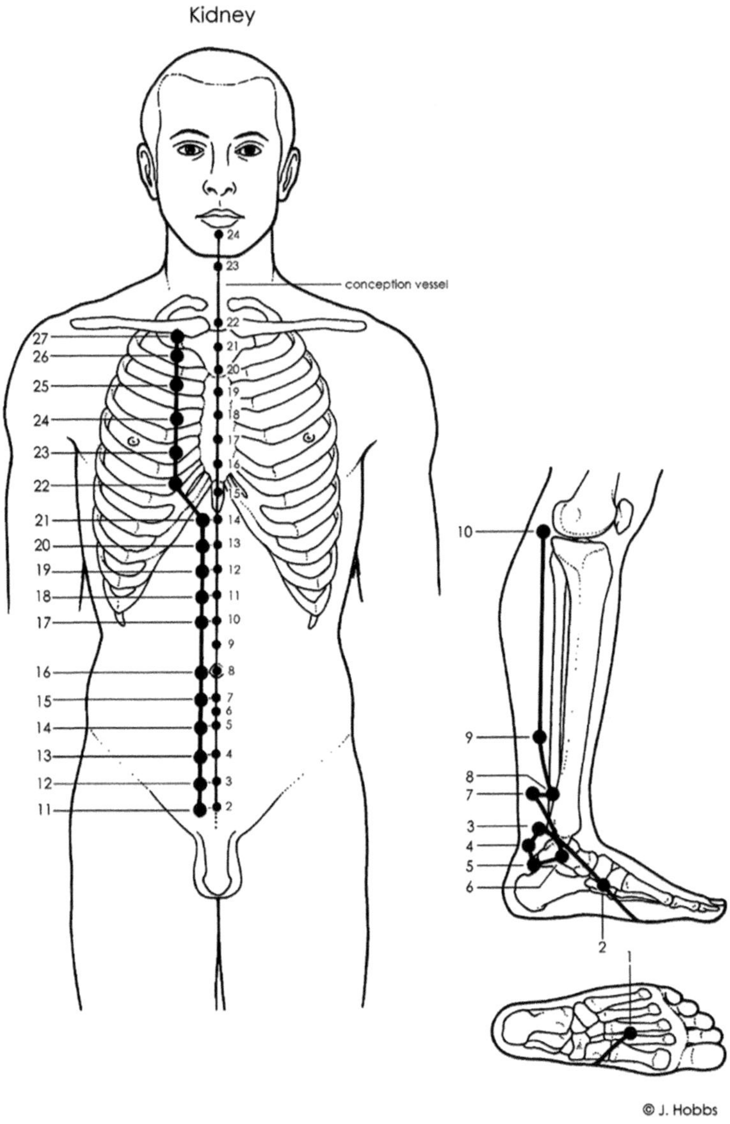

Fig. 6.7 Kidney (KI) Meridian—acupuncture points

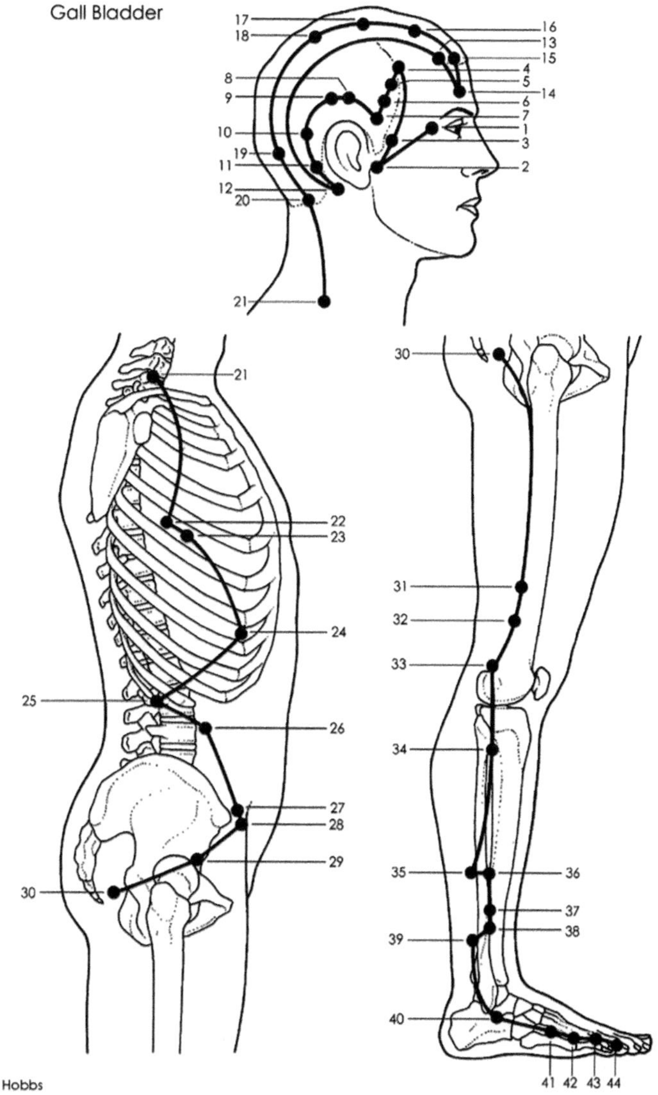

Fig. 6.8 Gallbladder (GB) Meridian—acupuncture points

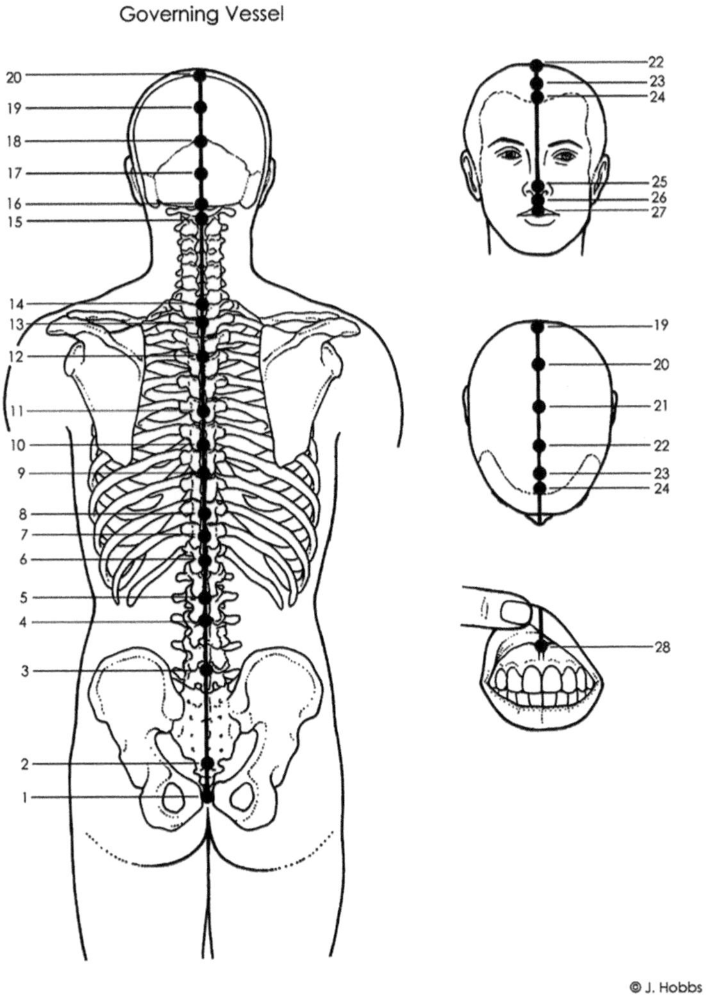

Fig. 6.9 Governing Vessel (GV) Meridian—acupuncture points

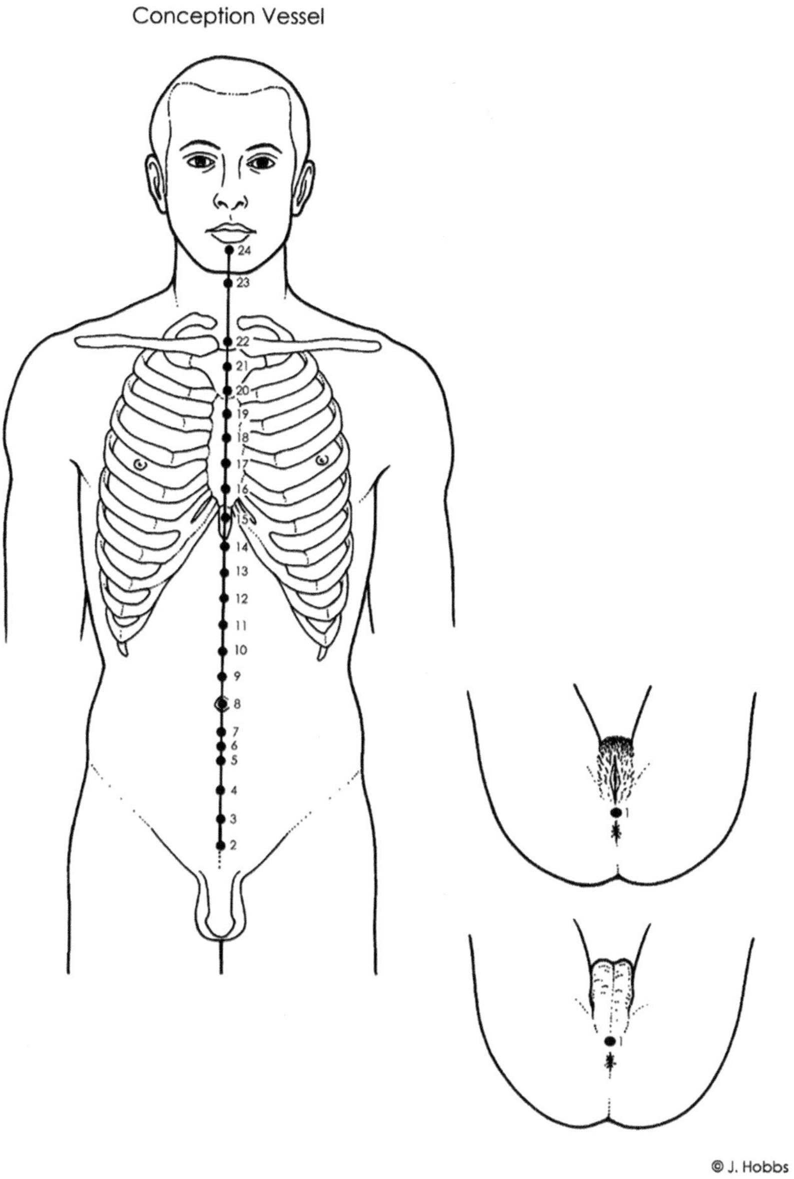

Fig. 6.10 Conception Vessel (CV) Meridian—acupuncture points

Fig. 6.11 Bladder (BL) Meridian—acupuncture points

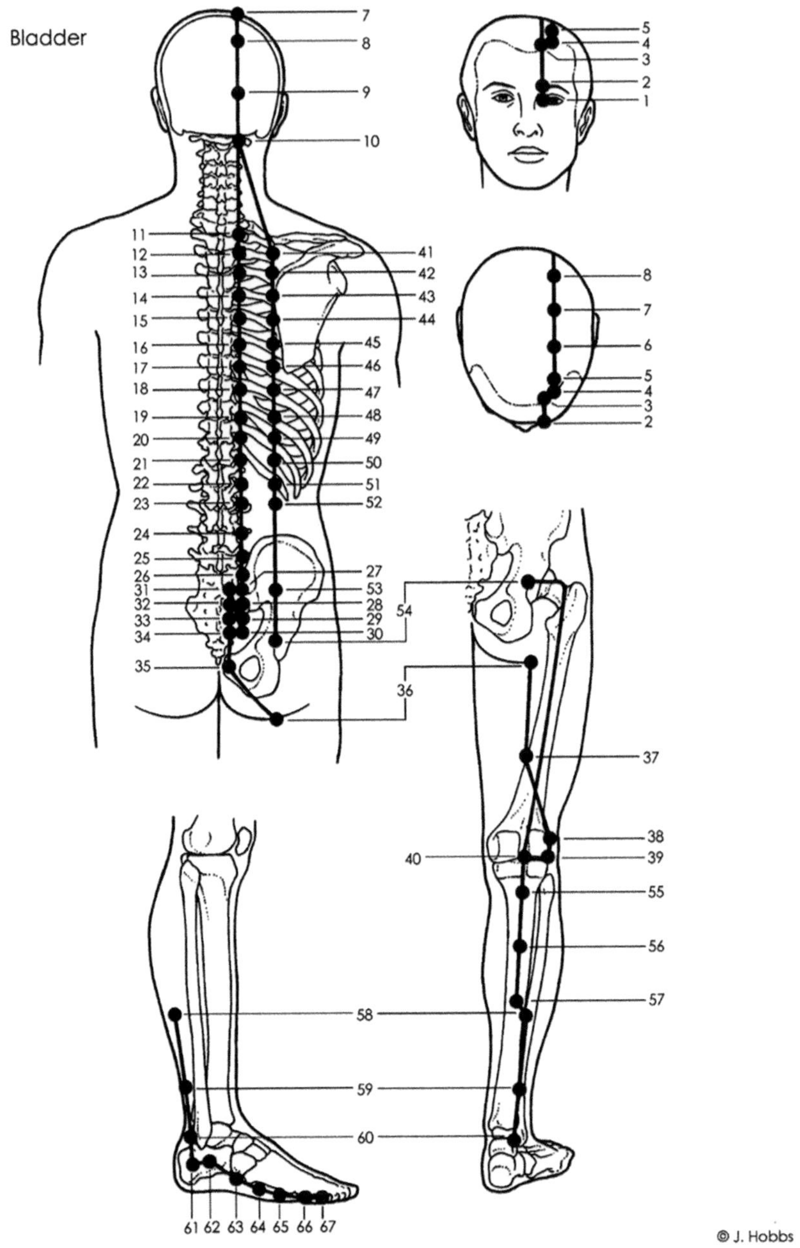

6.1.1.3 Dermomyotome

Dorsally, the somite now consists of segmental muscle precursors (myotome) and connective tissue related to the dermis (dermatome). Muscle cells are migratory and depending on their segment of origin can migrate a great distance. For example, myotomes originating from the lower cervical and upper thoracic regions of the embryo will eventually migrate down the developing upper limb to contribute distal musculature. Skeletal muscles are innervated by spinal nerves from the same segments of origin. If the interossei of the hand are used as an example, they will originate from the C8 and T1 myotomes and as they migrate down the developing limb, they will maintain their connection to the respective nerve roots. Thus, the nerve will lengthen to form the ulnar nerve and will be responsible for innervating the interossei.

6.2 A Note on the Autonomic Nervous System

As internal organs cannot be needled directly due to significant risk of functional damage, hemorrhage, and infection, it is likely that acupuncture affects the viscera indirectly, via the autonomic nervous system.

Needling in certain regions, particularly when close to the pelvis or sacrum, could cause direct damage to autonomic nerve fibers, leading to significant adverse effects. It is therefore important for an acupuncturist to understand the basic structure and function of the autonomic nervous system (ANS) before needling these points.

Due to the embryological derivation of somites, no spinal nerves reach the organs within the body cavities, therefore there is no somatic innervation of the internal viscera. Instead, organs are supplied by autonomic nerves, which are sympathetic or parasympathetic in function. Sympathetic nerves are present within all body regions, primarily targeting blood vessels to enable vasoconstriction. In contrast, parasympathetic nerves are more restricted in their distribution and do not extend into the limbs or body walls. They are contained only within the head, neck, and torso.

The primary function of the ANS is to maintain homeostasis, a function governed largely by the hypothalamus. One of the most basic principles of the ANS is to understand that sympathetic nerves trigger functions relating to the "fight or flight" response, while parasympathetic nerves trigger functions relating to "resting and digesting." These two nerve systems therefore exert reciprocal functions. A second key principle is to realize that autonomic efferent nerves have only three possible targets: smooth muscle, cardiac muscle, and glands. By contrast, all cutaneous sensation and innervation of skeletal muscle is via the somatic nervous system.

Autonomic nerves distribute to the major organs primarily via arteries; forming plexuses on their surfaces and using them as structural supports to reach their targets. There are exceptions to this pattern of vascular distribution, primarily concerning parasympathetic nerves superior to the diaphragm. In addition, somatic spinal nerves (e.g., the T6 intercostal nerve, the sciatic nerve, and the phrenic nerve) all have sympathetic nerve fibers traveling with them to reach peripheral targets such as blood vessels and sweat glands. Damage to autonomic nerves can therefore (generally) be prevented in acupuncture by avoiding needling of blood vessels and larger peripheral nerves.

When considering deeper needling, awareness of autonomic nerves that lie distinct from blood vessels and peripheral nerves is of particular importance. Within the posterior thorax and abdomen, the sympathetic chain and splanchnic nerves lie adjacent to the vertebral bodies and descending aorta. They are therefore not accessible when using safe needling approaches. Within the pelvis and neck, however, the autonomic nerves are less protected, and caution should be exercised when needling in these regions.

In the pelvis the autonomic nerves form an extensive plexus. This consists of the superior hypogastric plexus, formed by sympathetic nerves descending into the pelvis from the abdomen; and the inferior hypogastric plexus, formed by a mix of these sympathetic nerves and parasympathetic splanchnic nerves arising from S2, S3, and S4. These sacral nerve roots innervate the hindgut, pelvis, and perineum, and may be particularly vulnerable when needling the inner Bladder channel. Damage to them could have significant and life-altering consequences, including sexual dysfunction, and urinary or fecal incontinence. Note that the pudendal nerve provides somatic innervation to the pelvis and perineum and should not be confused with autonomic innervation to these regions.

In the neck, the sympathetic chain lies deeply but may be relevant to a multitude of needling points. Ascending bilaterally with only three cervical ganglia present, it is long and thin and particularly vulnerable. The cervical sympathetic chain is the only source of sympathetic innervation to the head, and damage to it may cause unilateral Horner's syndrome. The vagus nerve is also present within the neck, descending bilaterally within the carotid sheath. The vagus nerves are the source of all parasympathetic innervation to the heart, lungs, foregut, and midgut. As the other contents of the carotid sheath (the internal jugular vein and common carotid artery) should also be avoided, damage to the vagus nerve via needling of the neck should be avoidable.

6.2.1 Pain Referral Pathways

In addition to the efferent autonomic fibers which distribute toward smooth muscle, cardiac muscle, and glands, visceral afferent fibers are also sent back from these structures to update the central nervous system with information on the function and state of the internal organs. Visceral afferent fibers follow a similar course to efferent autonomic nerves and can therefore be found in the same regions.

When pain is detected in an organ, it is typically poorly localized and often referred to specific dermatomes of the skin. Within the thorax and abdomen, visceral pain sensations reach the spinal cord via sympathetic visceral afferents, while reflexive information travels via parasympathetic visceral afferents. In the pelvis, there is a theorized "pelvic pain-line" corresponding to the inferior limit of the peritoneum. Visceral pain from organs in contact with, or inferior to this "pain-line," will travel through pelvic splanchnic nerves, which are parasympathetic. Visceral pain from the thorax and abdomen therefore reaches the spinal cord at predictable levels between T1 and L2, while pain from pelvic organs primarily reaches the spinal cord at S2–S4. The levels at which visceral pain returns to the spinal cord determine which dermatomes this pain will be referred to.

6.3 A Note on Cun

For the non-acupuncturist, or those new to the field, descriptions of distance are traditionally measured in "cun"—equivalent to the length of the middle phalanx of the middle finger or the width of the thumb at the interphalangeal joint. These are used to provide quick reference points during palpation and thus should be used as a guide and not a definitive measurement. The word "cun," meaning "body inch" is a Mandarin term and when vocalized in the Pinyin (spelled sounds) System would be pronounced in English as either "chun" or "tsun" (Liao et al. 1994; White et al. 2018). The cun is a proportional measurement used as a guide to establish location of acupuncture points relative to palpable anatomical structures such as key bony landmarks or easily identifiable soft tissue structures. It is not an exact measurement due to the variability in measurement between individuals. It is however a useful approach to locating an approximated area prior to direct palpation when identifying an acupuncture point (Yin et al. 2005).

6.4 Applied Anatomy for Acupuncture Part 1: Upper Limb

Working from medial to lateral then anterior to posterior, the meridians that lay on the upper limb are the Heart, Pericardium, Lung, Large Intestine, Triple Energizer, and Small Intestine (Figs. 6.1, 6.4, 6.6, and 6.12). The most utilized classical acupuncture points of the upper extremity will be described anatomically, and the relative safety of needling these points in clinical practice are discussed. As in all sections of this chapter, location of these needlepoints relies on a thorough understanding of anatomy, accurate palpation skills, and an appreciation of variation between individuals.

6.4.1 Shoulder

LI16 (Ju Gu) is located on the superior aspect of the scapular region, in the depression posterior to the lateral end of the clavicle and anterior to the scapular spine, medial to the acromioclavicular joint. The needle is then inserted perpendicularly or obliquely 0.5–1.5 cun. The needle may first enter the most lateral fibers of the upper trapezius muscle or its fascia before entering supraspinatus at a deeper level. The lateral supraclavicular nerve is found subcutaneously in this region and should be avoided, though it does follow a highly variable course. Typically, it crosses the lateral

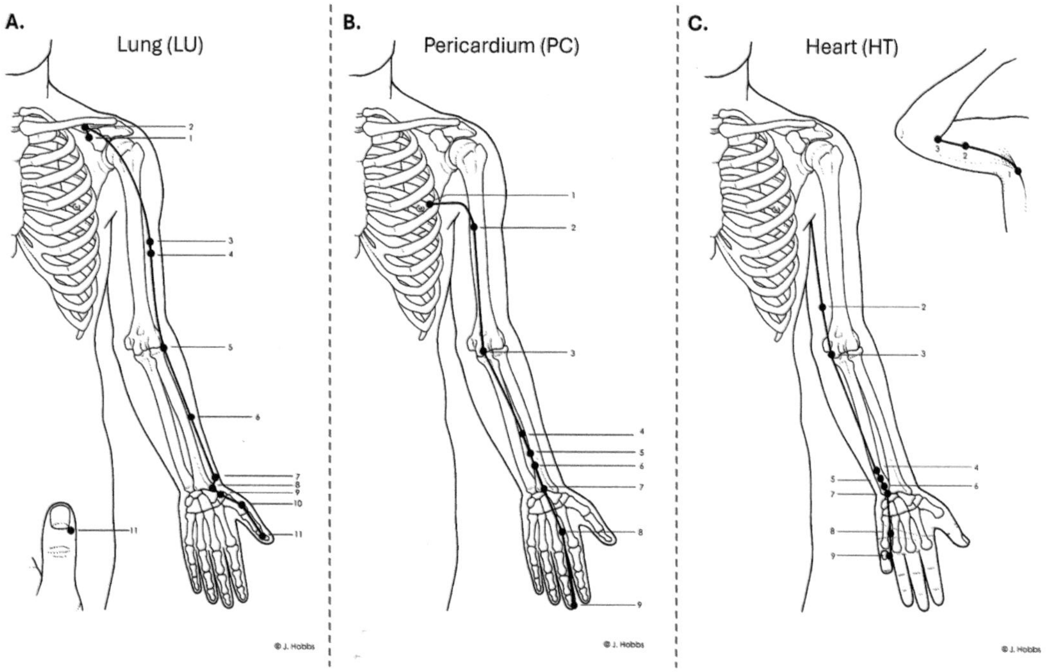

Fig. 6.12 Meridian acupuncture points specific to the upper extremity. (**a**) Lung (LU). (**b**) Pericardium (PC). (**c**) Heart (HT)

end of the clavicle, superficial to trapezius, and the acromion. Caution should also be exercised if deep needling occurs in an anterior or medial direction toward the thoracic cage, due to the potential risk of traumatic pneumothorax.

LI15 (Jian Yu) is located on the anterolateral aspect of the shoulder in the depression inferior to the anterior border of the acromial part of the deltoid muscle, between its anterior and middle fibers. This is best identified when the arm is abducted. This anatomical location may also be colloquially known as the anterior "eye of the shoulder." The needle may be inserted perpendicularly or obliquely 0.8–1.5 cun. Caution should be exercised not to damage the deltoid branch of the thoracoacromial artery and/or the branches of the lateral supraclavicular nerve. The axillary nerve is also at risk when needling deeply. If needling proximal to distal, care should also be taken not to injure the circumflex humeral arteries (anterior and posterior) which lie just distal to the head of the humerus. Deep to the deltoid and superficial to the greater tuberosity of the humerus lies the subdeltoid bursa which may also communicate with the subacromial bursa.

LI14 (Bi Nao) is located on the lateral arm, on the anterior margin of the insertion of the deltoid muscle. It follows the line connecting LI11 at the cubital crease and LI15 inferior to the acromion. The point is needled perpendicularly or obliquely 0.8–1.5 cun into the proximal attachment of the brachialis muscle. When the needle is inserted more anteriorly to the lateral border of the humerus it may also be inserted into the long head of biceps brachii. If the needle is located posterior to the lateral aspect of the humerus it will encounter the fibers of the lateral head of triceps brachii. Be aware that in smaller stature individuals, a perpendicular approach in the region of the deltoid insertion may lead to direct needling of the humerus with relatively superficial insertion.

LI13 (Shou Wu Li) is located on the lateral arm, 3 cun superior to LI11, on the line connecting LI11 at the cubital crease and LI15 inferior to the acromion. The needle is inserted perpendicularly 0.5–1 cun into the lateral aspect of the brachialis muscle. The cephalic vein should be avoided here as it runs lateral to the biceps brachii muscle.

TE14 (Jian Liao) is located on the posterolateral aspect of the shoulder in the depression inferior to the posterior border of the acromial part of the deltoid muscle, between its posterior and middle fibers. This is best identified when the arm is abducted. This anatomical location may also be colloquially known as the posterior "eye of the shoulder." The point is needled perpendicularly or obliquely 0.7–1 cun. The structures at risk when needling this point are like those of LI15, described above.

SI10 (Nao Shu) is located on the posterior shoulder, with the arm in adduction, in the depression inferior and lateral to the scapular spine at the acromion, on a line running from the acromion to the superior aspect of the posterior axillary fold. The point is needled perpendicularly 0.5–1 cun into the posterior deltoid initially and then the lateral aspect of infraspinatus at a deeper position. This point may be clinically recognized as the myofascial trigger point for the posterior fibers of the deltoid muscle. Caution should also be taken not to puncture the suprascapular vein, artery, or nerve, or the structures in the quadrangular space, which include the axillary nerve and posterior circumflex humeral artery and vein.

SI9 (Jian Zhen) is located on the posterior aspect of the shoulder. With the upper arm held in passive adduction, the point is identified approximately 1 cun superior to the posterior aspect of the axillary fold and inferior to the middle third fibers of the posterior deltoid. The point is needled perpendicularly 0.5–1 cun and may be inserted into either teres major or teres minor. This point may be clinically recognized as the myofascial trigger point for either the teres minor or teres major muscles dependent on location relative to palpation and depth of needling. The long head of triceps brachii may also be needled at SI9, as may latissimus dorsi at a significantly deeper location. Due to the proximity to the thoracic cage in this region, perpendicular insertion of the needle in relation to the posterior aspect of the upper limb is recommended. Deep needling toward teres major in a medial direction in this position is contraindicated due to the risk of traumatic pneumothorax. If teres major is the target tissue for this point, then the arm should be taken in abduction to ease identification of this muscle and reduce the risk of potential pneumothorax. Caution should also be taken not to puncture the circumflex scapular artery or vein, or the structures in the quadrangular space, which include the axillary nerve and posterior circumflex humeral artery and vein.

6.4.2 Extra Point

M-UE-48 (Jianneiling) is located in the depression between the coracoid process and the head of the humerus, approximately midway between the end of the anterior axillary fold and LI15 at the anterior "eye" of the shoulder. The needle is inserted perpendicularly 0.5–1 cun into the anterior deltoid muscle and progresses deeper to the lateral fibers of the pectoralis major muscle before overlying the tendon of the long head of biceps brachii muscle. Caution should be taken to avoid the underlying cephalic vein and middle supraclavicular nerve. When progressing deeper in this region, the anterior circumflex humeral artery and the deltoid branch of the thoracoacromial artery are also at risk.

6.4.3 Forbidden Point

HT1 (Ji Quan) is located in the axillary fossa when the arm is abducted and lies medial to the axillary artery. Traditionally this point is needled perpendicularly with caution due to its proximity to the axillary vessels and brachial plexus. For safety, the axillary artery can be identified via palpation and then the needle inserted medial and posterior to it. Caution should also be exercised to avoid the axillary vein, which adheres to the deep fascial layers within that region. Within the Acupuncture Association of Chartered Physiotherapists (AACP) foundation course in the UK, this point is considered to be a forbidden point for needling due to its proximity to the aforementioned artery.

6.4.4 Elbow and Forearm

TE10 (Tian Jing) is located in the posterior cubital region, in the depression 1 cun superior to the olecranon when the elbow is flexed. The needle is inserted perpendicularly 0.5–1 cun and passes into the tendon of the triceps brachii muscle and progresses into the olecranon fossa. Caution not to extend the elbow while the needle is in situ due to risk of trapping and damaging the needle and surrounding structures.

SI8 (Xiao Hai) is located with the elbow in flexion, in the depression between the olecranon and the medial epicondyle of the humerus, in the ulnar groove. Needle perpendicularly 0.3–0.5 cun from the posterior aspect of the elbow proximal to distal toward the wrist. In situ the needle may lie superficially along the course of the ulnar nerve which passes deep to it in the ulnar groove. Caution must be exercised to avoid traumatizing the underlying ulnar nerve. This can be achieved by directing the needle parallel to the ulnar nerve with a superficial proximal to distal insertion in the flexor carpi ulnaris muscle.

HT3 (Shao Hai) is located with the elbow flexed to 90° at the medial end of the transverse cubital crease at the midpoint between the medial epicondyle and the ulnar side of the tendon of biceps brachii. Needle perpendicularly exercising caution to avoid penetrating the basilic vein and medial cutaneous nerve of the forearm. If needling deeper, one should avoid the branches of the inferior ulnar collateral artery and vein and more inferiorly the anterior branches of the ulnar recurrent artery. The median nerve and brachial artery are also situated in the same region but tend to lie deeper and more lateral than the recommended needle insertion. The needle itself will be inserted initially into the pronator teres muscle superficially then if needled deeper it may fall within brachialis.

PC3 (Qu Ze) is located at the elbow, on the medial side of the biceps brachii tendon, medial to the brachial artery, and lateral to pronator teres. The needle is inserted perpendicularly 0.5–1 cun. Caution should be taken not to penetrate the medial cutaneous nerve of the forearm or the basilic vein. On deeper penetration, one should take care to avoid the brachial artery, vein, and median nerve which is covered superficially by the bicipital aponeurosis.

LU5 (Chi Ze) is located at the elbow, in the cubital crease, in the depression lateral to biceps brachii tendon and medial to the brachioradialis muscle. The needle is inserted perpendicularly 0.5–1 cun between these structures. Caution should be taken not to needle into the cephalic vein or lateral cutaneous nerve of the forearm. On deeper penetration, one should take care to avoid puncturing the collateral/recurrent radial artery or the radial nerve.

LI12 (Zhou Liao) is located on the lateral side of the cubital crease, in a depression 1 cun superior to LI11, at the junction of the lateral supracondylar ridge of the humerus with the lateral epicondyle. LI12 is best located with the elbow at 90 degrees of flexion. It is located on the lateral aspect of the triceps brachii muscle at the humeral attachment of the anconeus muscle. The needle is inserted perpendicularly 0.5–1 cun into the lateral aspect of the triceps brachii muscle and progresses medially to the humeral attachment of the anconeus muscle. The needle may also be directed into the brachioradialis or brachialis muscles from this position depending on the presentation of clinical symptoms and palpation. If needling deeper, caution should be taken as branches of the collateral/recurrent radial artery are found here.

LI11 (Qu Chi) is located on the lateral side of the cubital crease when the elbow is close to full flexion, midway between the tendon of biceps brachii and the lateral epicondyle, lateral to the brachioradialis muscle, in close proximity to the origin of the extensor carpi radialis longus muscle. The needle is inserted perpendicularly 1–1.5 cun and will penetrate the brachioradialis muscle and on deeper penetration may reach extensor carpi radialis longus and extensor carpi radialis brevis. Caution should be exercised not to damage the branches of the posterior cutaneous nerve of the forearm, and the deeper placed collateral/recurrent radial artery and the radial nerve. If the needle is directed medially and a deeper insertion is employed, then one should take care not to penetrate the brachial artery or the median nerve.

LI10 (Shou San Li) is located on the lateral side of the posterior forearm, 2 cun distal to the cubital crease, lateral to the brachioradialis muscle. On insertion the needle lies in the extensor carpi radialis longus muscle and deep to this in the extensor carpi radialis brevis muscle. On deeper insertion it may also enter the supinator muscle. Traditionally, it is referred to as laying on the line connecting LI5 at the wrist and LI11 at the lateral cubital crease (Fig. 6.6). The needle is inserted perpendicularly 0.5–1.5 cun into the muscle of choice. This point location is also clinically recognized as a significant myofascial trigger point in this region. Caution should be exercised not to damage the branches of the posterior or lateral cutaneous nerves of the forearm and at a deeper depth the recurrent radial artery and branches of the radial nerve.

LI9 (Shang Lian) is located on the lateral side of the posterior forearm, 3 cun distal to the cubital crease, on the line connecting LI5 at the wrist, and LI11 at the lateral cubital crease. The needle is inserted perpendicularly 0.5–1 cun initially into the brachioradialis muscle and then deeper into the tendon of the extensor carpi radialis longus muscle and then more deeply extensor carpi radialis brevis and the supinator muscle. The needle may also penetrate the extensor digitorum muscle and the abductor pollicis longus muscle depending on its angle and depth of insertion. Caution should be exercised not to damage the cephalic vein, branches of the lateral cutaneous nerves of the forearm and superficial branch of the radial nerve. At a deeper depth, the recurrent radial artery and deeper branches of the radial nerve are at risk.

LI8 (Xia Lian) is located on the lateral side of the posterior forearm, 4 cun distal to the cubital crease, on the line connecting LI5 at the wrist and LI11 at the lateral cubital crease. The needle is inserted perpendicularly 0.5–1 cun initially into the brachioradialis muscle and then deeper into the extensor carpi radialis longus muscle and then more deeply extensor carpi radialis brevis and the supinator muscle. The needle may also penetrate the extensor digitorum muscle and the abductor pollicis longus muscle depending on its angle and depth. Structures at risk are the same as the point above (LI9).

Note on Safe Practice: Nerves

In a study by Tang and Cheng (2019) which examined ultrasound visualization of radial nerve excursion during acupuncture, the authors found that a sharp electrical sensation may be elicited (that one may expect to be that of penetrating the nerve), well before the needle was actually close to the nerve. When the needle did encounter the nerve there was minimal or no sensation described by the subject. Ultrasound visualization demonstrated that as the needle encountered the nerve, the nerve deviated from the path of the needle and was observed to *"roll out of the way."* Following the study, some temporary aching sensation in the upper arm was present and continued for 2–3 hours, but there were no sensory or strength changes along the radial nerve distribution distal to the needle site. Tang and Cheng (2019) suggested that although some patients may describe sensations or paraesthesia post-needling, these subjective reports are not necessarily a confirmation of the needle contacting the nerve during treatment. That said, a clear understanding of the underlying anatomy surrounding the point is essential, as inadvertent contact or penetration of the nerve may occur without the patient being aware. While this is unlikely to cause any persistent nerve injury, it is best avoided if the nerve is not the desired target for needling.

PC5 (Jian Shi) is located on the anterior forearm, 3 cun superior to the transverse wrist crease, between the tendons of palmaris longus and the flexor carpi radialis muscle. The location is also approximately one-quarter of the distance between the transverse wrist crease and the cubital crease. The needle is inserted perpendicularly 0.5–1 cun medial to the tendon of the flexor carpi radialis muscle and will enter the flexor digitorum superficialis muscle and then progress to the flexor digitorum profundus tendons before moving deeper to penetrate the pronator quadratus muscle and then the interosseous membrane. Caution should be exercised if needling deeply in this area to avoid penetrating the median nerve or the anterior interosseous nerve.

PC6 (Nei Guan) is located on the anterior forearm, 2 cun superior to the transverse wrist crease, between the tendons of palmaris longus and the flexor carpi radialis muscle (Fig. 6.13). The median nerve lies lateral to this position and is located deep to flexor digitorum superficialis; between it and flexor digitorum profundus. Radial directing of the needle on deep insertion is therefore not recommended. In some individuals, palmaris longus may be absent and in this case the point is located adjacent to flexor carpi radialis on its ulnar aspect. In individuals where the two tendons are closely approximated, take care not to insert the needle into a tendon. Gentle digital pressure can be utilized to separate the tendons passively to permit easier insertion. The needle is inserted perpendicularly 0.5–1.5 cun toward TE5 and on deeper insertion will penetrate the pronator quadratus muscle and the interosseous membrane. When needling deeply, caution should be exercised to avoid penetrating the median nerve or the anterior interosseous nerve. Sharp pain on superficial insertion may signify needle contact to the cutaneous palmar branch of the median nerve or superficial branches of the cutaneous nerves of the forearm. These structures may lie immediately subcutaneously in these areas. If a strong "electric shock" sensation is felt the needle should be removed and repositioned on future insertions. PC6 is one of many common acupuncture points needled that lie directly over peripheral nerves and are needled due to their perineural location (Tang and Cheng 2019).

Note on Safe Practice: Nerve Damage

Although there is a significant frequency with which many of these acupuncture points are needled, there are only a small number of reports in the literature outlining nerve damage caused by acupuncture needling. There are two specific published case reports of peripheral nerve damage that exist within the literature. One report involves a broken needle in the carpal tunnel that caused neuropathy of the median nerve and one involves a needle inserted in the region of the fibular head, which led to fibular nerve palsy and subsequent foot drop (Peuker and Gronemeyer 2001). The direct needling of neural structures has been investigated in a number of prospective research studies. In a study from 2007 by Streitberger et al., PC6 was needled on 97 occasions bilaterally on 50 subjects and a De Qi sensation was achieved in 85 insertions. "De Qi" (De-chee) is the numbness, heaviness, and distension experienced during the mechanical stimulus of needling of deeper tissues and is thought to be attributed to the stimulation of a combination of A beta (Aβ) and A delta (Aδ) fibers (Zhang et al. 2012), and C fibers that are responsible for aching pain are also stimulated (Pomeranz 1987). C unmyelinated fibers have a slower conduction and are responsible for a dull, more burning, aching, throbbing pain, and are usually felt after the initial sharp stimulus of needle insertion (Wilkinson 2001). C fibers are thought to contribute to the overall analgesic effect generated by acupuncture and the associated phenomenon of De Qi as they give rise to a feeling of heaviness and soreness

(Wang et al. 1985). De Qi can be at the needle site and/or potentially referred locally or distally. In the study by Streitberger et al. 2007, the needle was inserted into the PC6 point and the location of the tip was then visualized using diagnostic ultrasound. The study reported that the median nerve was contacted 52 out of 97 times and penetrated on 14 out of 97 occasions. Immediately after needling, 2 patients reported having prolonged paraesthesia, and 1 patient needed to have the needle removed due to electric pain. During the week following needle insertion, 4 patients reported some transient unpleasant sensations; however, at 1 week post-needling, the neurologic examination yielded normal results for all the patients. The authors concluded: *"The rate of median nerve penetrations by the acupuncture needle at PC6 was surprisingly high, but these seemed to carry no risk of neurologic sequelae."* In a later study conducted by Kessler and Streitberger (2008), ultrasound was used to visualize an acupuncture needle completely perforating the median nerve at the acupuncture point PC6 on one of the authors. During this procedure, the author reported only a slight sensation occurring, and no pain. From this experiment, the authors concluded that *"in individual cases, it is possible for the median nerve to be perforated without causing pain or neurological problems."* Enblom and Johnsson (2017) reviewed 1298 PC6 acupuncture treatments, delivered to 221 participants (77% female, mean age 52.5, range 18–91 years). No serious complications occurred. Side effects during the 1298 acupuncture sessions included: minor bleeding in 5.0%, tiredness in 4.9%, numbness in 4.5%, and dizziness in 1.4%. After treatment, 4.3% reported soreness at the needling sites and 8.8% reported hematoma. Participants perceived the needling to be: not painful (47.4%), mildly painful (39.1%), moderately painful (11.6%), and very painful (1.4%). Few side effects occurred and those that did were mild. Nearly 90% found PC6 acupuncture to be not painful or only mildly painful. Nerve excursion has previously been described and documented in the literature in healthy median nerves at the wrist (Ellis et al. 2017; Hough et al. 2007). A lack of median-nerve excursion at the wrist has been linked with carpal tunnel syndrome and in pathological individuals the nerve may not *"roll"* as much as healthy nerves would during acupuncture needling. This may increase the potential risk of nerve penetration when needling near pathologic nerves when compared to needling near healthy nerves.

TE5 (Wai Guan) is located on the dorsal forearm, between the radius and ulna, 2 cun superior to the dorsal transverse wrist crease on the radial border of extensor digitorum. The needle is inserted perpendicularly 0.5–1.5 cun into extensor pollicis brevis. The needle can be progressed through deeper tissue anteriorly, passing through the interosseous membrane until it reaches the location of PC6. In this position, the needle may lie perineural to the median nerve and caution should be exercised to avoid needling the median nerve directly. One should also be aware not to damage the posterior cutaneous nerve of the forearm or the posterior or anterior interosseous nerves.

6.4.5 Wrist

HT4–HT6. As these three points lay in such proximity to each other they share the same surrounding anatomical structures. *HT5* (Tong Li) is located on the palmar surface of the forearm, 1 cun proximal to the transverse wrist crease. When inserted perpendicularly (0.3–0.5 cun), the needle lies on the radial side of the flexor carpi ulnaris tendon next to flexor digitorum superficialis with pronator quadratus laying deep to it. The

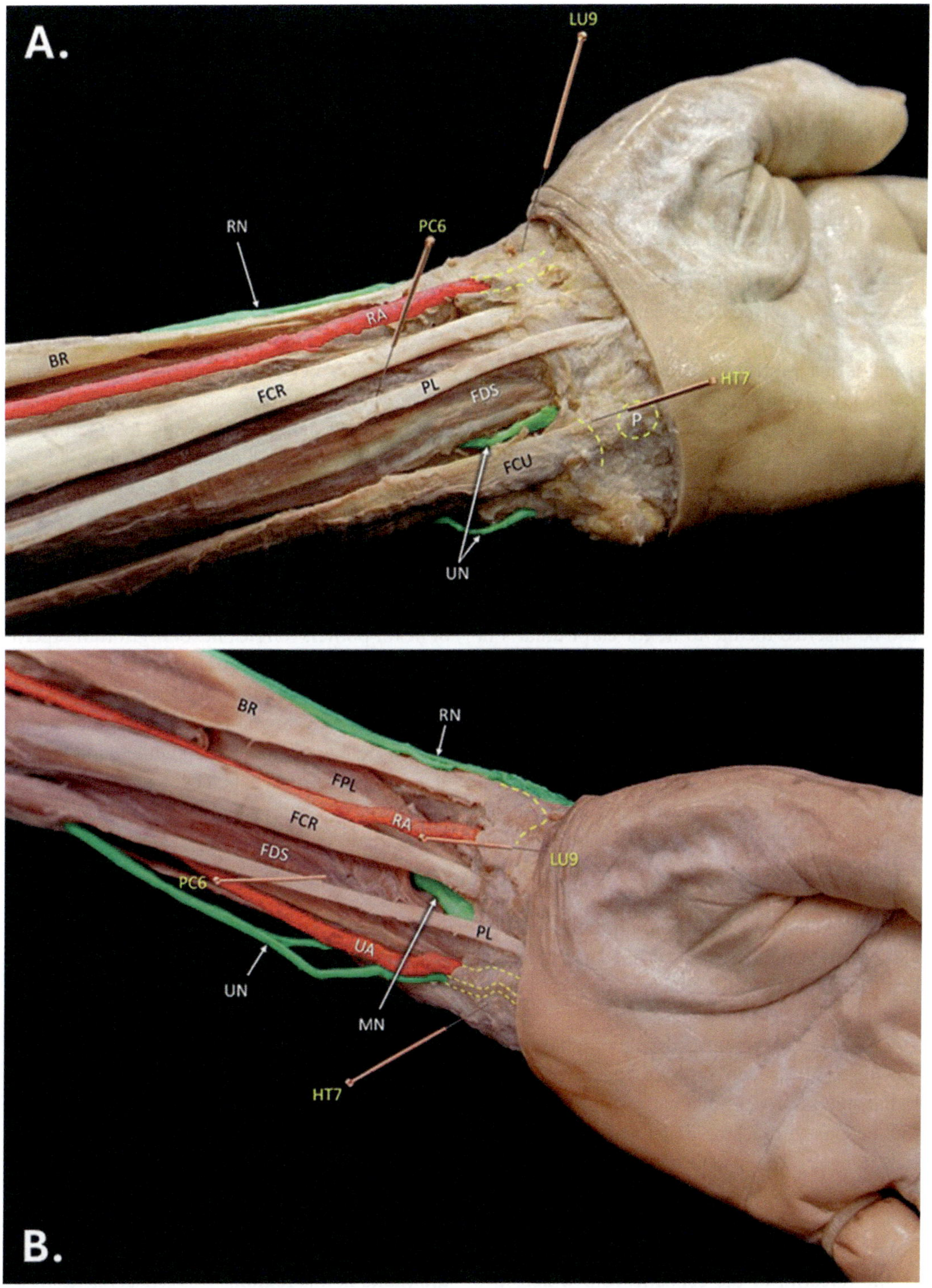

Fig. 6.13 *Left forearm and hand showing a comparison of needle positions PC6, LU9, and HT7*. Needle dimensions: 25 mm × 0.25 mm. (**a**) Dashed yellow lines demonstrate the continuation of the radial artery, the distal end of the ulna, and the outline of the pisiform bone. (**b**) Dashed yellow lines de monstrate the radial styloid process and the continuation of the ulnar artery and nerve. *BR* brachioradialis, *FCR* flexor carpi radialis, *FCU* flexor carpi ulnaris, *FDS* flexor digitorum superficialis, *FPL* flexor pollicis longus, *MN* median nerve (highlighted green), *P* pisiform bone (outlined in yellow), *PL* palmaris longus, *RA* radial artery (highlighted red), *RN* radial nerve (highlighted green), *UA* ulnar artery (highlighted red), *UN* ulnar nerve (highlighted green)

needle will also lie on the ulnar aspect of the ulnar artery, vein, and nerve, which should all be avoided. *HT4* (Ling Dao) is located 1.5 cun proximal to the transverse wrist crease whilst *HT6* (Yin Xi) lies just 0.5 cun proximal to the transverse wrist crease.

HT7 (Shen Men) is located on the transverse wrist crease, in the depression between the pisiform and ulna bones (Fig. 6.13). When inserted perpendicularly (0.3–0.5 cun) the needle lies on the radial side of the flexor carpi ulnaris tendon adjacent to flexor digitorum superficialis, with pronator quadratus lying deep to it. Lateral to the needle lays the ulnar artery, vein, and nerve. The ulnar nerve lies in the superficial layers of the adjacent tissue and caution should be exercised during insertion as placement toward the radius laterally may cause contact with the neural tissue. Contact with the ulnar nerve may be denoted by electric sensations radiating to the ulnar side of the hand affecting digits 4 and 5. The needle may also be inserted medial to flexor carpi ulnaris in line with the flexion crease and angled toward its target, passing deep to this tendon.

TE4 (Yang Chi) is located on the dorsal transverse wrist crease, ulnar to the central point, between the tendons of extensor digitorum and extensor digiti minimi muscles. The needle is inserted perpendicularly 0.3–0.5 cun into the extensor retinaculum of the wrist toward the junction of the triquetral and lunate bones with the ulna. Attention should be paid to the location and orientation of the related bony structures in this area to ensure success with needle placement.

LU7 (Lie Que) is located on the radial side of the forearm, 1.5 cun superior to the tip of the radial styloid process in a narrow space between the tendons of brachioradialis and abductor pollicis longus. The needle is inserted obliquely 0.3–0.5 cun and one should take caution not to penetrate the superficial branches of the radial artery, vein, or nerve, the lateral cutaneous nerve of the forearm, or the cephalic vein which lays superficial and passes through this location.

LU9 (Tai Yuan) is located on the lateral side of the anterior wrist crease, in the depression on the lateral side of the radial artery, lateral to the tendon of the flexor carpi radialis muscle, and medial to the tendon of the abductor pollicis longus muscle, distal to the radial styloid process (Figs. 6.13 and 6.14). It is also located level with HT7, which lies proximal to the pisiform bone. The needle is inserted perpendicularly 0.2–0.5 cun and directed between the radius and scaphoid bones into the underlying connective tissue overlying the joint space. Caution should be exercised when needling in this area to avoid puncturing the radial artery.

PC7 (Da Ling) is located at the transverse wrist crease, between the tendons of palmaris longus and flexor carpi radialis muscles (Fig. 6.15). If palmaris longus is absent, then the point is located medial to flexor carpi radialis. The needle is inserted perpendicularly 0.3–0.5 cun with attention paid to the median nerve which lies relatively superficially in this location. When needling deeply, the needle also passes through the flexor retinaculum, then between the tendons of flexor pollicis longus, flexor digitorum superficialis, and flexor digitorum profundus muscles. These muscles lie in the carpal tunnel deep to the median nerve. Care should be taken to avoid the palmar branch of the median nerve superficially. If a strong sensation is achieved in this area stimulation should be avoided, and if sharp *"electric shock"* pain is experienced the needle should be removed and relocated on future insertions.

6.4.6 Hand

SI3 (Hou Xi) is located at the medial end of the distal transverse palmar crease when a loose fist is formed, in the depression proximal to the head of the fifth metacarpal bone, at the junction of the dorsal and palmar surfaces, in the depression between associated muscle and bone. The point is needled perpendicularly (0.3–0.7 cun) initially into abductor digiti minimi and then deeper into flexor digiti minimi. Branches of the ulnar nerve are found in this region.

LI4 (He Gu) is located on the dorsum of the hand, approximately at the midpoint of the second metacarpal bone, in the belly of the first dorsal interosseus muscle (Fig. 6.16). The needle

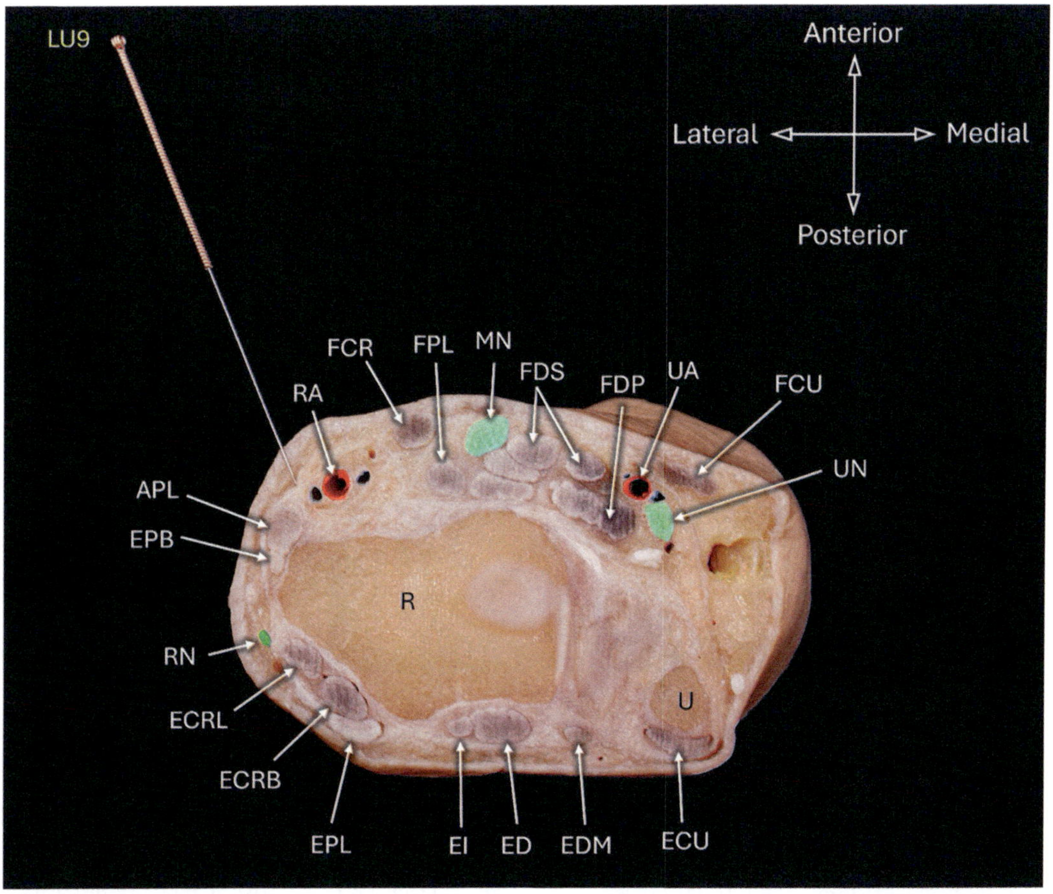

Fig. 6.14 Transverse slice through distal wrist showing needling of LU9. Needle dimensions: 25 mm × 0.25 mm. *APL* abductor pollicis longus, *BR* brachioradialis, *ECRB* extensor carpi radialis brevis, *ECRL* extensor carpi radialis longus, *ECU* extensor carpi ulnaris, *ED* extensor digitorum, *EDM* extensor digiti minimi, *EI* extensor indices, *EPB* extensor pollicis brevis, *EPL* extensor pollicis longus, *FCR* flexor carpi radialis, *FCU* flexor carpi ulnaris, *FDP* flexor digitorum profundus, *FDS* flexor digitorum superficialis, *FPL* flexor pollicis longus, *MN* median nerve (highlighted green), *R* radius, *RA* radial artery (highlighted red), *RN* radial nerve (highlighted green), *U* ulna, *UA* ulnar artery (highlighted red), *UN* ulnar nerve (highlighted green)

is inserted perpendicularly 0.5–1 cun. Caution not to penetrate the cephalic vein, cutaneous branches of the radial nerve, or the dorsal metacarpal artery and vein. When needling deeply into the adductor pollicis muscle take care to avoid the radial artery which is located distal to the junction of the base of the first and second metacarpals.

LI5 (Yang Xi) is located on the radial side of the wrist, distal to the tip of the radial styloid process, in the depression between the tendons of extensor pollicis longus and brevis, between the scaphoid and the radius, in the "anatomical snuff box" (Fig. 6.16). The needle is inserted perpendicularly 0.3–0.5 cun toward the joint space at the radial aspect of the wrist between the scaphoid and the radius. Take care not to penetrate the cephalic vein, cutaneous branches of the radial nerve, or the radial artery when needling deeper.

6.4.7 Summary

This chapter only covers the most common classical acupuncture points taught within foundation-level training courses within the fields of manual therapy or physical medicine. As a result, the point locations are specific and

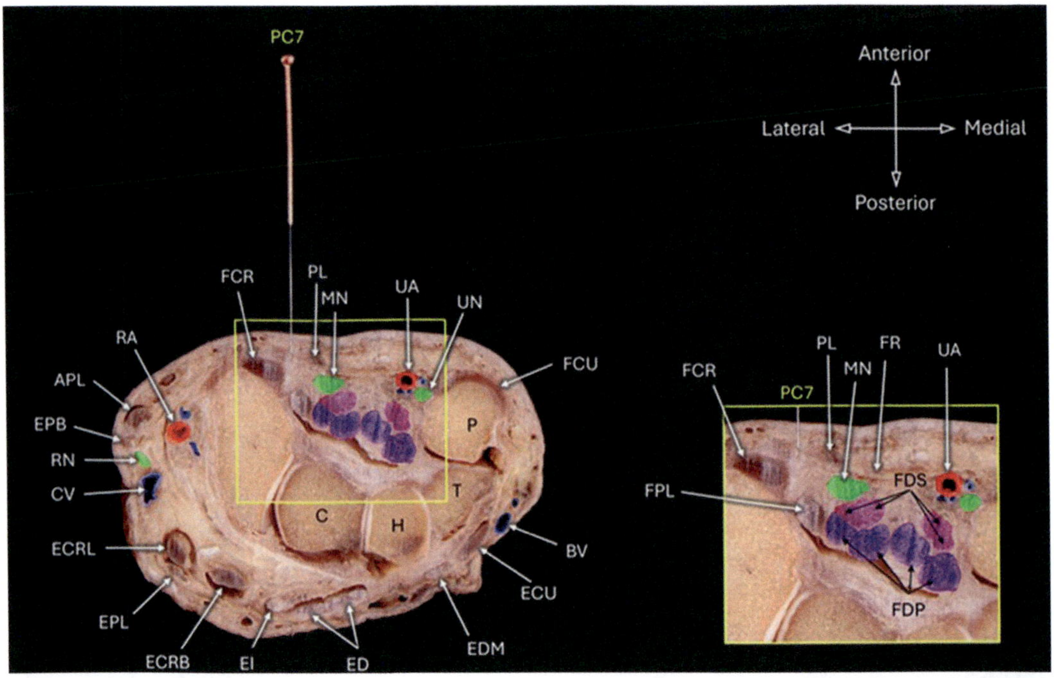

Fig. 6.15 Left-sided transverse section through the distal wrist, showing needling of PC7 within the carpal tunnel. Needle dimensions: 25 mm × 0.25 mm. *APL* abductor pollicis longus, *BV* basilic vein (highlighted blue), *C* capitate, *CV* cephalic vein (highlighted blue), *ECRB* extensor carpi radialis brevis, *ECRL* extensor carpi radialis longus, *ECU* extensor carpi ulnaris, *ED* extensor digitorum, *EDM* extensor digiti minimi, *EI* extensor indices, *EPB* extensor pollicis brevis, *EPL* extensor pollicis longus, *FCR* flexor carpi radialis, *FCU* flexor carpi ulnaris, *FDP* flexor digitorum profundus (highlighted purple), *FDS* flexor digitorum superficialis (highlighted pink), *FPL* flexor pollicis longus, *FR* flexor retinaculum, *H* hamate, *MN* median nerve (highlighted green), *P* pisiform, *PL* palmaris longus, *RA* radial artery (highlighted red), *RN* radial nerve (highlighted green), *S* scaphoid, *T* triquetrum, *UA* ulnar artery (highlighted red), *UN* ulnar nerve (highlighted green)

do not permit for generalization of the associated anatomy outside of these areas or points. Whenever needling into purely tender, "Ah Shi," trigger or motor points of the musculoskeletal system that are not within the boundaries of these classical points, the onus is upon the practitioner to educate themselves on the underlying risk factors related to any invasive procedure. The term "Ah Shi" refers to the patient's acknowledgment that the point reproduces their pain on palpation or needling. "Ah Shi" literally translates to "Oh yes" in Mandarin. An Ah Shi point is therefore a term used to describe points on the body that when palpated or needled reproduce a recognized pain presentation for the patient. These points can be a classical traditional acupuncture point, an extra point, or a myofascial pain point, which may also be referred to as a "trigger point" as popularized by Travell and Simons (Shah et al. 2015).

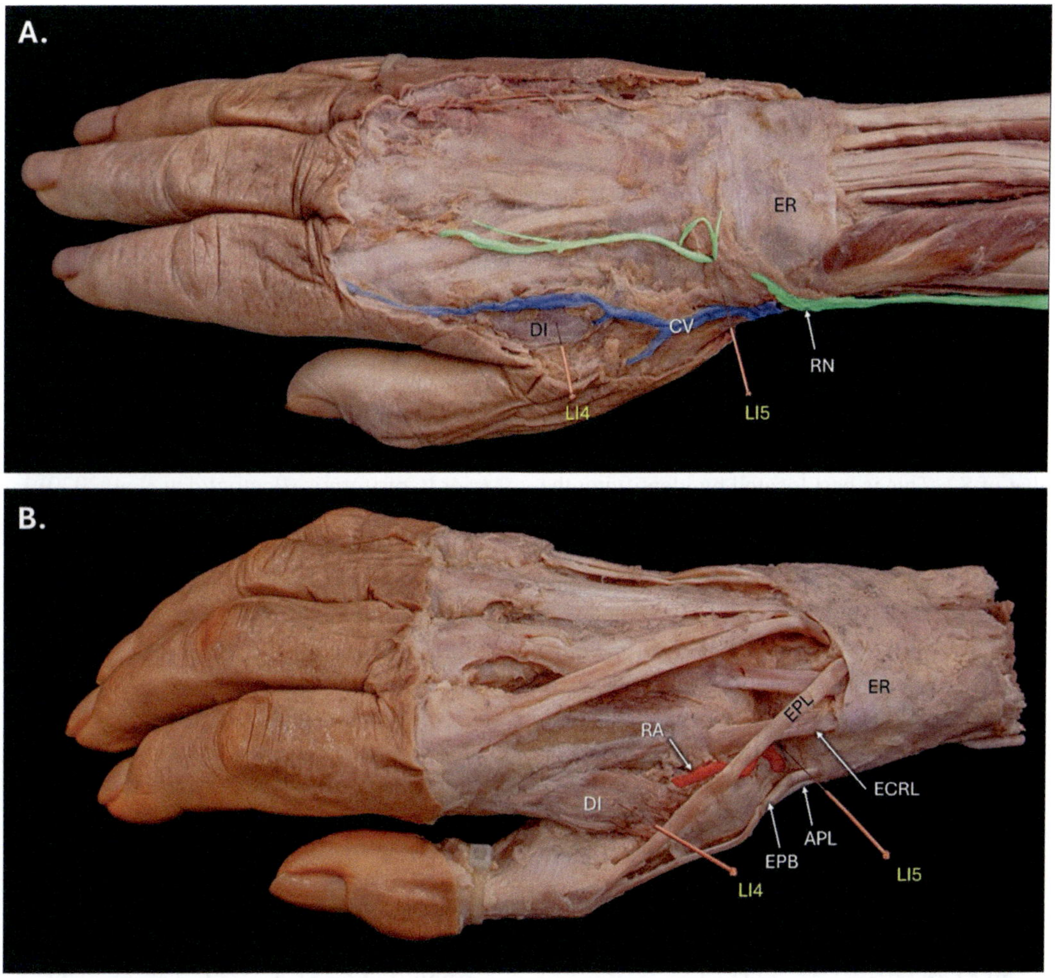

Fig. 6.16 Right-sided dorsal hand showing needling of LI4 and LI5. Needle dimensions: 25 mm × 0.25 mm. (**a**) superficial dissection showing subcutaneous structures of the dorsal hand. (**b**) deeper dissection exposing muscle tendons and radial artery within the anatomical snuffbox. *APL* abductor pollicis longus, *CV* cephalic vein (highlighted blue), *DI* first dorsal interosseus muscle, *ECRL* extensor carpi radialis longus, *EPB* extensor pollicis brevis, *EPL* extensor pollicis longus, *ER* extensor retinaculum, *RA* radial artery (highlighted red), *RN* superficial branch of radial nerve (highlighted green)

6.5 Applied Anatomy for Acupuncture Part 2: Lower Limb

This section will provide information on the most common acupuncture points on the lower limb, as defined and taught by a cross-section of acupuncture courses delivered within physical therapy. Key to the safe practice of any acupuncture or dry needling procedure is adherence to the triumvirate of respect; that being, respect the patient, respect the anatomy, and respect the technique. The potential incidence of adverse events can be significantly reduced when a solid knowledge of the underlying anatomy is present. Familiarization with the anatomy as described in this chapter can help support that knowledge. Working from medial to lateral then anterior to posterior, the meridians that lay on the lower limb are the Kidney, Liver, Spleen, Stomach, Gallbladder, and Bladder (Figs. 6.2, 6.3, 6.5, 6.7, 6.8, and 6.11). Needle point locations will be described anatomically, and the relative safety of needling these points in clinical practice will be discussed. As in all sections of this chapter, location of these needle points relies on a thorough understanding of anatomy, accurate palpation skills, and an appreciation of variation between individuals.

6.5.1 Hip and Gluteal Region

BL53 (Bao Huang) is located in the sacral region, 3 cun lateral to the posterior midline, at the level of the second sacral foramen. The most easily identified foramen of the sacrum via palpation is that of S2. This is best located by identifying the posterior superior iliac sine (PSIS) and then palpating inferiorly and medially to the bony prominence. S2 is located adjacent to the PSIS in this region. BL53 can also be located on palpation of the most tender point of the gluteus maximus muscle inferior and lateral to the posterior superior iliac spine. The needle is inserted perpendicularly 0.5–1.5 cun. On oblique lateral insertion, the needle may also penetrate the posterior aspect of the gluteus medius muscle. BL53 is recognized clinically as a common myofascial trigger point within the gluteus maximus and medius muscles. The superior gluteal nerve and vessels are located deep in the musculature in this region.

BL54 (Zhi Bian) is located in the sacral region, 3 cun lateral to the posterior midline, at the level of the fourth sacral foramen and sacral hiatus in the gluteus maximus muscle (Fig. 6.17). The needle is inserted perpendicularly 1.5–2 cun into gluteus maximus and may pass into gluteus medius. Progressing deeper to 2–3 cun in the direction of the pubic symphysis, the needle can reach the piriformis muscle. Exercise caution when needling in this region as the superior cluneal nerve and inferior gluteal vessels and nerve are present. The sciatic nerve also lies in this region within close proximity to the piriformis muscle, so it too should be considered when needling deeply here. BL54 is recognized clinically as a common myofascial trigger point within both the gluteus maximus and piriformis muscles that refer pain to the posterior hip and gluteal region.

Note on Safe Practice: Piriformis

Formed by the union of the ventral rami of the L4-S3 spinal cord segments, the sciatic nerve has a very close relationship with the piriformis muscle in the gluteal region. Most commonly, the sciatic nerve exits the greater sciatic foramen inferior to piriformis. Significant variations to this course do exist, however. The most frequently reported variation (5–10%) sees the common fibular nerve emerge through piriformis, splitting the muscle, with the tibial nerve following the expected course inferior to piriformis (Poutoglidou et al. 2020). Other variations are less frequent but may involve the whole nerve or its divisions passing above, though, or below piriformis. Anecdotally, the authors based at KASTC have observed a higher proportion of variations than is reported in the literature (Tomaszewski et al. 2016). These variations alter the course of the nerve and may increase the risk of accidental nerve damage in this region. Awareness of the existence of these variations is therefore important.

GB29 (Ju Liao) is located at the hip, with the hip joint in flexion, in the depression midway between the anterior superior iliac spine and the superior aspect of the greater trochanter. GB29 is located while the patient is in the lateral recumbent position with the thigh flexed slightly. The needle is inserted perpendicularly 1–3 cun. In its most superficial position, it will lie in the tensor fasciae latae muscle but travelling deeper it may enter the gluteus medius and gluteus minimus muscles. In its deepest position, it may lie in the attachments of the rectus femoris muscle. In this location caution should be exercised to avoid puncturing the superficial circumflex iliac artery or vein or the lateral femoral cutaneous nerve. At deeper levels avoid needling into the lateral circumflex femoral artery and vein and the branches of the superior gluteal artery, vein, and nerve.

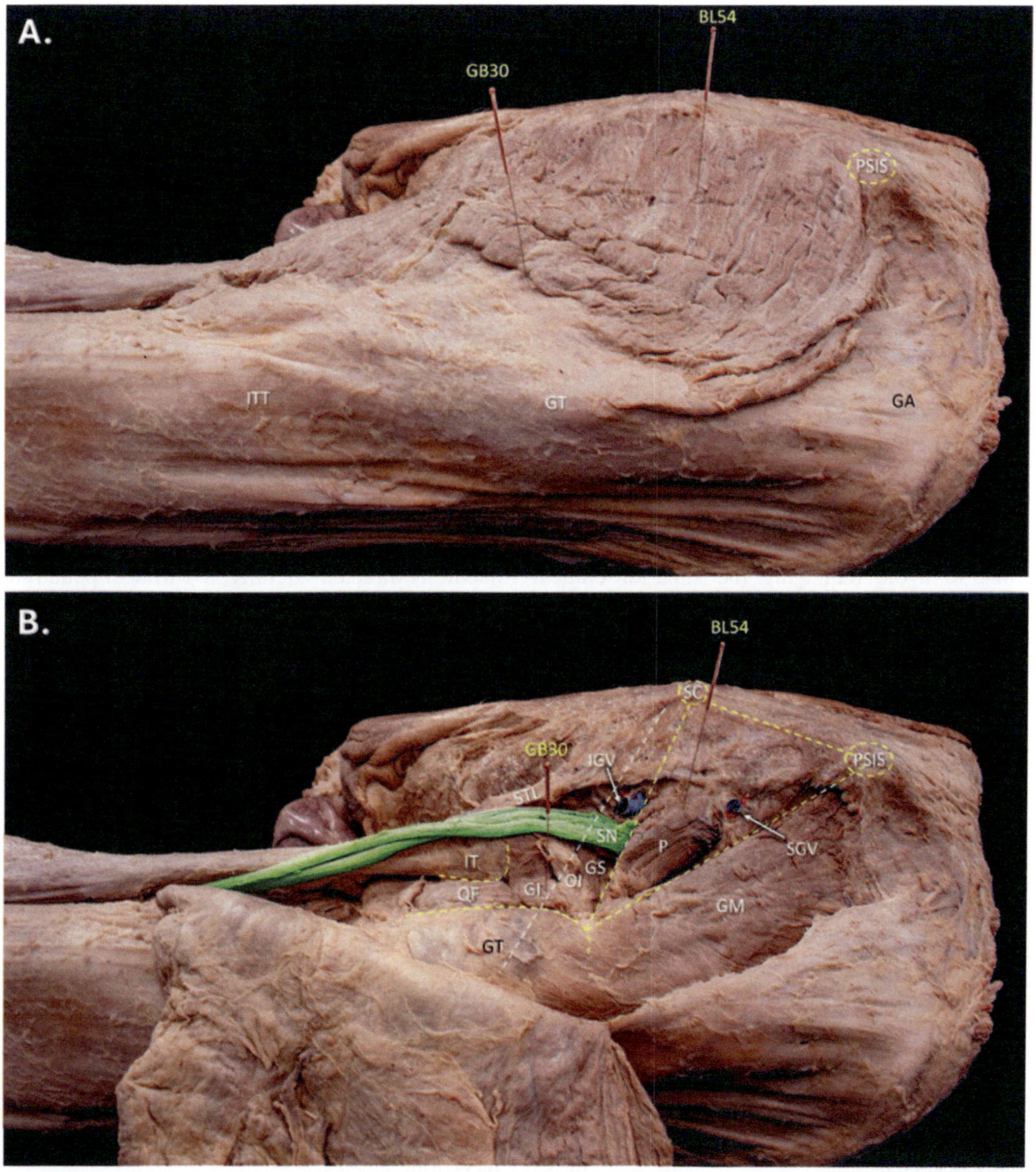

Fig. 6.17 Right-sided gluteal region (prone position) showing needling of BL54 and GB30. Needle dimensions 90 mm × 0.30 mm. (**a**) Superficial insertion into gluteus maximus. (**b**) Deeper dissection with gluteus maximus reflected to show deeper needle insertions into piriformis (BL54) and gemellus inferior (GB30). The sciatic nerve has been reflected medially to expose the lateral rotator muscles. Dashed white line extends from the sacral hiatus to the most lateral aspect of the greater trochanter. GB30 is positioned along the lateral third of this line. Dashed yellow lines represent a triangle formed between the posterior superior iliac spine, greater trochanter, and sacral cornu. Piriformis can be identified within this triangle. *GA* gluteal aponeurosis, *GI* gemellus inferior, *GM* gluteus medius, *GS* gemellus superior, *GT* greater trochanter (medial border represented by dashed yellow curve), *IGV* inferior gluteal vessels (highlighted blue and red), *IT* ischial tuberosity (superolateral border represented by dashed yellow curve), *ITT* iliotibial tract, *OI* obturator internus, *P* piriformis, *PSIS* posterior superior iliac spine (outlined by dashed yellow circle), *QF* quadratus femoris, *SC* sacral cornu (outlined by dashed yellow circle), *SGV* superior gluteal vessels (highlighted blue and red), *SN* sciatic nerve (highlighted green), *STL* sacrotuberous ligament

GB30 (Huan Tiao) is located in the gluteal region, one-third of the way from the most lateral aspect of the greater trochanter to the sacral hiatus (Fig. 6.17). GB30 is located while the patient is in the lateral recumbent position with the thigh flexed slightly. The needle is inserted perpendicularly 1.5–2.5 cun into the gluteus maximus muscle and may travel deeply between the inferior margin of the piriformis muscle and the superior margin of the quadratus femoris muscle, terminating within the gemellus superior or inferior muscles, obturator internus muscle, or quadratus femoris muscle dependent on the position of the hip and the angulation of the needle. The needle should lie in a position midway between the ischial tuberosity and the greater trochanter, lateral to the sciatic nerve. Take caution when needling to avoid damaging the superior cluneal nerve or the inferior gluteal vessels and nerve. When needling deeper, care needs to be taken to avoid the sciatic nerve.

BL36 (Cheng Fu) is located in the gluteal region, inferior to gluteus maximus in the middle of the transverse gluteal fold, between the tendons of the semitendinosus and biceps femoris muscles. The needle is inserted perpendicularly 0.5–1.5 cun toward the ischial tuberosity and may pass through the lower fibers of gluteus maximus and can be directed between or through the tendons of the semitendinosus muscle and the long head of the biceps femoris muscle. In this position, the needle passes medial to the path of the sciatic nerve. Be cautious not to penetrate the posterior femoral cutaneous nerve superficially in this region, or the sciatic nerve when needling more deeply.

6.5.2 Thigh

ST31 (Bi Guan) is located on the anterolateral aspect of the thigh, directly inferior to the anterior superior iliac spine, level with the lower border of the pubic symphysis, in the depression lateral to the sartorius and medial to tensor fasciae latae muscles. The needle is inserted perpendicularly 1–2.5 cun between sartorius and medial to tensor fasciae latae muscles and directed toward the rectus femoris muscle. Clinically this point overlies the myofascial trigger point in the proximal aspect of rectus femoris that may refer pain to the knee. Caution should be taken as the lateral femoral cutaneous nerve lies superficially in this region.

ST32 (Fu Tu) is located on the anterior thigh, on a line between the anterior superior iliac spine and the superior lateral border of the patella, 6 cun (or two hands widths) superior to lateral patella. Its position is also identifiable by dividing the distance between the prominence of the greater trochanter and the superior lateral patella into thirds, as ST32 lies just superior to the junction of the lower and middle thirds. The needle is inserted perpendicularly 1–1.5 cun into the lateral aspect of the rectus femoris muscle. If the needle is directed laterally, it will also penetrate the vastus lateralis muscle. The needle may be directed deeper and medially into the vastus intermedius muscle. At its terminal point of full penetration, the needle will contact the anterior aspect of the femur. This depth of penetration is not often required for treatment effects to be realized.

ST34 (Liang Qiu) is located on the anterior thigh, on a line between the anterior superior iliac spine and the superior lateral aspect of the patella, 2 cun proximal to lateral patella between the vastus lateralis and rectus femoris muscles. To aid location the height of the patella is considered to be equivalent to 2 cun. The point is needled perpendicularly at a depth of 0.5–1 cun. The needle may enter both vastus lateralis and rectus femoris muscles, dependent on angle and depth of insertion.

SP10 (Xue Hai) is located on the medial thigh, with the knee in flexion, 2 cun proximal to the superior medial angle of the patella, on vastus medialis muscle in line with ST34 in the vastus lateralis muscle. To aid location the height of the patella is considered to be equivalent to 2 cun. The needle is inserted perpendicularly 0.5–1.5 cun into the vastus medialis muscle. Caution should be taken in this area to avoid the penetration of anterior femoral cutaneous nerve and if needling deeper the branches of the femoral vessels and nerve. Clinically SP10 overlies the myofascial trigger point in the proximal aspect of vastus medialis muscle that may refer pain to the knee. The myofascial trigger point in the region of SP10 in adductor magnus can also refer pain from the medial knee to the groin.

BL37 (Yin Men) is located in the posterior thigh between the heads of the semitendinosus and biceps femoris muscles, 6 cun inferior to the transverse gluteal fold, on a line joining BL36 and BL40 at the popliteal fossa. The needle is inserted perpendicularly 0.5–1 cun between the heads of the semitendinosus and biceps femoris muscles. Caution should be exercised to avoid penetration of the posterior femoral cutaneous nerve or the deep femoral artery or vein. Underlying this point is the sciatic nerve and this should also be avoided when needling more deeply. A slight oblique angle laterally may help to avoid the sciatic nerve.

GB31 (Feng Shi) is located on the lateral thigh, 7 cun superior to the popliteal crease at the posterior border of the iliotibial tract. GB31 can also be located directly inferior to the tip of the patient's middle finger when they are standing with their arm extended along the thigh. The needle is inserted perpendicularly 0.5–1.5 cun into or posterior to the border of the iliotibial tract, and will enter the vastus lateralis muscle. Dependent on angle of insertion and depth of needling, the vastus intermedius muscle or the short head of the biceps femoris muscles may also be reached from this point. This point is clinically recognized as being in the region of common myofascial trigger points within the vastus lateralis muscle. Take caution to avoid damaging branches of the lateral femoral cutaneous nerve and branches of the femoral nerve.

6.5.3 Knee

ST35 (Du Bi) is located at the knee, in the depression below the lateral side of the patella when the knee is flexed slightly. This location is also referred to as the lateral *"eye of the knee"* and correlates with the lateral arthroscopic portal to the knee. The needle is inserted between the patella ligament and the lateral patella retinaculum perpendicularly 0.7–1 cun, dependent on the thickness of the underlying tissue. Caution should be taken not to enter the knee joint without consideration for appropriate levels of sterilization and hygiene to avoid joint infection.

Xiyan is paired with ST35 and is located at the knee, in the depression below the medial side of the patella when the knee is flexed slightly. This location is also referred to as the medial *"eye of the knee"* and correlates with the medial arthroscopic portal to the knee. The needle is inserted between the patella ligament and the medial patella retinaculum perpendicularly, with the same precautions as applied to ST35.

LR8 (Qu Quan) is located on the medial side of the knee, in the depression of the transverse popliteal crease between the upper border of the medial epicondyle of the femur and semitendinosus and semimembranosus tendons. The needle is inserted perpendicularly 1–1.5 cun posterior to the sartorius muscle and the tendon of the gracilis muscle and anterior to the semimembranosus muscle. Deeply the needle may be located in the medial head of the gastrocnemius muscle. Take care not to penetrate the great saphenous vein, the anterior branch of the femoral cutaneous nerve or the saphenous nerve. When needling deeply caution is required not to puncture the popliteal vessels or the tibial nerve in the region of the medial popliteal fossa.

KI10 (Yin Gu) is located on the medial side of the popliteal fossa, in the depression between tendons of the semitendinosus and gracilis muscles when the knee is flexed. The needle is inserted perpendicularly 0.5–1.5 cun into the semimembranosus muscle or in a deeper position into the medial head of the gastrocnemius muscle. In this region be cautious not to damage the great saphenous vein or saphenous nerve that lay anterior to this point. If needling more deeply also take care not to penetrate the medial superior genicular artery and vein.

BL40 (Wei Zhong) is located at the midpoint of the popliteal fossa, midway between the tendons of the semitendinosus and biceps femoris muscles. The needle is inserted perpendicularly 0.5–1.5 cun. Caution when needling to avoid the posterior femoral cutaneous nerve superficially.

At a deeper level, consideration should be made to avoid the tibial nerve, the popliteal artery, and vein.

6.5.4 Leg

SP9 (Yin Ling Quan) is located proximally on the medial leg in the region of the conjoined tendinous insertion of the pes anserinus (Fig. 6.18). It lies on the inferior border of the medial tibial condyle, in the depression between the posterior border of the tibia and gastrocnemius muscle. The needle is inserted perpendicularly 0.5–1 cun and is directed toward the popliteus muscle. The needle will pass posterior to the insertion of the semimembranosus muscle at the medial condyle of the tibia. Caution should be exercised if needling in an anterior direction so as not to penetrate the great saphenous vein or saphenous nerve. If needling deeper one should be aware of the medial inferior genicular artery and vein and the popliteal artery and veins. The posterior tibial artery and veins and tibial nerve are also of concern if the needle is directed posteriorly and deeply. An alternative approach in needling anterolaterally at this location will direct the needle toward the popliteus muscle at its medial tibial attachment.

ST36 (Zu San Li) is located on the leg in the tibialis anterior muscle, one fingers' breadth lateral to the tibia's anterior crest, 3 cun inferior to ST35 and level with the inferior aspect of the tibial tuberosity. The needle is inserted perpendicularly 0.5–1.5 cun. Caution should be exercised on deeper needling or when needling this point too laterally to avoid the deep fibular nerve, anterior tibial artery, and veins. A branch of the lateral sural cutaneous nerve also lies in this region. Deeper penetration in this region may permit the needling of tibialis posterior which lies posterior to the interosseous membrane that separates tibialis anterior from tibialis posterior. The tibial nerve, posterior tibial artery, and vein also lie in this region posterior to tibialis anterior and caution should be taken when needling deeply in this area.

ST38 (Tiao Kou) is located on the leg, one cun lateral to the tibia's anterior crest, at the midpoint of a line between the lateral malleolus and ST35 at the lateral patella. The needle is inserted perpendicularly into tibialis anterior to a depth of 0.5–1.5 cun. Deeper penetration in this region may also permit the needling of tibialis posterior which lies posterior to the interosseous membrane. The tibial nerve, posterior tibial artery, and vein also lie in this region and caution should be taken when needling deeply in this area.

GB34 (Yang Ling Quan) is located on the lateral aspect of the leg, in the depression approximately 1 cun inferior and 1 cun anterior to the head of the fibula. The needle is inserted perpendicularly 0.5–1.5 cun at the anterior border of the fibularis longus muscle and enters the extensor digitorum longus muscle. This point may also be clinically recognized as being in the region of common myofascial trigger points within the extensor digitorum longus muscle. Some traditional needling techniques also involve directing the needle 2–3 cun deep through the interosseous membrane toward SP9 on the medial aspect of the tibia. When needling in this region take care not to damage the superficial or deep fibular nerves, lateral branches of the sural nerve, or the fibular artery. If needling in a more anterior direction caution should be exercised to avoid the anterior tibial artery and veins. If needling deeper then avoid the fibular artery and veins.

GB39 (Xuan Zhong) is located on the lateral side of the leg, 3 cun superior to the prominence of the lateral malleolus between or through the tendons of the fibularis longus and fibularis brevis muscles. The needle is inserted perpendicularly 0.5–1.5 cun into the fibers of the fibularis brevis muscle. This point may be clinically recognized as being in the region of common myofascial trigger points within the fibularis brevis muscle. GB39 may also be found in an alternate location between the posterior border of the fibula and the tendon of the fibularis brevis muscle. It may also be situated on the anterior border of the fibula if insertion between the tendons in the original position is not permissible due to myofascial tightness. Be cautious not to penetrate

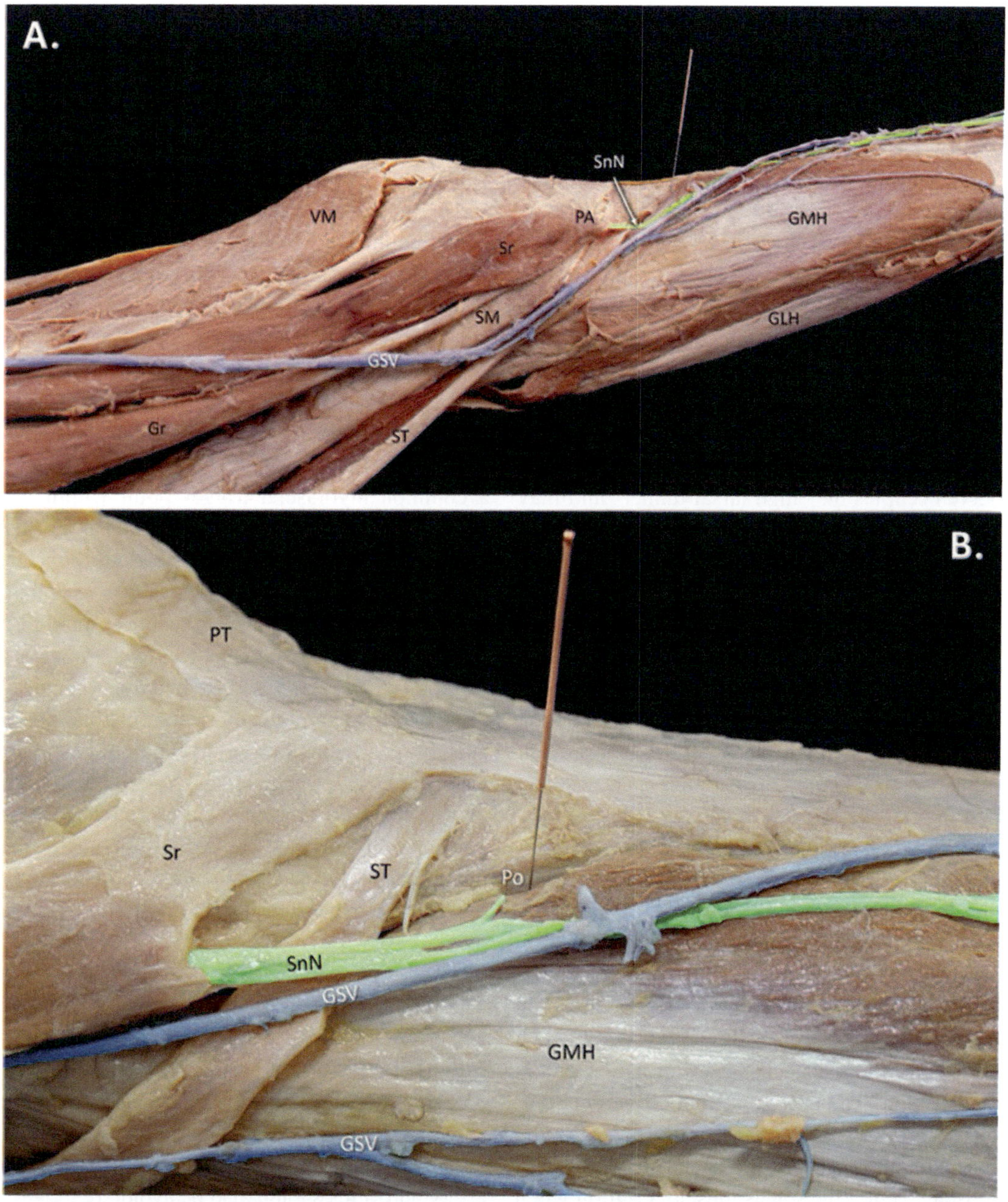

Fig. 6.18 Medial view of left lower limb showing needling of SP9. Needle dimensions 40 mm × 0.25 mm (**a**) view of lower limb musculature and course of great saphenous vein. (**b**) Zoomed in view of needling at SP9 showing insertion into popliteus muscle and close relationship with great saphenous vein and saphenous nerve. *GLH* gastrocnemius lateral head, *Gr* gracilis, *GSV* great saphenous vein (highlighted blue). *PA* pes anserinus, *GMH* gastrocnemius medial head, *Po* popliteus, *PT* patellar tendon, *SM* semimembranosus, *SnN* saphenous nerve (highlighted green), *Sr* sartorius, *ST* semitendinosus, *VM* vastus medialis

branches of the sural nerve or the superficial fibular nerve. If needling more deeply and anteriorly then caution should be exercised to avoid the anterior tibial artery and veins and the deep fibular nerve. If needling deeper still, avoid the fibular artery and veins.

SP6 (San Yin Jiao) is located on the medial leg, 3 cun superior to the medial malleolus on the posterior border of the tibia (Fig. 6.19). The needle is inserted perpendicularly 0.5–1.5 cun posterior to the tibia. Depending on the depth of insertion, the needle can penetrate the flexor digitorum longus and soleus muscles. If it is inserted too deeply it can penetrate the posterior tibial artery or vein. If it is inserted in a posterior oblique direction, it can penetrate the posterior tibial artery and vein or the tibial nerve. If the needle is inserted too anteriorly the great saphenous vein may be damaged, though the vein's course is variable particularly if varicosities are noted. At the region of SP6, the tibial nerve occupies the "high tarsal tunnel" and may generate symptoms related to the compression of the surrounding flexor digitorum longus, flexor hallucis longus, soleus, and tibialis posterior muscles. Myofascial trigger points can be identified within both flexor hallucis longus and soleus muscles in the region of SP6. Clinically, this point may be needled for its perineural location to the posterior tibial nerve and its subsequent influence on sacral plexus innervated visceral presentations.

BL57 (Cheng Shan) is located on the posterior leg, 8 cun inferior to BL40, directly between and inferior to the two heads of the gastrocnemius muscle. This is approximately the location where the two heads blend to become the calcaneal tendon. The tibia is approximately 16 cun in length and BL57 also lies midway between BL40 and the calcaneal insertion of the calcaneal tendon. The needle is inserted perpendicularly 0.8–1.2 cun and passes through the musculotendinous junction of the gastrocnemius muscle to enter the soleus muscle. Caution when needling deeply as the tibial nerve lies deep to this location, overlying tibialis posterior. BL57 is recognized clinically as a common myofascial trigger point within the soleus muscle that can refer pain to the ipsilateral sacroiliac joint. Clinically this point may be needled for its perineural location to the sural nerve and its subsequent influence on sacral plexus innervated visceral presentations.

KI9 (Zhu Bin) is located on the medial leg, 5 cun superior to KI3, anterior to the calcaneal tendon, at the lower end of the gastrocnemius muscle. The needle is inserted perpendicularly 0.5–1.5 cun into the distal aspect of the soleus muscle. This point is also clinically recognized as a common myofascial trigger point within the soleus muscle that can refer to pain along the medial aspect of the calcaneal tendon, the heel, and plantar aspect of the foot toward the metatarsal heads. When needling KI9 caution should be exercised not to penetrate the great saphenous vein or saphenous nerve superficially. When needling deeper and more anteriorly the tibial nerve and the posterior tibial artery and veins should also be avoided.

KI7 (Fu Liu) is located on the medial leg, 2 cun superior to KI3 in the depression anterior to the border of the calcaneal tendon and inferior to the soleus muscle. The needle is inserted perpendicularly 0.5–1 cun anterior to the calcaneal tendon and the tendon of the plantaris muscle. The needle passes posterior to the tibial nerve and posterior tibial artery and vein. Caution should be exercised when needling in this region to avoid the great saphenous vein and saphenous nerve superficially. When needling deeper and more anteriorly the tibial nerve and the posterior tibial artery and veins should also be avoided.

Fig. 6.19 *Needling of SP6.* Needle dimensions 40 mm × 0.25 mm. (**a**) medial view of left leg showing position of the needle relative to neurovascular structures. (**b**) Transverse section of distal leg showing insertion of needle into tibialis posterior. *ATA* anterior tibial artery (highlighted red), *CT* calcaneal tendon, *DFN* deep fibular nerve (highlighted green), *EDL* extensor digitorum longus, *EHL* extensor hallucis longus, *F* fibula, *FA* fibular artery (highlighted red), *FB* fibularis brevis, *FDL* flexor digitorum longus, *FHL* flexor hallucis longus, *FL* fibularis longus, *GSV* great saphenous vein (highlighted blue), *MM* medial malleolus, *PT* plantaris tendon, *PTA* posterior tibial artery (highlighted red), *PTV* posterior tibial vein (highlighted blue), *SN* sural nerve (highlighted green), *SnN* saphenous nerve (highlighted green), *T* tibia, *TA* tibialis anterior, *TN* tibial nerve (highlighted green), *TP* tibialis posterior, *TS* tibial shaft

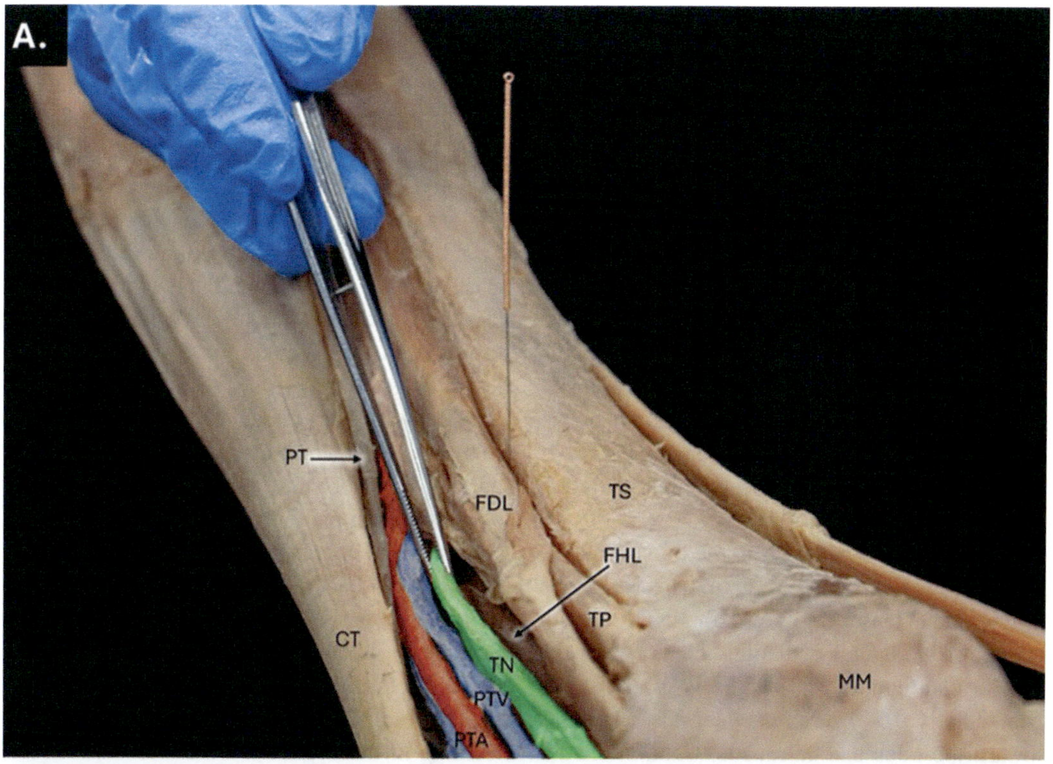
A.
PT
FDL
TS
FHL
TP
CT
TN
PTV
PTA
MM

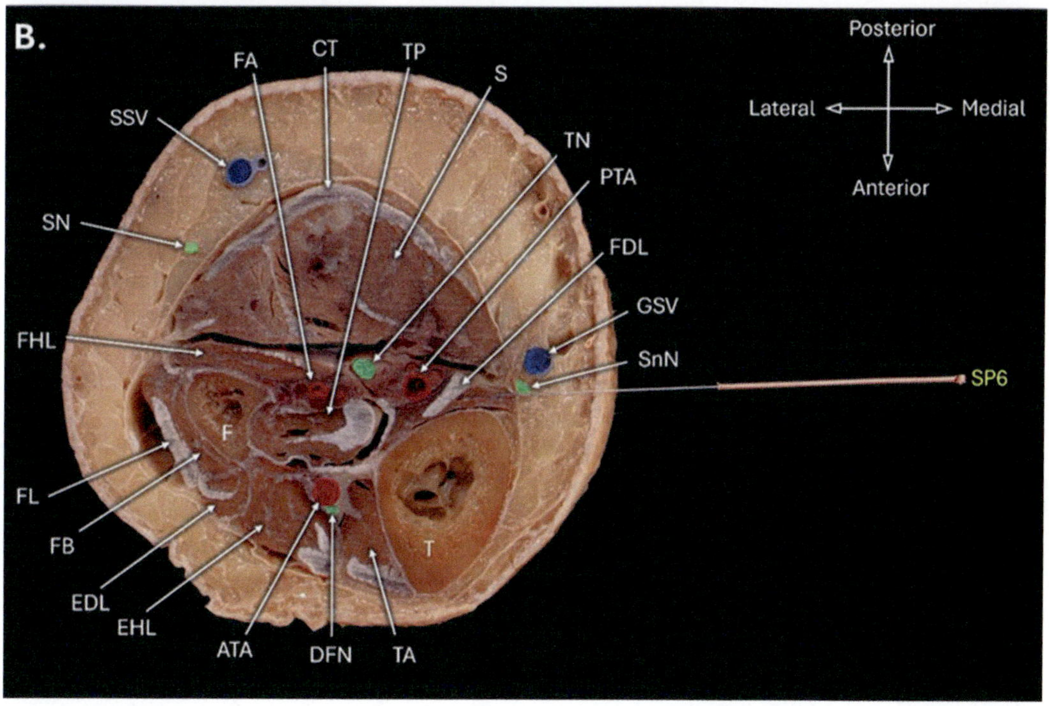
B.
Posterior
Lateral
Medial
Anterior
FA
CT
TP
S
SSV
TN
PTA
SN
FDL
GSV
FHL
SnN
SP6
F
FL
FB
EDL
EHL
ATA
DFN
TA
T

6.5.5 Ankle

LR4 (Zhong Feng) is located on the dorsum of the foot, 1 cun anterior to the medial malleolus, in the depression medial to the tibialis anterior tendon with the ankle in plantigrade. The needle is inserted perpendicularly 0.3–1 cun anterior to the great saphenous vein and posterior to the tendon of the tibialis anterior muscle. The needle can also be directed obliquely in a distal and lateral direction beneath the tendon of the tibialis anterior muscle. Do not needle deeply in a posterior direction due to the relative location of the great saphenous vein and saphenous nerve. If needling deeply, avoid the local medial malleolar network of arteries and veins.

ST41 (Jie Xi) is located on the dorsum of the foot, between the tendons of extensor hallucis longus and extensor digitorum longus, level with the tip of the lateral malleolus, ideally with the foot in plantigrade. The needle is inserted perpendicularly 0.5–0.7 cun between these tendons toward the talus and calcaneus. Caution should be exercised to avoid puncturing the dorsalis pedis artery which is a continuation of the anterior tibial artery in this region. Other neurovascular structures of note in this area include the deep fibular nerve and the medial dorsal cutaneous branch of the superficial fibular nerve.

GB40 (Qiu Xu) is located at the lateral ankle, in the depression anterior and inferior to the lateral malleolus. It is also lateral to the extensor digitorum longus tendon and anterior to the tendon of the fibularis longus muscle and posterior to the fibularis tertius tendon, if present. The needle is inserted perpendicularly 0.5–1.5 cun. This point overlies the inferior extensor retinaculum and is in the region of the anterior talofibular ligament. It also overlies the interosseus talocalcaneal ligament more deeply. When needling in this region avoid the branches of the small saphenous vein and the sural nerve. The deep fibular nerve, lateral tarsal artery, and the anterior lateral malleolar artery should also be avoided when needling more deeply in this location.

BL62 (Shen Mai) is located on the lateral foot, in the depression directly inferior to the lateral malleolus, caudal to or between the tendons of the fibularis longus and brevis muscles at the dorsal-plantar skin junction. The needle is inserted perpendicularly 0.5 cun between the tendons of the fibularis longus and brevis muscles. Alternatively, if the space between the tendons is restricted the needle can be inserted obliquely 1 cun inferior and posterior to the tendons. Caution should be exercised when needling in this region to avoid the small saphenous vein, the sural nerve, and the lateral malleolar artery. Clinically this point may be needled for its perineural location to the sural nerve.

BL60 (Kun Lun) is located on the lateral ankle, in the depression midway between the lateral malleolus and the calcaneal tendon when the foot is positioned in plantigrade (Fig. 6.20). The needle is inserted perpendicularly 0.5 cun anterior to the calcaneal tendon. From this location, the needle can also pass to a depth of 1 cun or greater through the precalcaneal (Kager's) fat pad in the direction of KI3, located from the medial aspect of the ankle. Caution should be exercised when needling in this region to avoid the sural nerve, the small saphenous vein, and the lateral malleolar artery and vein. When needling deeper, care should also be taken to avoid the fibular artery and veins. If deeper needling to KI3 is employed, then the tibial nerve and the posterior tibial artery and vein are also a consideration. If the foot is not in plantigrade but in plantarflexion, then the position of the calcaneus is altered and sits in a relatively superior location. This requires relocation of BL60 needle entry to a more superior location to avoid insertion into the underlying bone. Clinically, this point may be needled for its perineural location to the sural nerve.

KI3 (Tai Xi) is located on the medial ankle, at the midpoint between the prominence of the medial malleolus and the calcaneal tendon (Fig. 6.21). This point should be identified via palpation of the posterior tibial artery pulse. The needle is inserted perpendicularly 0.5 cun immediately posterior to the posterior tibial artery and passes posterior to the tendons of the flexor digitorum longus and tibialis posterior muscles. From this location, the needle can also be passed to a depth of 1 cun or greater through Kager's fat pad in the direction of BL60, located from the

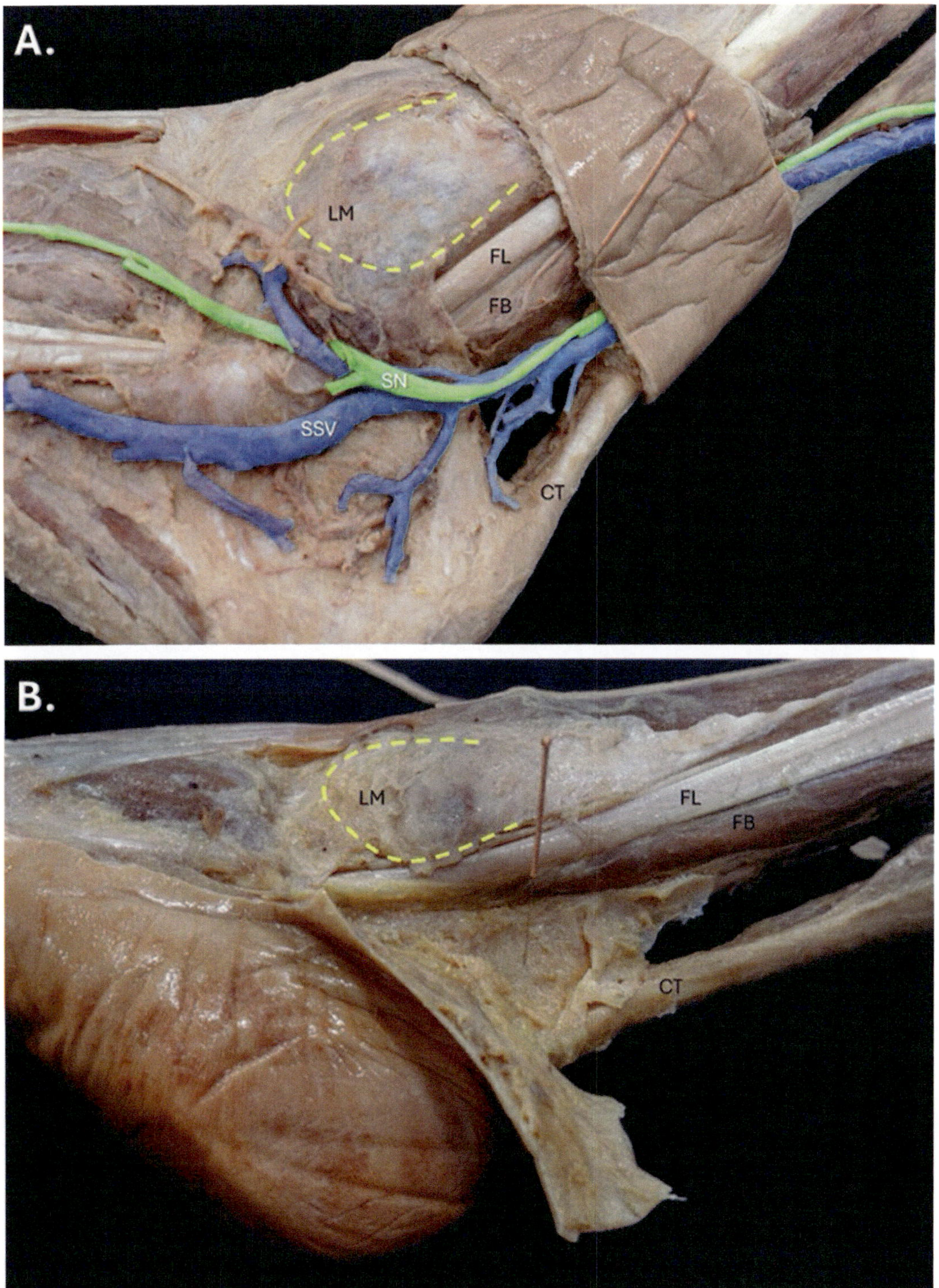

Fig. 6.20 Lateral view of left ankle showing needling of BL60. Needle dimensions 25 mm × 0.25 mm. (**a**) Superficial dissection showing subcutaneous structures. (**b**) Deeper dissection showing insertion of needle into the precalcaneal (Kager's) fat pad. *CT* calcaneal tendon, *FL* fibularis longus, *FB* fibularis brevis, *LM* lateral malleolus (outlined in yellow), *SN* sural nerve (highlighted green), *SSV* small saphenous vein (highlighted blue)

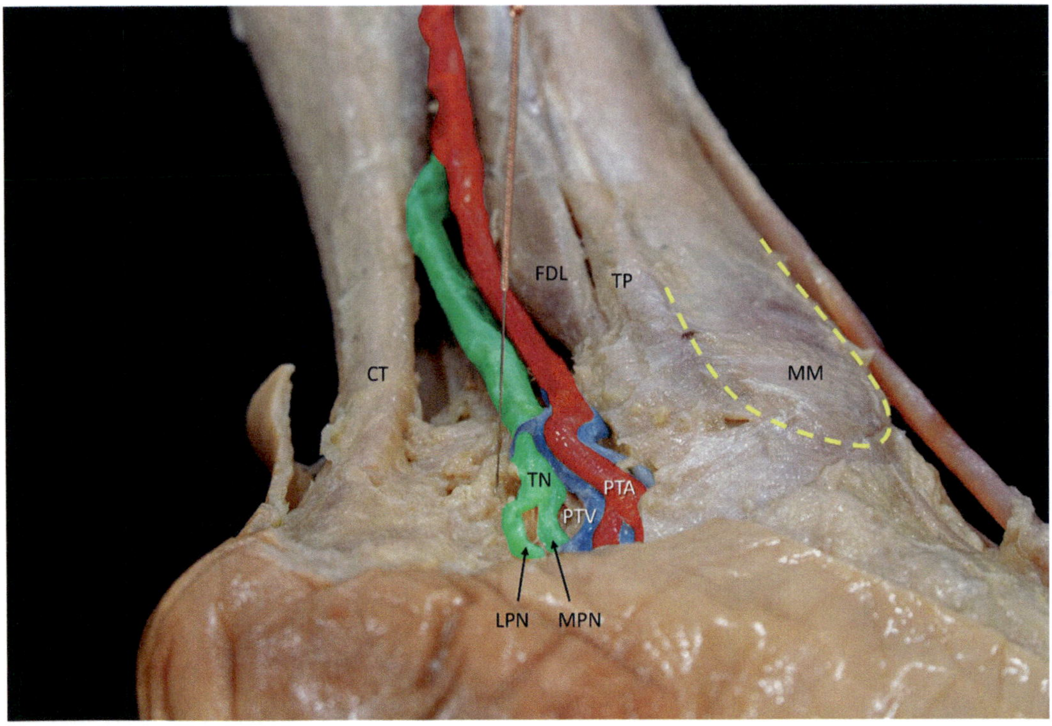

Fig. 6.21 Medial view of left ankle showing needling of KI3. Needle dimensions 25 mm × 0.25 mm. *CT* calcaneal tendon, *FDL* flexor digitorum longus, *LPN* lateral plantar nerve (highlighted green), *MM* medial malleolus (outlined in yellow), *MPN* medial plantar nerve (highlighted green), *PTA* posterior tibial artery (highlighted red), *PTV* posterior tibial vein (highlighted blue), *TN* tibial nerve (highlighted green), *TP* tibialis posterior

lateral aspect of the ankle. Caution should be exercised when needling in this region to avoid the great saphenous vein and saphenous nerve superficially, and the tibial nerve and posterior tibial vein deeply. Needling too anteriorly in this position can lead to penetration of the tibial nerve or the posterior tibial artery or vein. Clinically this point may be needled for its perineural location to the tibial nerve as it lies immediately proximal to the branching of the medial and lateral plantar nerves.

KI6 (Zhao Hai) is located on the medial ankle, in the depression 1 cun inferior to the medial malleolus at the joint space between the calcaneus and the talus between the tendons of the tibialis posterior and flexor digitorum longus muscles. The needle is inserted perpendicularly or obliquely in a slightly superior direction 0.3–0.5 cun in between these tendons. If the space between the tendons is restricted, then the needle can be inserted in a depression inferior to the sustentaculum tali. Caution should be exercised when needling in this region to avoid the great saphenous vein and saphenous nerve. If needling posteriorly and deeply, avoid the tibial nerve, the posterior tibial artery, and vein. Clinically this point may be needled for its perineural location to the tibial nerve adjacent to the branching of the medial and lateral plantar nerves.

6.5.6 Foot

SP3 (Tai Bai) is located on the medial aspect of the foot, in the depression proximal to the first metatarsophalangeal joint, between the inferior surface of the shaft of the first metatarsal and the tendon of abductor hallucis, at the dorsal–plantar junction. The needle is directed perpendicularly 0.3–1 cun and in its deeper position inserts into flexor hallucis brevis. Insertion of the needle too far inferiorly into the tendon of abductor hallucis or deeply into the tendon of flexor hallucis longus may cause discomfort and should be avoided. Caution should be taken not to penetrate the medial tarsal artery, the medial plantar artery, or the medial plantar nerve.

LR2 (Xing Jian) is located on the dorsum of the foot, between the first and second metatarsophalangeal joints, 0.5 cun proximal to the web margin. The needle is inserted perpendicularly 0.3–0.5 cun or obliquely toward the calcaneus up to 1 cun. The needle is directed into the depression between the first and second proximal phalanges between the tendons of the extensor digitorum longus and extensor digitorum brevis muscles and the tendon of the extensor hallucis longus muscle. Take care not to puncture the dorsal digital vessels or nerves.

LR3 (Tai Chong) is located on the dorsum of the foot, between the first and second metatarsal bones, approximately 2 cun superior to the web margin (Fig. 6.22). The needle is inserted perpendicularly 0.3–1 cun and is directed into the depression between the first and second metatarsal bones, between the tendons of the extensor digitorum longus muscle and extensor hallucis longus muscle, lateral to the tendon of the extensor hallucis brevis. The needle enters the belly of the first dorsal interosseous muscle. Caution should be taken as LR3 is extremely close to the digital cutaneous branch of the deep fibular nerve, the first dorsal metatarsal artery, and the dorsal venous arch, and the. Take care not to puncture any of these associated structures when needing this point.

ST44 (Nei Ting) is located on the dorsum of the foot, at the proximal end of the web space in the fossa between the second and third proximal phalanges in the region of the metatarsophalangeal joints. The needle is inserted perpendicu-

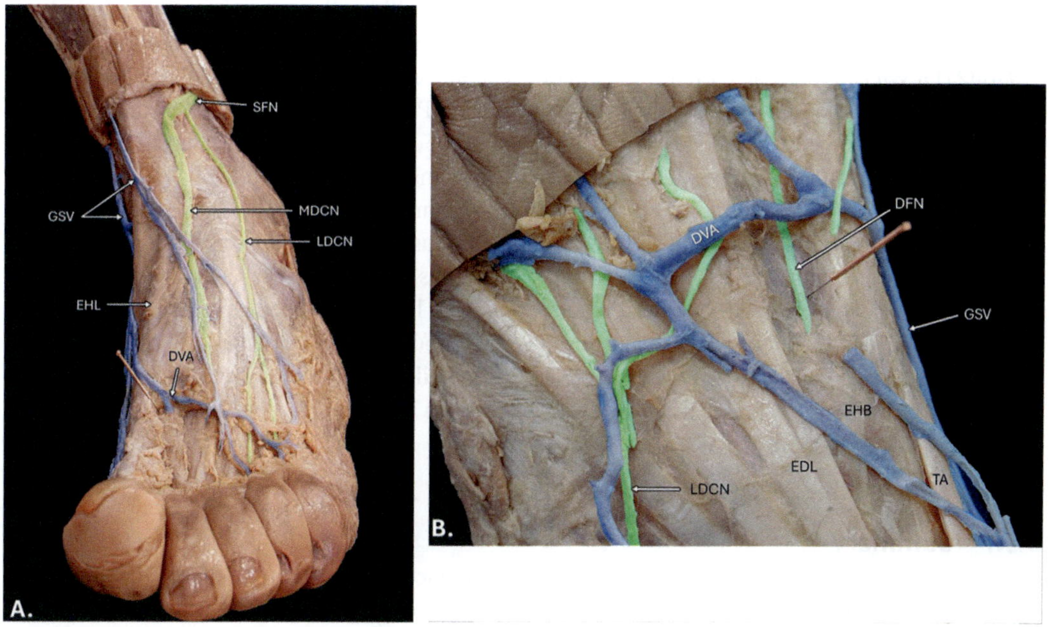

Fig. 6.22 Left-sided dorsum of foot showing needling of LR3. Needle dimensions 25 mm × 0.25 mm. (**a**) Cutaneous nerves and superficial vasculature on the dorsal foot and ankle. (**b**) Zoomed in view of dorsum of foot to show proximity of needle to the digital cutaneous branch of the deep fibular nerve. Note that the first dorsal metatarsal artery (not visible in this dissection) is normally also present within this region. *DFN* digital cutaneous branch of deep fibular nerve (highlighted green), *DVA* dorsal venous arch (highlighted blue), *EDL* extensor digitorum longus, *EHB* extensor hallucis brevis, *EHL* extensor hallucis longus, *GSV* great saphenous vein (highlighted blue), *LDCN* lateral dorsal cutaneous branch of superficial fibular nerve (highlighted green), *MDCN* medial dorsal cutaneous branch of superficial fibular nerve (highlighted green), *SFN* superficial fibular nerve (highlighted green), *TA* tibialis anterior

larly 0.3–0.5 cun into the skin and connective tissue between the tendons of extensor digitorum longus and extensor digitorum brevis muscles. Caution as branches of the dorsal venous arch and a branch of the lateral dorsal cutaneous branch of the superficial fibular nerve lie in this region.

GB41 (Zu Lin Qi) is located on the dorsum of the foot, in the depression between the fourth and fifth metatarsal bones, between the tendons of the extensor digitorum longus and fibularis tertius muscles. The needle is inserted perpendicularly 0.3–0.5 cun between the fourth and fifth metatarsal bones into the fourth dorsal interosseus muscle and to the third plantar interosseous muscle. On deeper insertion it may also enter the transverse head of the adductor hallucis muscle. Caution should be exercised as branches of the dorsal venous arch and a branch of the lateral cutaneous branch of the superficial fibular nerve lie in this region. When needling deeper, avoid inserting the needle into the associated bones or the dorsal metatarsal artery.

GB42 (Di Wu Hui) is located on the dorsum of the foot, in a depression proximal to the heads of the fourth and fifth metatarsal bones. This point lies on the medial aspect of the extensor digitorum longus tendon of the fifth digit. The needle is inserted perpendicularly 0.3–0.5 cun between the fourth and fifth metatarsal bones into the fourth dorsal interosseus muscle and to the third plantar interosseous muscle. Caution should be exercised as this point shares similar risks to the previous point (GB41), with an increased risk of hitting the dorsal metatarsal artery.

6.5.7 Summary

As with all sections of this chapter, the focus is on the most common classical acupuncture points taught within foundation-level training courses within the fields of manual therapy or physical medicine. As a result, the point locations are specific and do not permit for generalization of the associated anatomy outside of these areas or points. Whenever needling into purely tender, "Ah Shi," trigger or motor points of the musculoskeletal system that are not within the boundaries of these classical points, the onus is upon the practitioner to educate themselves to the underlying risk factors related to any invasive procedure.

6.6 Applied Anatomy for Acupuncture Part 3: The Trunk

This is the third of four sections that will provide information on anatomy relative to common acupuncture points. The meridians that lay on the trunk are the Conception Vessel, Governing Vessel, Bladder, Stomach, Small Intestine, Liver, Kidney, and Gallbladder (Figs. 6.2, 6.4, 6.5, 6.7, 6.8, 6.9, 6.10, and 6.11). Needle point locations in the thoracic, abdominal, and lumbar regions of the torso will be described anatomically, and the relative safety of needling these points in clinical practice will be discussed. Due to the underlying viscera in these regions, there are some unique challenges and considerations when needling the trunk. Poor awareness of the anatomy can increase the risk of incorrect needle placement, which may result in severe adverse effects, such as pneumothorax. Emphasis is therefore placed on discussing needle points with specific safety concerns, such as GB21. As in all sections of this chapter, location of needle points relies on a thorough understanding of anatomy, accurate palpation skills, and an appreciation of variation between individuals.

6.6.1 The Thorax

Acupuncture points often fall at the convergence or divergence of myofascial structures, at separations of muscle or muscle bellies. The points of the inner Bladder channel are no exception, being located approximately at the midpoint (or crest) of the erector spinae group in the thoracic region bilaterally (Fig. 6.11). In the thoracic region their paraspinal location lies 1.5 cun lateral to the spinous processes of each vertebra. The needle is inserted obliquely to avoid the underlying intercostal spaces in this region and so therefore falls predominantly into the longissimus thoracis muscle throughout its length. Between T1 and L2, needles should not be inserted at a perpendicular angle due to the risk of pneumothorax, as these points overly the intercostal spaces throughout the thoracic region. Insertion between these spinal levels should always be in an oblique medial direction (Fig. 6.23).

BL11 (Da Zhu) is located on the upper back, 1.5 cun lateral to the lower border of the spinous process of the first thoracic vertebra (T1). When standing in the anatomical position, the distance between the medial border of the scapula and the spinous process of T1 is approximately the equivalent of four fingers' breath or 3 cun. This is the lateral position of the outer Bladder channel. The inner Bladder channel is located half this distance at 1.5 cun or two fingers' breadth. The inner and outer Bladder channel points are innervated by the lateral branches of the dorsal rami of spinal nerves, which include both sensory and motor fibers. The dorsal rami emerges through the intervertebral foramina, turning posterolaterally to supply the erector spinae muscles: longissimus, iliocostalis, and spinalis. Medially, the dorsal rami innervates the transversospinalis muscle group: multifidus, semispinalis, and rotatores. Further supply is given to the zygapophysial joints, interspinous ligament(s), and the interspinales muscles. BL11 is at this location and the needle is inserted obliquely medially 0.5–0.7 cun initially into the trapezius muscle then into the rhomboid minor muscle. Deeper insertion may take the needle into the serratus posterior superior muscle and the longissimus thoracis. Deeper insertion with an oblique angle directed at the articular pillar of the thoracic spine, dependent

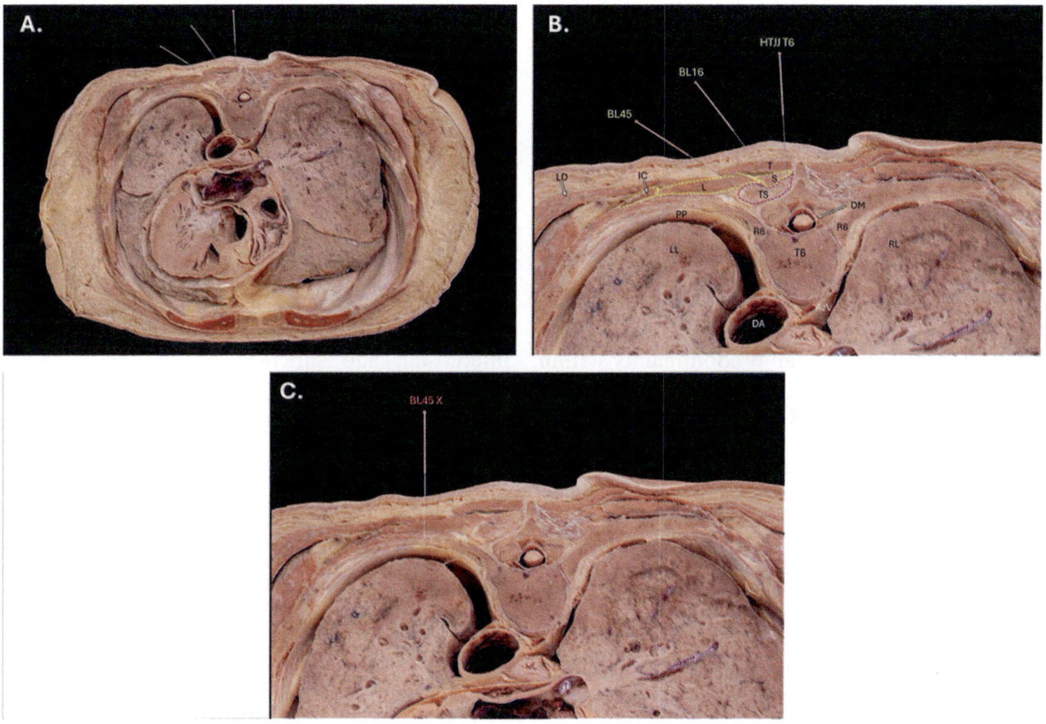

Fig. 6.23 Transverse thoracic slice at the level of T6 showing needling of BL45, BL16, and HTJJ T6. Needle dimensions: 40 mm × 0.25 mm. (**a**) A complete view of thoracic slice at T6. (**b**) Close-up view with annotations and highlighted erector spinae and transversospinalis muscle groups. (**c**) Perpendicular angling of needle leading to penetration of the pleural cavity. *AzV* azygos vein, *DA* descending aorta, *DM* dura mater, *IC* iliocostalis, *L* longissimus, *LD* latissimus dorsi, *LL* left lung, *PP* parietal pleura, *R6* rib 6, *RL* right lung, *S* spinalis, *T* trapezius, *T6* T6 vertebral body. *TS* transversospinalis muscle group (outlined in purple)

on angulation may also take the needle to the splenius cervicis, splenius capitis, semispinalis cervicis, semispinalis capitis, and the multifidus muscles. The needle may pass through several myofascial trigger points in the underlying muscles, any of which may elicit pain referral to the neck and medial scapula. Caution when needling deeply as this point overlies the lateral aspect of the transverse process of the second thoracic vertebrae (T2) and second intercostal space, therefore perpendicular or oblique lateral needling may create a risk of pneumothorax.

B12–B21. The points that follow BL11 in an ascending numerical sequence progress caudal from T1 to L1 and all lie 1.5 cun lateral to the lower border of the corresponding spinous process of their adjacent thoracic and lumbar vertebra. All utilize the same needling technique of oblique, medial insertion 0.5–0.7 cun as they carry the same risk of pneumothorax. BL12 (Feng Men) is located in the upper thoracic region, 1.5 cun lateral to the lower border of the spinous process of the second thoracic vertebra (T2), BL13 at T3, and so on. Due to the proximity of the points, needles inserted at BL12–BL15 will pass through the same layers as indicated within the description of BL11 (above). Here too the needle may pass through several myofascial trigger points in the underlying muscles which can influence pain and dysfunction in the scapular and thoracic regions. Those from BL16 to BL21 (Fig. 6.23b) will fall predominantly within the longissimus thoracis, spinalis thoracis, and the multifidus muscles at their deepest level once they have passed through the more superficial structures of the trapezius muscle or the latissimus dorsi muscle.

BL41 (Fu Fen) is located on the upper back, 3 cun lateral to the lower border of the spinous process of the second thoracic vertebra (T2), level with BL12. When standing in the anatomical position the distance between the scapula and the spinous process of T1 is approximately the equivalent of four fingers' breath or 3 cun. This is the lateral position of the outer Bladder channel. Anatomically this point lies along the lateral border of the iliocostalis muscle overlying the second intercostal space. Care should be taken to palpate the underlying ribs in this region and identify the posterior rib angle. When needling the outer Bladder channel, the direction of needle insertion should give due consideration to the angle of the underlying rib. If the point falls medial to the posterior rib angle, then the needle is inserted obliquely and medially 0.5–0.7 cun initially into the trapezius muscle then into the rhomboid muscles. Deeper insertion may take the needle into the iliocostalis cervicis muscle. Caution should be exercised when needling deeply at thoracic outer Bladder points (BL41–BL50) as these points overlie the intercostal spaces and therefore perpendicular or oblique lateral needling may create a risk of pneumothorax.

BL42–BL46 are located caudal to BL41 along the back, 3 cun lateral to the lower border of the thoracic spinous processes. BL42 (Po Hu) therefore lies immediately caudal to BL41, and level with BL13, at the level of the third thoracic vertebra (T3). The same precautions and considerations for needling are exercised at BL42 that are employed at BL41. The Bladder points from BL41 to BL46 follow this same anatomical location sequence over the second to seventh intercostal spaces. There is no inner or outer Bladder point at the level of the eighth thoracic vertebra, although the extra point of Weiwanxiashu lies 1.5 cun lateral to the inferior aspect of the spinous process of T8.

BL47–BL50 relate to the ninth, tenth, eleventh intercostal, and subcostal spaces, respectively (Fig. 6.11). Specifically, BL50 is 3 cun lateral to the lower border of the spinous process of the 12th thoracic vertebra. For these points the needle is inserted medially obliquely 0.5–1 cun initially into the latissimus dorsi and then deeper into the serratus posterior inferior and iliocostalis thoracis muscles. All three of these muscles may generate myofascial pain in the region of these points. The pain may refer to a paraspinal or ventral distribution in a quasi-dermatomal pattern and may potentially confuse diagnosis of visceral pain. Caution should be exercised when needling deeply in a perpendicular direction at any of the points in this region due to the proximity of the underlying aspect of the pleura, the kidneys, and on the right, the liver.

SI11 (Tian Zong) is located on the scapula and targets the infraspinatus muscle. The needle is inserted perpendicularly 0.5–1 cun, one-third of the distance between the mid-point of the scapular spine and the inferior angle of the scapula. Close attention should be paid to the relative depth and size of infraspinatus which occupies this fossa. Clinically, several myofascial trigger points may be identified in the region of SI11. Caution should be exercised when needling deeply to the infraspinous fossa surface of the scapula as some studies have identified scapular defects presenting as foramina in these underling regions (Pate et al. 1985). Posteriorly, branches of the suprascapular and circumflex scapular arteries and veins are found, as well as the suprascapular nerve.

SI12 (Bing Feng) is located on the upper back, with the arm in slight abduction, in the middle of the supraspinous fossa directly superior to SI11. Superficially the needle will pass through the upper fibers of trapezius and then deeper into supraspinatus (Fig. 6.24). The point is needled either medially along the supraspinous fossa or perpendicularly 0.5–1 cun to the belly of supraspinatus within the supraspinous fossa. Caution should be exercised with this technique however, with initial entry being adjacent to the spine of the scapula and needle direction being posterior to identify the supraspinous fossa with the needle. Once the bony landmark of the spine of the scapula has been identified deeper needling into supraspinatus may be pursued. Needling into the anterior direction is contraindicated due to the risk of traumatic pneumothorax. Consideration should also be given to the suprascapular nerve, artery, and vein in this location too. This point may be clinically recognized as the myofascial trigger point for the belly of the supraspinatus muscle.

SI13 (Qu Yaun) is located on the upper back, in the depression on the medial end of the supraspinous fossa. This point is needled perpendicularly 0.3–0.5 cun. Superficially, the needle will pass through the upper fibers of trapezius and then deeper into supraspinatus. Identification of the surrounding anatomy is essential when needling in this area. Palpation of the spine of the scapula and the medial border of the scapula is recommended, as is the identification of the adjacent ribs. Once orientation to the thoracic cage and bony demarcation of the medial aspect of the scapula has been achieved, safe needling can progress. Caution should be exercised with deep medially oblique needling of this point due to the potential risk of traumatic pneumothorax.

SI14 (Jian Wai Shu) is situated 3 cun lateral to the lower border of the spinous process of the first thoracic vertebra (T1). This places it in the region of the superior angle of the scapula. The needle will enter trapezius and serratus posterior superior which is overly iliocostalis and the scapular attachment of levator scapulae (Jarmey et al. 2021). Innervation in this region is via the dorsal scapular nerve which innervates levator scapulae and rhomboids minor and major. Clinically, this point may often be tender in those with cervical and upper limb disorders due to the relationship of the three muscles associated with the dorsal scapular nerve and myofascial trigger points in the upper fibers of trapezius and levator scapulae. Caution should be exercised on needle insertion as SI14 overlies the thoracic cage and subsequently the pleura and lungs. It is, therefore, needled obliquely (0.3–0.7 cun) either from the spine toward the shoulder or from the shoulder toward the spine, dependent on the position of the patient (prone, side lying, or sitting). As with all points in this region, palpation of the scapular spine, superior angle, and medial border of the scapula are recommended, as is the identification of the adjacent underlying ribs and intercostal spaces. Once orientation to the thoracic cage and bony demarcation of the scapula has been achieved, safe needling can progress. Oblique needling of SI14 in some directions may also penetrate rhomboid minor whose fibers may cross those of levator scapulae at their attachments to the medial border and superior angle of the scapula, respectively. White et al. (2018) warn against the deep needling of SI14 unless one is confident of the angle and depth of the needle in relation to the scapula. Deep perpendicular needling of SI14 is to be avoided in all instances due to the inherent risk of pneumothorax in this region (Jarmey et al. 2021). Consideration should also be paid to the other

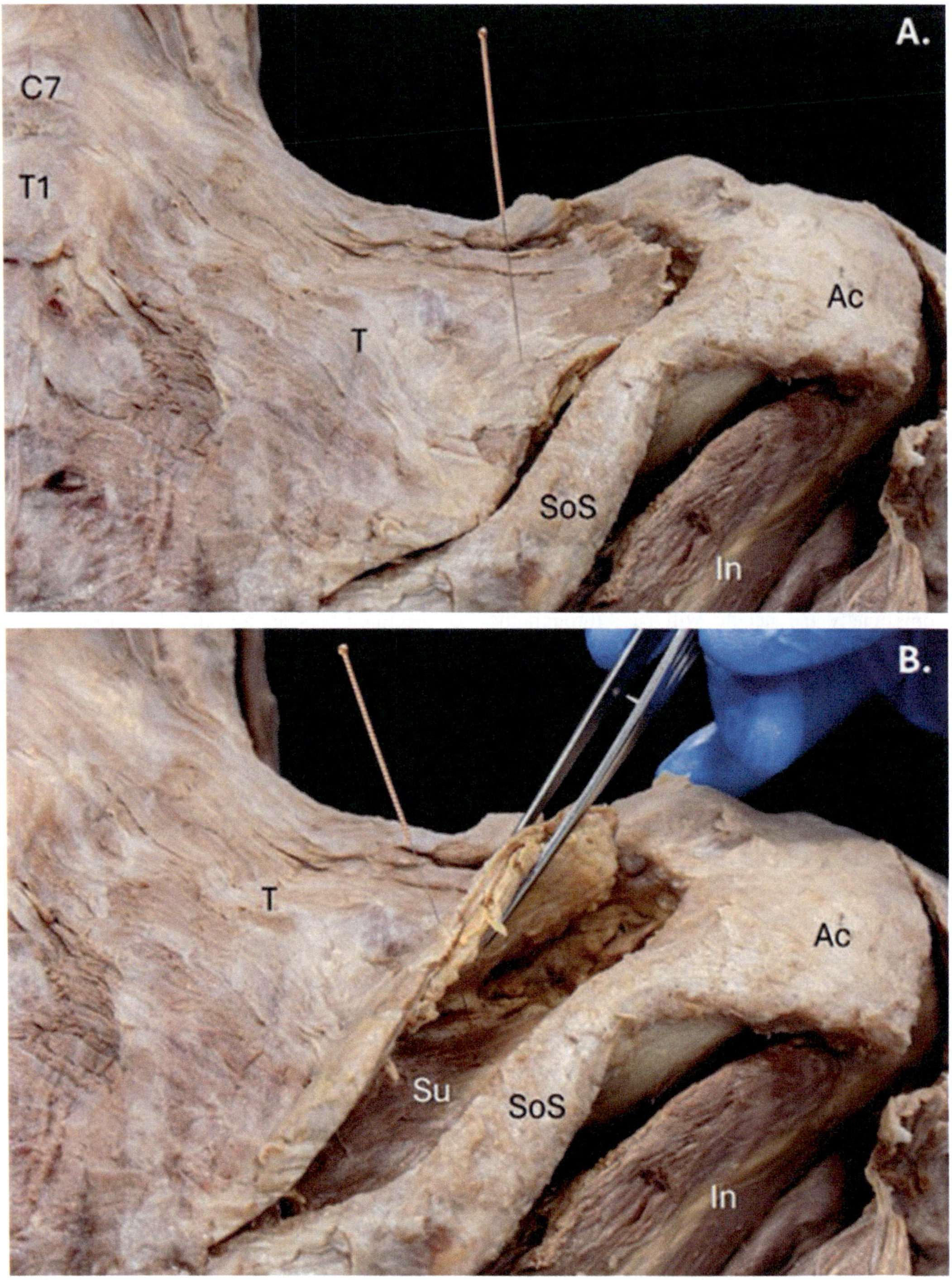

Fig. 6.24 Posterior view of the neck and right shoulder showing needling of SI12. Needle dimensions: 40 mm × 0.25 mm. (**a**) superficial insertion into trapezius. (**b**) Needling of SI12 with trapezius partially reflected to see insertion into supraspinatus. *Ac* acromion, *C7* C7 spinous process, *In* infraspinatus, *T* trapezius, *T1* T1 spinous process, *SoS* spine of scapula, *Su* supraspinatus

underlying structures such as the spinal accessory nerve and the medial cutaneous branches of the posterior rami of the first and second thoracic nerves. The transverse cervical artery and vein and the dorsal scapular nerve also lie deep in this area and caution should be taken when needling.

SI15 (Jian Zhong Zhu) is located on the upper back, 2 cun lateral to the lower border of the spinous process of the seventh cervical vertebra (C7). Like SI14, SI15 overlies the levator scapulae muscle and is needled obliquely 0.3–0.6 cun due to its proximity to the apices of the pleura and lungs. As discussed above (SI14), palpation of the scapular spine, superior angle, and medial border of the scapula are always recommended, as is the identification of the adjacent ribs and their associated intercostal spaces. Incorrect deep needling at this point carries the risk of pneumothorax if needled incorrectly. This point is also recognized clinically as a myofascial trigger point, and pain from it can often extend distally along the posteromedial aspect of the upper arm and medial forearm to the fourth and fifth digits of the hand (Baldry 2004). Deep needling of SI15 not only carries a risk of traumatic pneumothorax but the transverse cervical artery and vein and branches of the third and fourth cervical nerves also lay in this area, as does the spinal accessory nerve. Additional caution should therefore be taken so as not to compromise any of these adjacent neurovascular structures.

GB21 (Jian Jing) is located in the suprascapular region, midway between the tip of the acromion process and below the spinous process of the seventh cervical vertebra (C7) (Fig. 6.25a). It can also be located on the dorsal vertical extension of the mammary line. In this location, GB21 should lie lateral to the first rib and superior to the second rib and therefore lay outside of the lung fields. The needle is inserted obliquely 0.5–1.5 cun posterior to anterior into the upper fibers of the trapezius muscle. Clinically this point may overlie a myofascial trigger point in the upper fibers of trapezius that can refer pain to the ipsilateral side of the head. Accurate palpation and point identification are essential in this region and caution should be exercised at all times. Avoid deep perpendicular needling or deep anteromedial oblique needling due to the ever-present risk of pneumothorax (Fig. 6.25a, b). Care should also be taken not to damage the suprascapular artery, vein, and nerve, as well as the spinal accessory nerve or the deeper transverse cervical artery and vein (Jarmey et al. 2021).

Note on Safe Practice: Pneumothorax

There are numerous trials that have explored the effectiveness of acupuncture and dry-needling interventions at the location of GB21 in the upper fibers of trapezius (Ziaeifar et al. 2014; Gerber et al. 2015; Cerezo-Tellez et al. 2016; Wu et al. 2016; Abbaszadeh-Amirdehi et al. 2017). When needling a selection of 57 cadavers, Zhang et al. (2002) reported that the anatomical structures encountered beneath GB21 perpendicularly were (from superficial to deep); skin, superficial fascia, deep fascia, trapezius muscle, lateral part of the levator scapulae muscle, serratus anterior muscle, external intercostal muscle, internal intercostal muscle and parietal pleura, visceral pleura, and lung. In a systematic review from Zhang et al. (2010), reported that the most frequently needled acupuncture point recorded for pneumothorax was GB21, with it being implicated in 30% of the 201 cases documented. In an MRI-based study from by Chou et al. (2015), the depth of the lungs directly perpendicular beneath GB21 was measured via MRI in 394 subjects. It was determined that depth to the lungs went from a minimum of 37 mm in underweight males to 75 mm in obese males. In the female subjects evaluated it was between a minimum of 31 mm to a maximum of 72 mm, dependent on body composition and deposition of adipose tissue. In 16–18 year olds across both sexes the depth was 32.45 mm (± 10.06 mm) Ma et al. (2016). Chu et al. (2018) conducted a study to examine whether there was a change in safe needling depth at acupoint GB21 according to posture and breathing. They noted that needling depth changed significantly with a change in posture from sitting to prone lying, with prone having a greater safe needling depth. They also found that although a difference in depth was observed between maximal inspiration and expiration it was not clinically significant, with an excursion of less than 1 mm in most subjects, although outliers could reach up to 4.5 mm. From the result of this study, the mean safe needling depth of GB21 in individuals in their twenties is less than 35 mm in males and 30 mm in females. The authors also concluded that needling depths can easily deviate from the mean and standard deviation in many subjects and so simply applying needling depth on a sex or BMI basis is not sufficient or recommended. A better practice would be to also consider the individual's anatomical characteristics when deciding on needling depth. Finally of note, Wang et al. (2018) compared superficial and traditional acupuncture in the treatment of trigger points in the upper trapezius and found significant improvements in pain symptoms in both groups. This led the authors to conclude that because superficial acupuncture was associated with less pain and less potential adverse effects than deep needling, while still producing immediate pain relief, it could be recommended for treating myofascial pain syndrome in this region.

CV17 (Shan Zhong) is located on the sternal midline, level with the fourth intercostal space, between the nipples. Due to the location of CV17 directly overlying the sternum it is needled subcutaneously only 0.3–0.5 cun. Caution should be exercised when needling over the sternum as in 4% of females and 10% of males, sternal foramina are considered to remain patent into adulthood, with these structures not being effectively identified on standard thoracic radiographs (McCormick 1981). More recently, Choi et al. (2017) highlighted the risk of cardiac tamponade resulting from sternal foramina. The remaining CV points between CV18 and CV21 also lay on the sternal midline and the same needling technique and precautions should be employed.

CV22 (Tian Tu) is in the suprasternal fossa, on the anterior midline, just above the jugular notch. The needle is inserted retrosternally 0.5–1 cun. This point carries a significant risk when needling due to the underlying anatomical structures. Potential adverse events that can arise from needling CV22 incorrectly include tracheal or aortic arch puncture, venous injury (left brachiocephalic and inferior thyroid), or pneumothorax. The needle is inserted perpendicularly initially then directed in an oblique caudal direction retrosternally, avoiding a posterior placement or deviation to the left or right. Caution should be exercised to avoid the trachea and the brachiocephalic trunk which is located to its right. The thymus gland (more relevant in younger individuals) also lays in the region of the suprasternal notch in the superior mediastinum between the trachea and the manubrium. This point is rarely needled but it is imperative to understand the underlying anatomy and associated risk factors that can allow a clinician to make a reasoned informed decision based on risk–benefit ratio relating to any treatment.

6.6.2 Extra Points

The *HTJJ* (Hua Tuo Jia Ji) points are a series of 34 or 48 paired paraspinal points dependent on which acupuncture text they are described in. If they are referred to as 34 points, then these are located 0.5 cun lateral to either the lower border or the tips of the spinous processes of T1-L5. This approximates with the facet joints or the level of the spinal nerve roots. When 48 points are described, these are located 0.5 cun lateral to the spinous processes of C1-L5. The sacral region does not have HTJJ points as the Bladder points from 31 to 34 (also known as "Baliao"), overlie the sacral foramina in this region (Fig. 6.26). The relative location of the

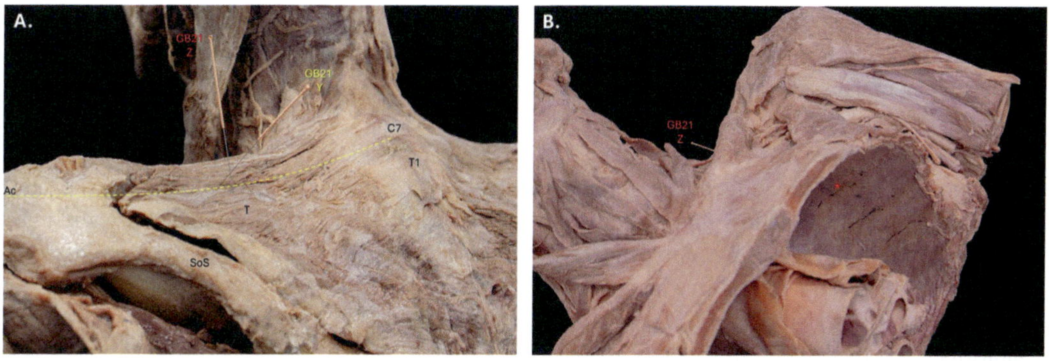

Fig. 6.25 *Needling of GB21.* Needle dimensions: GB21 Y = 40 mm × 0.25 mm. GB21 Z = 90 mm × 0.3 mm. (**a**) Posterior view of the neck and left shoulder showing a direct comparison of the safe needling orientation at GB21 with the improper technique likely to penetrate the pleural cavity. Y = safe orientation at GB21. Z = unsafe orientation at GB21. The yellow dashed line is drawn between the tip of the acromion and the inferior margin of the C7 spinous process. (**b**) Antero-inferior view showing penetration of the pleural cavity via incorrect positioning of GB21. When pushed into the hilt, 20 mm of the 90 mm needle projected into the third intercostal space (marked with a red asterisk). It should be noted that this cadaver had very little adipose tissue present, and all superficial tissues at the needle entry site have been removed. *Ac* acromion, *C7* C7 spinous process, *SoS* spine of scapula, *T* trapezius, *T1* T1 spinous process

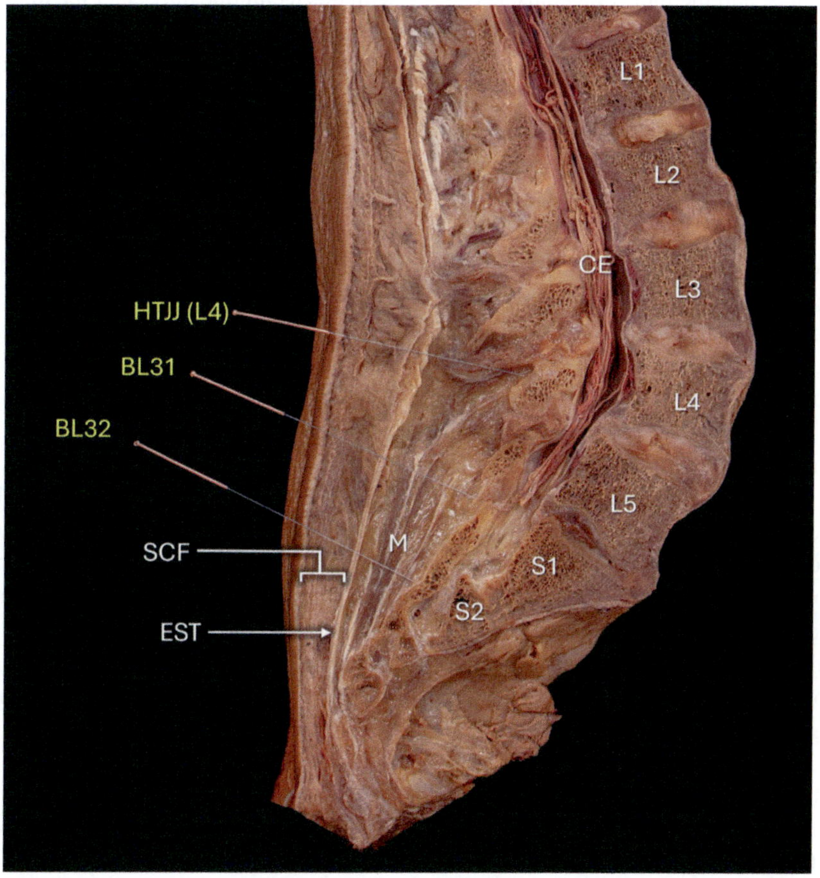

Fig. 6.26 *Sagittal section of the lower vertebral column shows needling of HTJJ (L4), BL31, BL32.* Needle dimensions: 70 mm × 0.25 mm. *CE* cauda equina, *EST* erector spinae tendon, *M* multifidus, *SCF* subcutaneous fat

HTJJ points to the spinous processes may vary between individuals from approximately 0.3 to 1 cun dependent on anatomy and palpation of reactive tissue. Due to their anatomical location, the points are employed to treat local spinal pain and dysfunctions, as well as influencing segmental response for dermatomes, myotomes, and their pain referral distributions. If needling the HTJJ between 0.5 and 1 cun lateral to the spinous processes within the cervical, thoracic, and/or lumbar regions then a perpendicular insertion of the needle is recommended. The tissues underlying the HTJJ points are those innervated by the medial branches of the dorsal rami (see BL11).

In the cervical spine, when needling the HTJJ points the needle will pass initially through the superficial trapezius and then dependent on their level and depth may pass through the semispinalis capitis, splenius capitis, semispinalis cervicis, spinalis cervicis, and the multifidus muscles.

In the thoracic region when needling the HTJJ points, dependent on the spinal level and depth the needle will pass initially through the trapezius or the medial musculotendinous aspect of latissimus dorsi and then deeper to the rhomboid major or minor muscles between C6 and T5. Passing deeper between C7 and T3 the needle will enter the serratus posterior superior muscle before progressing to the splenius capitis, semispinalis thoracis or spinalis thoracis, and finally the multifidus muscle.

In the lumbar region when needling the HTJJ points the needle will pass initially through the medial musculotendinous aspect of latissimus dorsi at its insertion onto the thoracolumbar fascia. Between T11 and L2 it will pass through the serratus posterior inferior muscle before progressing to the longissimus thoracis muscle and finally reaching the multifidus muscle. If needling within the sacral region into the Baliao points known as BL31–BL34, the needle will pass through thoracolumbar fascia before entering into the longissimus muscle, then ultimately reaching the multifidus muscles overlaying the sacral foramina. The Baliao points are described in further detail below.

In numerous classic acupuncture texts, the HTJJ point at C7 will be independently referenced as Dingchuan, an extra point utilized in the treatment of respiratory disorders. It is needled to a depth of 0.5–1 cun, 0.5 cun lateral to the spinous process of C7. The needle will first entertrapezius muscle and then progress to the rhomboid minor, the serratus posterior superior, the splenius capitis, and the semispinalis cervicis muscles and finally end in the multifidus muscle, if needled deeply enough.

6.6.3 The Lumbar Region

After the spinal level of L2, the needles of the inner Bladder channel may be directed perpendicular into the longissimus muscle to a depth of 1.5 cun. Throughout the lumbar region the needle may also be directed medially obliquely toward the articular pillar of the lumbar vertebrae and with this technique may enter the multifidus muscle too at a deeper level.

BL23–BL26 These points share a common target tissue, method, and direction of needling. BL23 (Shen Shu) is located on the lower back, 1.5 cun lateral to the lower border of the spinous process of the second lumbar vertebra (L2). The needle is inserted perpendicularly 1–1.5 cun initially into the thoracolumbar fascia and the tendinous attachments of the latissimus dorsi muscle before entering the longissimus muscle. If the needle is directed medially obliquely toward the articular pillar of the lumbar spine it may also pass into the multifidus muscle at that level. BL24–BL26 are needled in the same plane from L3 to L5. Caution to avoid needling beyond the depth of the transverse process anteriorly in this region due to the risk of reaching the kidneys. The lower borders of the kidneys are located approximately at L2 on the left side and L3 on the right, with the right's lower position being dictated by the mass of the liver which is located superior to the kidney on that side.

BL27–BL30 These points are in the sacral region between S1 and S4. Similar to the lumbar Bladder points they will enter the thoracolumbar fascia initially but will then pass into the medial aspect of the gluteus maximus muscle. On deeper insertion, when directed medially obliquely toward the posterior surface of the sacrum, they may also enter the multifidus muscle.

BL31 (Shang Liao) is in the sacral region, overlying the first posterior sacral foramen approximately one finger's breadth from the posterior midline (Fig. 6.26). The most easily identified foramen of the sacrum via palpation is that of S2. This is best located by identifying the posterior superior iliac sine (PSIS) and then palpating inferiorly and medially to the bony prominence. S2 is located adjacent to the PSIS in this region. Located superior to the foramen of S2 is the foramen of S1, and inferior is the foramen of S3. The needle is inserted perpendicularly 0.5–1.5 cun into the thoracolumbar fascia initially and will then pass into the multifidus muscle if needled deeply. BL27 is adjacent to BL31 but lies 1.5 cun lateral to the sacral foramen of S1 within the inner Bladder channel. When needling within the sacral region into the Baliao points (BL31–34), the needle will pass through thoracolumbar fascia before entering into the longissimus muscle then ultimately reach the multifidus muscle overlaying the sacral foramen. Caution should be exercised not to needle deeply into the sacral foramina themselves as this may damage the sacral nerve trunks of that level.

BL32 (Ci Liao) is located in the sacral region, overlying the second posterior sacral foramen approximately one finger's breadth from the posterior midline (Fig. 6.26). As previously stated, S2 is the most easily identified foramen of the sacrum via palpation and lies inferiorly and medially to the bony prominence of the PSIS. The needle is inserted perpendicularly 0.5–1.5 cun into the thoracolumbar fascia initially, then it will pass into the multifidus muscle if needled deeply here. The S2, S3, and S4 nerve roots form the pelvic splanchnic nerves which provide parasympathetic supply to the smooth muscle and glands of the hindgut, pelvis, and perineum. The pudendal nerve also arises from these nerve roots and is responsible for somatic supply of the perineum and associated structures. This includes sensation to the external genitalia and innervation of the perineal muscles, including the external urethral sphincter and the external anal sphincter. S2 also contributes to the sciatic nerve (L4–S3), the inferior gluteal nerve and innervation to piriformis (S1, S2), obturator internus, and the superior gemellus (L5–S2), the posterior femoral cutaneous nerve (S2, S3) and the perforating cutaneous nerve (S2, S3). Damage to these sacral nerve roots can therefore have significant and life-altering consequences, including lower limb motor weakness, sexual dysfunction, and urinary or fecal incontinence.

BL33 (Zhang Liao) is located in the sacral region, overlying the third posterior sacral foramen approximately one finger's breadth from the posterior midline. To locate S3, first identify the PSIS and then the S2 foramen as described above. S3 is the next palpable foramen located immediately inferior to S2. The needle is inserted perpendicularly 0.5–1.5 cun into the thoracolumbar fascia initially, then it will pass into the multifidus muscle if needled deeply here.

BL34 (Xia Liao) is located in the sacral region, overlying the fourth posterior sacral foramen approximately one finger's breadth from the posterior midline and adjacent to the sacral hiatus. The S4 foramen can be found by palpating and identifying the PSIS, then the S2 and S3 foramina as described above. S4 is the next palpable foramen located immediately inferior to S3. To confirm this location, one may also identify the inferior aspect of the sacrum and progress superiorly onto the dorsal aspect of the sacrum. The foramen of S4 will be the first palpable recess located lateral to the sacral hiatus. The needle is inserted perpendicularly 0.5–1.5 cun into the thoracolumbar fascia initially, then it will pass into the multifidus muscle if needled deeply here.

BL51 (Huang Men) is located on the lower back, 3 cun lateral to the lower border of the spinous process of the first lumbar vertebra (L1), at the level of BL22. The needle is inserted medially obliquely 0.5–1 cun initially into the latissimus dorsi muscle and the thoracolumbar fascia, then deeper into the serratus posterior inferior and iliocostalis muscles. When needled beyond these structures the needle may also reach the lumbar interfascial triangle and the quadratus lumborum muscle in this location. Caution should be exercised when needling deeply in a perpendicular or superior direction in this region due to the proximity of the underlying kidneys.

BL52 (Zhi Shi) (Fig. 6.27) is located inferior to BL51, 3 cun lateral to the lower border of the spinous process of the second lumbar vertebra (L2), at the level of BL23. As with BL51, the needle is inserted medially obliquely 0.5–1 cun initially into the latissimus dorsi muscle and the thoracolumbar fascia, and then deeper into the serratus posterior inferior and iliocostalis muscles. When needled beyond these structures, the needle may also reach the lumbar interfascial triangle and the quadratus lumborum muscle in this location. Clinically this point may overly myofascial trigger points in the iliocostalis and quadratus lumborum muscles. Caution should be exercised when needling deeply in a perpendicular or superior direction in this region due to the proximity of the underlying kidneys.

6.6.4 The Abdominal Region

ST25 (Tian Shu) is located on the abdomen, 2 cun lateral to the umbilicus, on the rectus abdominis muscle. The needle is inserted perpendicularly or obliquely 0.7–1.2 cun into the rectus abdominis muscle and may also pass into the aponeuroses of the internal and external oblique muscles and then the transverse abdominis muscle in a deeper location. Clinically this point may overly myofascial trigger points in the rectus abdominis and the transverse abdominis muscles. Care should be taken, however, not to needle deeply in a perpendicular direction as the potential to puncture the peritoneum and underlying viscera exists, particularly in those with a low level of abdominal adipose tissue. The distance from the pubic symphysis to the umbilicus is divided into fifths with each subdivision being considered equivalent proportionally to one cun.

ST26–ST29 One cun caudal to ST25 lies ST26, then one cun caudal to that ST27 and so on until ST30. The same technique for needing and same considerations for underlying viscera and exercised for all this sequence of points. Additional care should be exercised in the region of ST29 when needling male subjects as the spermatic cord lies in the underlying tissue. The superficial inguinal ring, an opening in the external oblique fascia, can be palpated just superolateral to the pubic tubercle. The spermatic cord traverses this opening and courses through the inguinal canal. In female subjects, the main structure located here is the round ligament of the uterus and may also be at risk.

ST30 (Qi Chong) is located at the superior border of the pubic symphysis, 2 cun lateral to the

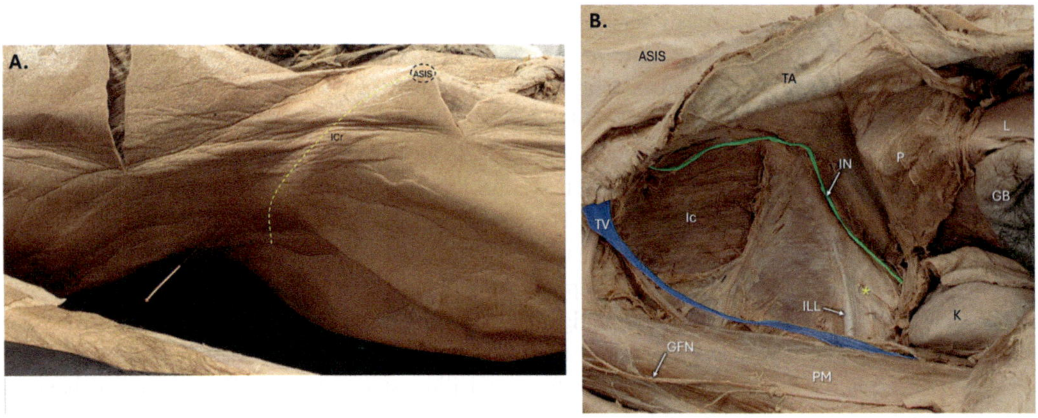

Fig. 6.27 *Needling of BL52.* Needle dimensions 90 mm × 0.30 mm. (**a**) superficial insertion on right side. (**b**) Internal view of penetration through quadratus lumborum. Note that this cadaver had very little adipose tissue. *Yellow asterisk* needle position within quadratus lumborum, *ASIS* anterior superior iliac spine, *GB* gallbladder, *GFN* genitofemoral nerve, *Ic* iliacus, *ICr* iliac crest (yellow dashed line in A), *ILL* iliolumbar ligament, *IN* ilioinguinal nerve (highlighted green), *K* kidney (with renal fat removed), *L* liver, *P* peritoneum (reflected), *PM* psoas major, *TA* transversus abdominis, *TV* testicular vein (highlighted blue)

anterior midline, at the level of CV2, superior to the inguinal ligament on the medial aspect of the external iliac artery. The needle is inserted perpendicularly at a depth of 0.5–1 cun. Care should be taken not to damage the associated neurovascular structures in the area. These are more vulnerable to deep needling in individuals with less adipose tissue and include the anterior cutaneous branch of the iliohypogastric nerve and the superficial epigastric artery and vein. Caution should always be exercised with deep needling to avoid the external iliac vein, artery, spermatic cord in males, and round ligament of the uterus in females.

CV2 (Qu Gu) is located in the pubic region on the anterior midline, at the superior border of the pubic symphysis. Needle perpendicularly 0.5–1 cun. The depth of needling over the abdomen will be influenced by the level of adipose tissue present. This may vary greatly between individuals and will also vary depending on the location of the abdomen. On initial insertion, the needle will pass through the skin and subcutaneous adipose tissue before progressing to the target tissue of the linea alba of the rectus abdominis muscle. Proceed with caution and do not needle deeply beyond the rectus sheath as deeper adipose tissue and peritoneum directly overlie the bladder at this point. When needling in this region it is advisable to empty the bladder prior to treatment. Between the acupuncture points of CV2–CV4 is the location of the initial incision site for the insertion of a suprapubic catheter (Corder and LaGrange 2024).

CV3–CV8 The distance from the pubic symphysis to the umbilicus is divided into fifths with each subdivision being considered equivalent proportionally to 1 cun. CV2 is located at the superior border of the pubic symphysis whilst CV7 lies 1 cun caudal to the umbilicus. CV3 is 1 cun superior to CV2, then CV4 is 1 cun superior to CV3, and so on until CV8 which overlies the umbilicus. CV6 is omitted from this 1 cun step sequence however as it lies 0.5 cun between CV5 and CV7. As before, the relationship of the underlying viscera should be considered for safe needling of these points. Differences in deposition of adipose tissue should also be expected. This may lead to a potential deeper insertion of the needle required to reach the target tissue of the linea alba and rectus abdominis muscle. This consideration should be taken into account with all individuals based on examination and palpation of the abdominal tissue.

Note on Safe Practice: Diastasis Recti

Prior to needling the abdominal midline, careful palpation should be undertaken to ascertain the integrity of the underlying tissue and evaluate the presence of a potential diastasis of the rectus abdominis muscle. This condition may exist in a significant percentage of the population with 33% of females still experiencing this 12 months postpartum (Sperstad et al. 2016). A diastasis of the rectus abdominis muscles is a thinning and widening of the linea alba, combined with laxity of the abdominal musculature of the anterior trunk. If a diastasis is identified, caution should be exercised when needling Conception Vessel points on the abdomen due to the paucity and thinning of the underlying tissue. In such cases, the points of the Kidney channel from KI11 to KI21 may be needled as an alternative (Fig. 6.7). The Kidney channel in the abdominal region commences from KI11 which is located in the pubic region, immediately superior to the pubic symphysis, 0.5 cun lateral to the anterior midline, at the level of CV2. It travels cranially to KI21 (You Men) located in the epigastric region, 6 cun superior to the umbilicus and 0.5 cun lateral to the midline at the level of CV14. These series of Kidney points may be needled perpendicularly 0.5–1 cun as an alternative to the corresponding CV points which they lay lateral to. The same caution should be exercised at these points with an additional consideration of the liver on the right side, laying deep to the rectus abdominis muscle, rectus sheath, deeper adipose tissue, and parietal peritoneum.

CV12 (Zhon Guan) is located in the upper umbilical region on the anterior midline, 4 cun superior to the umbilicus, midway between it and the xiphisternal joint. The needle is inserted perpendicularly 0.5–1.5 cun through skin and superficial adipose tissue to the target tissue of the linea alba of the rectus abdominis muscle and transversalis fascia. Caution should be exercised to avoid needling deeply beyond the rectus sheath into the deeper adipose tissue, the peritoneal cavity, or the falciform ligament.

CV14 (Juque) is located in the epigastric region on the anterior midline, 2 cun inferior to the xiphoid process, and 6 cun superior to the umbilicus.

6.6.5 Summary

This third section focuses on the most common classical acupuncture points found within the thoracic, abdominal, and lumbar regions as taught within foundation-level training courses within the fields of manual therapy or physical medicine. Due to the underlying structures in these areas, particularly the viscera, there are some unique challenges and considerations to needling. The risk of incorrect placement of a needle in the thoracic region can result in severe (but ultimately avoidable) sequelae. The point locations described are specific to the underlying tissues and do not permit for generalization of the associated anatomy outside of these areas or points. An exception here is when reference is made to a series of points such as the inner Bladder channel that all utilize the same angle of entry for safety reasons. Whenever needling into purely tender "Ah Shi" trigger or motor points of the musculoskeletal system beyond the boundaries of these classical points, one must pay particular attention to the location and angle of the ribs as well as any underlying viscera to avoid potential adverse events.

6.7 Applied Anatomy for Acupuncture Part 4: The Head and Neck

In this final section, the most common classical acupuncture points of the head, neck, and face will be described anatomically, and the relative safety of needling these points in clinical practice are discussed. There is a high volume of neurovascular structures in the head and neck that require a particularly detailed understanding of anatomy to comprehend in full. As many of these structures are relatively superficial, incorrect needle placement can lead to serious adverse effects such as cranial nerve palsy and hematoma. The meridians that lay on the head, neck, and face are Gallbladder, Triple Energizer, Conception Vessel, Governing Vessel, Bladder, Stomach, Small Intestine, and Large Intestine (Figs. 6.1, 6.2, 6.4, 6.6, 6.8, 6.9, 6.10, 6.11, and 6.28). As in all sections of this chapter, location of these needle points relies on a thorough understanding of anatomy, accurate palpation skills, and an appreciation of variation between individuals.

6.7.1 The Head

ST8 (Tou Wei) is located at the corner of the forehead, at the coronal suture, 0.5 cun posterior to the anterior hairline, 4.5 cun lateral to the anterior midline. The needle is inserted obliquely and subcutaneously 0.5–1 cun into the superior margin of the temporal muscle. Clinically this point overlies the myofascial trigger point in the superior margin of the temporalis muscle and may contribute to headache symptoms. Caution should be exercised to avoid damaging the superficial temporal artery and vein.

GB8 (Shuai Gu) is located 1.5 cun superior to the auricular apex and directly above TE 20 at the midpoint of the apex of the ear and the tubercle of the parietal bone. The needle is inserted subcuta-

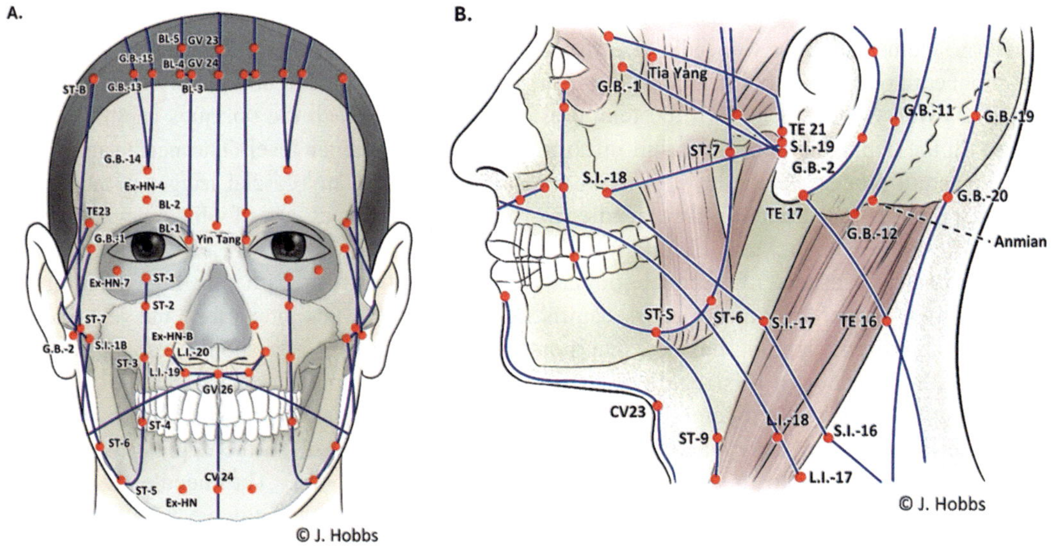

Fig. 6.28 Schematic diagrams of acupuncture points (**a**) of the head and face and (**b**) of the lateral head and neck

neously 0.3–0.5 cun, initially into the superior auricular muscle and then deeper into the temporalis muscle.

GB19 (Nao Kong) is located in the occipital region, lateral to the posterior midline, at the level of the upper border of the external occipital protuberance, superior to GB20. The needle is inserted subcutaneously 0.3–0.5 cun into the depression between the origins of sternocleidomastoid and trapezius muscles and into the occipital belly of the occipitofrontalis muscle. Clinically this point overlies the myofascial trigger point in the occipital belly of the occipitofrontalis muscle that can refer to pain across the ipsilateral aspect of the head forward toward the eye.

GV20 (Bai Hui) is located in the midline, intersecting a line connecting the right and left ear apices overlying the sagittal suture of the parietal bones. A small depression may be palpable in this region. The needle is inserted subcutaneously 0.3–0.5 cun into the epicranial aponeurosis and the superficial connective tissue. The periosteum of the parietal bone may be encountered with deeper needling at this point. Be aware that the subcutaneous tissue in this region is heavily vascularized and includes the anastomotic branches of the superficial temporal and occipital artery and veins so vigorous needling should therefore be avoided to reduce the risk of subsequent bleeding.

6.7.2 The Neck

GB20 (Feng Chi) is located on the posterior head, at the junction of the occipital and nuchal regions. The needle is inserted into a depression between the origins of the sternocleidomastoid muscle laterally and the trapezius muscle medially. The needle is inserted obliquely 0.5–1.5 cun toward the opposite eye or apex of the nose. The needle will pass between the attachments of the sternocleidomastoid and trapezius muscles into the splenius capitis muscle initially and then the semispinalis muscle and then may enter the obliquus capitis superior and the rectus capitis posterior major muscles dependent on angle and depth of insertion. Clinically the myofascial trigger points that lay in any of the muscles in the region of this point may contribute to head and neck pain. Caution when needling deeply in this area to avoid the suboccipital vessels and the greater and lesser occipital nerves. Also, avoid the suboccipital nerve (dorsal ramus of C1) and vertebral artery which lie deeper still in this region. Care should also be taken not to needle deeply in another angle or to needle directly to the spinal cord or medulla oblongata.

GB12 (Wan Gu) is located in the temporal region, in the depression posterior and inferior to the mastoid process, at the level of the inferior margin of the auricle. The needle is inserted obliquely 0.3–0.5 cun initially into the superior posterior aspect of sternocleidomastoid and then deeper into the splenius capitis and the longissimus capitus muscles. Caution when needling deeper in this region to avoid greater auricular or lesser occipital nerves or the posterior auricular artery and vein. Also consider the proximity to the vertebral artery and vein.

BL10 (Tian Zhu) is located in the nuchal region, on the lateral border of the trapezius muscle at the level of the C1/C2 joint, 1.3 cun lateral to the midline (Fig. 6.29). The needle is inserted perpendicularly 0.5–1 cun into the lateral border of the trapezius muscle. Deep to the trapezius muscle the needle may pass into the splenius capitis muscle and then the semispinalis capitis muscle. With a medial oblique insertion, the needle may also reach the obliquus capitis inferior muscle at a deeper level. The needle may pass through several myofascial trigger points in the underlying muscles, any of which may elicit pain referral to the neck and head. Caution: do not needle upward toward medulla oblongata. Branches of the occipital artery and vein and the greater occipital nerve should also be avoided here when needling deeply.

CV23 (Lian Quan) is located on the anterior midline of the neck, in the depression superior to the hyoid bone. The needle is inserted obliquely (toward the tongue root) 0.5–1 cun into the superficial platysma muscle initially and then the digastric muscle before it passes deeper into the mylohyoid muscle. Deeper to this lay the geniohyoid muscles. Caution when needling here is not to damage the cutaneous branches of the transverse cervical nerve or the

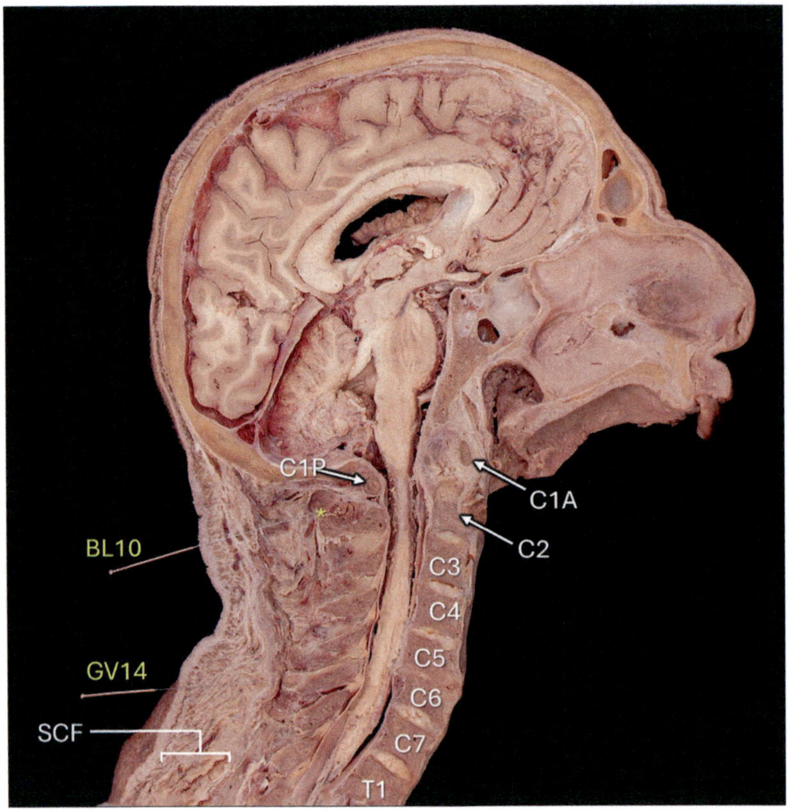

Fig. 6.29 *Sagittal head showing needling of BL10 and GV14*. Needle dimensions 25 mm × 0.25 mm. *C1A* anterior arch of C1 vertebra, *C1P* posterior arch of C1 vertebra, *C2–C7* vertebral bodies of cervical vertebrae, *T1* vertebral body of T1, *SFC* subcutaneous fat. Yellow asterisk shows tip of needle at BL10

branches of the hypoglossal nerve sublingual artery and nerve that lay in the deeper tissues. Also take care to avoid the anterior jugular vein in the region.

6.7.3 The Face

LI19 (He Liao) is located in the oral region, directly inferior to the lateral margin of the nostril, level with GV26 (at the junction of the upper third with the lower two-thirds of the philtrum). The needle is inserted obliquely 0.2–0.3 cun into the orbicularis oris muscle. Care should be taken as branches of the facial vessels are present here.

LI20 (Ying Xiang) is located in the nasolabial groove, 0.5 cun lateral to the nostril, level with the midpoint of the lateral border of the nostril. The needle is inserted obliquely, horizontally, and medially or subcutaneously 0.3–0.5 cun into the levator labii superioris muscle. The maxillary bone may be reached with deep needling in this location. Care should be taken as branches of the facial vessels are present here.

ST2 (Si Bai) is located on the face, superficial to the infraorbital foramen, 0.5 cun inferior to the midpoint of the orbit. The needle is inserted perpendicularly or obliquely transversely, cranially, and laterally 0.3–0.5 cun. The needle is inserted into the orbicularis oculi muscle and may reach the levator labii superioris muscle. Care should be taken not to needle too deeply due to the proximity of the underlying infraorbital foramen or to damage the neurovascular bundle in this area.

ST3 (Ju Liao) is located on the face, level with the nostrils, inferior to ST2. The needle is inserted perpendicularly 0.3–0.5 cun into the zygomaticus minor muscle and if directed deeper then it enters the levator anguli oris muscle. Care should be taken as branches of the facial vessels are present here.

ST4 (Di Cang) is located on the face, in the oral region, 0.5 cun lateral to the corner of the mouth, in the continuation of the nasal labial groove, in line with the midpoint of the orbit. The needle is inserted perpendicularly or obliquely transverse and laterally 1–1.5 cun. The needle may pass into the risorius muscle, the marginal aspect of the orbicularis oris muscle, the zygomaticus major muscle, or at a deeper level into the buccinator muscle. Caution not to damage the parotid duct, facial nerve or facial artery and vein which are located lateral to this point.

ST5 (Da Ying) is located on the cheek, anterior to the angle of the mandible, on the anterior border of masseter muscle. The needle is inserted obliquely 0.3–0.5 cun into the platysma and masseter muscles. Care should be taken not to penetrate the facial artery, vein, or the facial nerve which lies in this region. Clinically this point overlies the myofascial trigger point in the anterior inferior aspect of the superficial masseter.

ST6 (Jia Che) is located on the cheek, 1 cun anterior and superior to the angle of the mandible, in the posterior inferior aspect of the belly of masseter muscle. The needle is inserted perpendicularly 0.3–0.5 cun in the masseter muscle, which has both superficial and deep fibers. With a deep perpendicular insertion, the needle will reach the ramus of the mandible. The needle may also be inserted obliquely 0.5–2 cun anteriorly obliquely toward ST4. When the needle is directed toward ST4 the needle may pass through the risorius, depressor anguli oris, orbicularis oris, and/or zygomaticus major muscles dependent on its angle and path. Caution should be exercised not to damage the buccal branches of the facial nerve. Clinically this point overlies the myofascial trigger point in the posterior inferior aspect of the superficial masseter.

ST7 (Xia Guan) is located on the lateral face, in the depression at the lower border of the zygomatic arch, anterior to the condyloid process of the mandible at the approximate level of the tragus of the ear (Fig. 6.30). ST7 is best located with the mouth open but then the point is subsequently needled with the mouth closed in a relaxed state. The needle is inserted perpendicularly 0.3–0.5 cun, initially through the parotid gland and into the fibers of the deep masseter. Deeper insertion, 1–1.5 cun, however, will direct the needle to the lateral pterygoid muscles. Clinically this point overlies the myofascial trigger point in the deep masseter and at a deeper level, the myofascial trigger points of the lateral pterygoid muscle. The

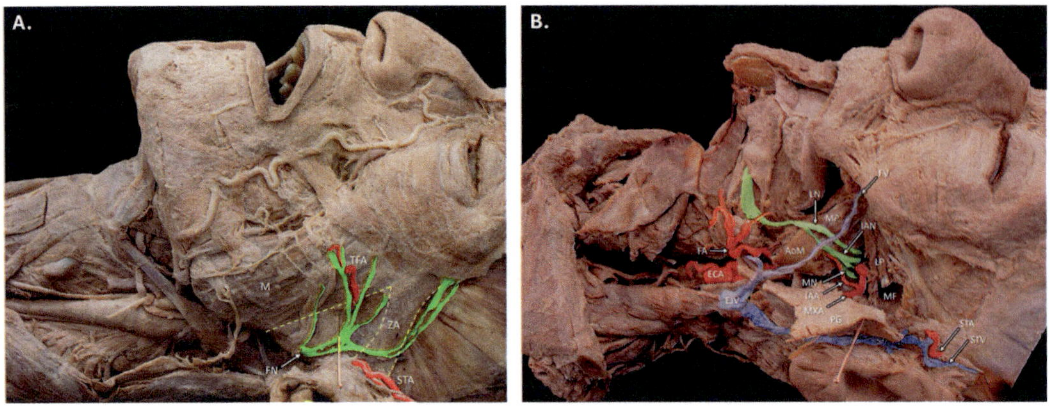

Fig. 6.30 *Lateral head and neck showing superficial and deep needling of ST7.* Needle dimensions 25 mm × 0.25 mm. (**a**) Lateral face with parotid gland removed to reveal facial muscles and branches of the facial nerve. The posterior border of the superficial head of the masseter has been outlined in yellow. (**b**) Deeper dissection with reflected parotid gland to show deeper insertion of needle into lateral pterygoid. The zygomatic arch, condyloid, and coronoid processes of the mandible have been removed to expose the structures in the infratemporal fossa. *AoM* angle of mandible, *ECA* external carotid artery (highlighted red), *EJV* external jugular vein (highlighted blue), *FA* facial artery (highlighted red), *FN* facial nerve (highlighted green), *FV*, facial vein (highlighted blue), *IAA* inferior alveolar artery (highlighted red), *IAN* inferior alveolar nerve (cut, highlighted green), *LN* lingual nerve (highlighted green), *LP* lateral pterygoid, *M* masseter, *MF* mandibular fossa (of temporal bone) *MN* mylohyoid nerve (highlighted green), *MP* medial pterygoid, *MXA* maxillary artery (highlighted red), *PG* parotid gland, *STA* superficial temporal artery (highlighted red), *STV* superficial temporal vein (highlighted blue), *TFA* transverse facial artery (highlighted red), *ZA* Zygomatic arch (outlined in yellow)

lateral pterygoids are involved in protrusion, depression, and unilateral movement of the mandible and may relate to temporomandibular dysfunction and pain in some individuals. Caution should be exercised when needling in this region to avoid the parotid duct, the zygomatic branches of the facial nerve and the transverse facial artery and vein as well as the branches of the mandibular division of the trigeminal nerve.

SI18 (Quan Liao) is located on the face, in the depression inferior to the lower border of the zygomatic bone, inferior to the lateral canthus of the eye on the anterosuperior border of the masseter muscle. The needle is inserted perpendicularly 0.5–0.8 cun into the zygomaticus major and masseter muscles. Be aware of the transverse facial artery and vein, zygomatic branches of the facial nerve, and infraorbital nerve. Also, exercise caution to avoid the buccal artery, vein, and nerve as well as the parotid duct.

SI19 (Ting Gong) is located on the anterior to the tragus of the ear, in the depression between the tragus and the temporomandibular joint when the mouth is open slightly. This point lies inferior to TE21 and superior to GB2. The needle is inserted in a slightly posteriorly oblique direction 0.5–1 cun into the superficial cartilaginous portion of the external acoustic meatus. Take care not to penetrate the anterior auricular nerve or the superficial temporal artery and vein.

BL2 (Zan Zhu) is located on the face, toward the medial end of the eyebrow, on the supraorbital notch. The needle is inserted subcutaneously medially or caudally 0.3–0.5 cun into the occipitofrontalis, depressor supercilii, and orbicularis oculi muscles and more deeply into the corrugator supercilii muscle. Further insertion will bring the needle to the periosteum of the frontal bone. Clinically this point overlies the myofascial trigger points in the occipitofrontalis muscle which may be associated with headache symptoms. Care should be taken not to puncture the supratrochlear and supraorbital neurovascular bundles.

TE21 (Er Men) is located anterior to the ear, in the depression anterior to the supratragic notch and posterosuperior to the condyloid process of the mandible when the mouth is open. The needle is inserted perpendicularly 0.3–0.5 cun or inferior and obliquely into the underlying subcutaneous tissues toward SI19 and then subsequently GB2. Be aware not to penetrate the superficial temporal artery which can be identified with light palpation in this region.

TE23 (Si Zhu Kong) is located in the depression at the lateral end of the eyebrow, along the supraorbital margin in line with the lateral canthus of the eye. The needle is inserted in a horizontal posterior direction laterally subcutaneously into the orbicularis oculi muscle or medially toward Yu Yao at the center of the supraorbital ridge. Take care not to penetrate the temporal branch of the facial nerve, the zygomaticofacial nerve, and branches of the superficial temporal artery and vein. Clinically this point overlies the myofascial trigger point in the orbicularis oculi muscle which can refer pain to the cheek, and upper lip ipsilaterally.

GB2 (Ting Hui) is located anterior to the ear, in the depression anterior to the supratragic notch and posterosuperior to the condyloid process of the mandible when the mouth is open. The needle is inserted perpendicularly 0.3–0.5 cun into the underlying subcutaneous tissues. Avoid needling the great auricular nerve or the superficial temporal artery and vein.

GB3 (Shang Guan) is located on the lateral face, anterior to the ear, on the upper border of the zygomatic arch, directly superior to ST7. The needle is inserted perpendicularly 0.3–0.5 cun into the temporal muscle. Take care not to damage the superficial temporal artery and vein, the auriculotemporal nerve, and the temporal branches of the facial nerve. The deep temporal artery vein and nerve should also be avoided in this region.

GB14 (Yang Bai) is located in the frontal region, in the depression 1 cun superior to the eyebrow, in line with the midpoint of the orbit. The needle is inserted subcutaneously 0.3–0.5 cun into the occipitofrontalis muscle. Clinically this point overlies the myofascial trigger point in the occipitofrontalis muscle that can refer pain from the midline to the superior temporal region. Deeper insertion will take the needle to the periosteum of the frontal bone.

CV24 (Cheng Jiang) is located on the chin, below the middle of the lower lip, in the depression in the center of the mentolabial groove created by the mentalis and orbicularis oris muscles. The needle is inserted obliquely 0.2–0.3 cun into the connective tissue between the left and right mentalis muscles and the left and right depressor labii inferioris muscles and into the inferior aspect of the orbicularis oris muscle. Exercise caution not to penetrate the mental nerve of the inferior labial artery and vein.

GV26 (Shui Gou) is located in the upper philtrum, one-third of the distance from the nose and the top of the lip. The needle is inserted obliquely 0.3–0.5 cun into the orbicularis oris and depressor septi nasi muscles. Take care not to damage the superior labial artery and vein in this region.

6.7.4 Extra Points

EX-HN 3 (Yintang) is located on the anterior midline in the glabella region, midway between the medial ends of the eyebrows. The needle is inserted obliquely 0.5 cun in an inferior direction toward the nasal tip into the procerus muscle.

EX-HN 4 (Yuyao) is located at the midpoint of the orbit within the middle of the eyebrow. The needle is inserted transversely 0.5 cun, either medially or laterally along the eyebrow. The needle passes into the orbicularis oculi muscle initially and may also enter the occipitofrontalis muscle. Care should be taken not to puncture the supratrochlear and supraorbital neurovascular bundles.

EX-HN 5 (Taiyang) is located on the temple, in a depression about 1 cun posterior to the midpoint between the lateral end of the eyebrow and the lateral canthus of the eye. The needle is inserted obliquely 0.5 cun in a lateral direction. The needle will pass into the lateral aspect of the orbicularis oculi and temporalis muscles. Take care not to damage the superficial temporal artery, or the temporal branches of the facial nerve.

6.7.5 Summary

This final section focused on the most common classical acupuncture points found on the head, neck, and face. Due to the underlying anatomy in these areas, particularly the neurovascular structures that can lay relatively superficial beneath the overlying tissues, there are some specific considerations to needling. The risk of incorrect needle placement in the face or neck can result in severe (but ultimately avoidable) neurovascular injury. The point locations described are specific to the underlying tissues and do not permit for generalization of the associated anatomy outside of these areas or points. Whenever needling into purely tender, "Ah Shi," trigger or motor points of the musculoskeletal system that are not within the boundaries of these classical points, one must pay particular attention to the location, particularly in the cervical region, to avoid the underlying neurovascular structures and avoid potential adverse events.

6.8 Conclusion

This chapter is the first text of its kind to clearly illustrate key acupuncture points on human cadavers. Utilized as a cadaveric atlas, it offers a unique insight into the anatomical structures that lie deep or adjacent to common needle points. The cadaveric images highlight palpable landmarks used to identify key points and demonstrate their close relationships to neurovascular structures or underlying viscera.

It can be noted that the key to the safe practice of any acupuncture or dry needling procedure is adherence to the triumvirate of respect: respect the patient, respect the anatomy, and respect the technique (Hobbs 2020). If this triumvirate is adhered to then the potential incidence of adverse events can be significantly reduced. It must also be recognized that technique as it relates to needling is completely reliant on anatomy for the optimization of effectiveness and the maintenance of safety. Therefore, a profound and practical understanding of anatomy of the area a clinician is needling must be in effect. This can be considered particularly important when needling in regions with vulnerable underlying viscera located at relatively shallow depths or in regions of neurovascular structures (Peuker et al. 1999). Peuker and Gronemeyer (2001) noted that all the traumatic adverse events recorded in the literature that they examined could have been avoided if there had been a better application of the knowledge of anatomy. A review in 2003 (Grant and Ma 2003) expanded on this and stated that poor knowledge of anatomy, physiology, pathology, histology, individual abnormalities, and variances and past medical history were likely to lead to the occurrence of avoidable adverse events. This chapter may therefore be considered a resource for clinicians studying acupuncture or dry needling or for those already in practice. Readers can utilize the cadaveric images and extensive discussion of acupuncture points to review and refresh their anatomical knowledge with a view to maintaining their practice of nonmaleficence and beneficence.

References

Abbaszadeh-Amirdehi M, Ansari NN, Naghdi S, Olyaei G, Nourbakhsh MR (2017) Therapeutic effects of dry needling in patients with upper trapezius myofascial trigger points. Acupunct Med 35:85–92

Baldry P (2004) Acupuncture, trigger points and musculoskeletal pain, 3rd edn. Churchill Livingstone, London

Buddle S (2023) Time-lapse dissection videos: traditional practice in a new, digital format. J Vis Commun Med 46(4):197–208. https://doi.org/10.1080/17453054.2024.2304838

Cerezo-Tellez E, Torres-Lacomba M, Fuenes-Gallardo I et al (2016) Effectiveness of dry needling for chronic nonspecific neck pain: a randomized, single-blinded, clinical trial. Pain 157(9):1905–1917

Choi PJ, Iwanaga J, Tubbs RS (2017) A comprehensive review of the sternal foramina and its clinical significance. Cureus 9(12):e1929. https://doi.org/10.7759/cureus.1929

Chou PC, Huang YC, Hsueh CJ, Lin JG, Chu HY (2015) Retrospective study using MRI to measure depths of acupuncture points in neck and shoulder region. BMJ Open 5(7):e007819

Chu H, Kim J, Ha W et al (2018) Change of safe needling depth at Acupoint GB21 according to posture and breathing. Evid Based Complement Alternat Med 2018:230810

Chung KS (2016) Wade-Giles Romanization system. In: The Routledge encyclopedia of the Chinese language, 1st edn. Routledge, pp 756–757

Corder CJ, LaGrange CA (2024) Suprapubic bladder catheterization [Updated 2022 Nov 7]. In: StatPearls [Internet]. StatPearls Publishing, Treasure Island, FL. Available from: https://www.ncbi.nlm.nih.gov/books/NBK482179/

Ellis R, Blyth R, Arnold N, Miner-Williams W (2017) Is there a relationship between impaired median nerve excursion and carpal tunnel syndrome? A systematic review. J Hand Ther (America) 30(1):3–12

Enblom A, Johnsson A (2017) Type and frequency of side effects during PC6 acupuncture: observations from therapists and patients participating in clinical efficacy trials of acupuncture. Acupunct Med 35(6):421–429

Gerber LH, Shah J, Rosenberger W et al (2015) Dry needling alters trigger points in the upper trapezius muscle and reduces pain in subjects with chronic myofascial pain. PM & R 7(7):711–718

Grant A, Ma B-Y (2003) The safe use of difficult & dangerous acupuncture points. J Chin Med 72:1

Hobbs J (2020) The implications and considerations of perineural acupuncture and dry needling with respect to the median and radial nerves at PC 6 and LI 13. Acupunct Physiother 32(2)

Hough AD, Moore AP, Jones MP (2007) Reduced longitudinal excursion of the median nerve in carpal tunnel syndrome. Arch Phys Med Rehabil 88(5):569–576

Jarmey C, Bouratinos I, Pearce L (2021) The definitive guide to acupuncture points: a practical approach, 3rd edn. Lotus Publishing, Chichester

Kessler J, Streitberger K (2008) Perforation of the median nerve with an acupuncture needle guided by ultrasound. Acupunct Med 26(4):231–233

Liao SJ, Lee MHM, Ng LKY (1994) Principles and practice of contemporary acupuncture. Marcel Dekker Inc. Also in White, Cummings & Filshie

Ma YC, Peng CT, Huang YC, Lin HY, Lin JG (2016) Safe needling depths of upper back acupoints in children: a retrospective study. BMC Complement Altern Med 16(1):85

McCormick WF (1981) Sternal foramina in man. Am J Med Pathol 2(3):249–252

Pate D, Kursunoglu S, Resnick D et al (1985) Scapular foramina. Skeletal Radiol 14:270–275. https://doi.org/10.1007/BF00352618

Peuker E, Gronemeyer D (2001) Rare but serious complications of acupuncture: traumatic lesions. Acupunct Med 19(2):103–108

Peuker E, White A, Ernst E, Pera F, Filler TJ (1999) Traumatic complications of acupuncture: therapists need to know human anatomy. Arch Fam Med 8:553–558

Pomeranz B (1987) Scientific basis of acupuncture. In: Stux G, Pomeranz B (eds) Acupuncture textbook and atlas. Springer-Verla, Heidelberg, pp 1–18

Poutoglidou F, Piagkou M, Totlis T, Tzika M, Natsis K (2020) Sciatic nerve variants and the piriformis muscle: a systematic review and meta-analysis. Cureus 12(11):e11531. https://doi.org/10.7759/cureus.11531

Shah JP, Thaker N, Heimur J, Aredo JV, Sikdar S, Gerber LH (2015) Myofascial trigger points then and now: a historical and scientific perspective. PM R 7:746. https://doi.org/10.1016/j.pmrj.2015.01.024

Sperstad JB, Kolberg M, Hilde G, Ellstrom-Engh M, Kari B (2016) Diastasis recti abdominis during pregnancy and 12 months after childbirth: prevalence, risk factors and report of lumbopelvic pain. Br J Sports Med 50(17):1092–1096

Streitberger K, Eichenberger U, Schneider A et al (2007) Ultrasound measurements of the distance between acupuncture needle tip at P6 and the median nerve. J Altern Complement Med 13(5):585–591

Tang CT, Cheng AL (2019) Ultrasound visualization of radial nerve excursion during acupuncture. Med Acupunct 31(3):185–188

Tomaszewski KA, Graves MJ, Henry BM, Popieluszko P, Roy J, Pękala PA, Hsieh WC, Vikse J, Walocha JA (2016) Surgical anatomy of the sciatic nerve: a meta-analysis. J Orthop Res 34:1820–1827. https://doi.org/10.1002/jor.23186

Wang KM, Yao SM, Xian YL, Hou ZL (1985) A study on the receptive field of acupoints and the relationship between characteristics of needling sensation and groups of afferent fibres. Sci Sin B 28(9):963–971

Wang CC, Huang TH, Chiou KC, Chang ZY (2018) Therapeutic effect of superficial acupuncture in treating myofascial pain of the upper trapezius muscle: a randomized controlled trial. Evid Based Complement Alternat Med 2018:9125746

White A, Cummings M, Filshie J (2018) An introduction to Western medical acupuncture, 2nd edn. Churchill Livingstone, Edinburgh

Wilkinson PR (2001) Neurophysiology of pain. Part 1: mechanisms of pain in the peripheral nervous system. CPD Anaesthesia 3(3):103–108

Wu CL, Liao LL, Yang SW (2016) Efficacy of acupuncture treatment on myofascial pain syndrome in upper trapezius. J Med Biol Eng 36(5):10

Yin SC, Park H-J, Seo J-C, Lim S, Koh H-G (2005) An evaluation of the Cun measurement system of acupuncture point location. Am J Chin Med 33(5):729–735

Zhang JH, Yu AS, Zhao YX et al (2002) Study on the anatomic structure and safe needling depth of Jianjing (GB 21) point. Acupunct Res 27(2):152–154

Zhang J, Shang H, Gao X, Ernst E (2010) Acupuncture related adverse events: a systematic review of the Chinese literature. Bull World Health Organ 88:915–921

Zhang ZJ, Wang XM, McAlonan GM (2012) Neural acupuncture unit: a new concept for interpreting effects and mechanisms of acupuncture. Evid Based Complement Alternat Med 2012. https://doi.org/10.1155/2012/429412. [WWW document]

Ziaeifar M, Arab AM, Karimi N, Mosallanejad Z (2014) The effect of dry needling on range of motion of neck lateral flexion in subjects with active trigger point in upper trapezius muscle. Jentashapir J Health Res 5(6):e26659

Morphological Variations of the Vertebral Artery: Clinical Implications

7

Bukola Omotoso, Rohen Harrichandparsad, Indheresan Govindsamy Moodley, Nalini Govender, and Lelika Lazarus

Abstract

The vertebrobasilar system is associated with morphological variations such as hypoplasia, duplication, tortuosity, fenestration, asymmetry of the vessels, and a variable posterior inferior cerebellar artery (PICA) origin. Deviation from typical morphologic configuration can predispose or aggravate the development of cerebrovascular diseases such as vertebrobasilar insufficiency, posterior circulation stroke, and aneurysms. In addition, morphologic variations can predispose the arterial system to iatrogenic injuries during endovascular or surgical intervention. This chapter examined the anatomical features of the extracranial (V1–V3) and intracranial (V4) components of the vertebral artery (VA), highlighting the clinical management of pathologies associated with anatomical variations. Using computed tomography and digital subtraction angiography images, we illustrated five clinical cases of VA morphologic variations and associated pathologies. For each type of variation, we describe the clinical management of associated pathologies. Due to the unpredictable characteristics of vascular variations, this information will contribute significantly to the management of VA pathology associated with variation and help in decision-making for surgical intervention around each segment of the VA and the principal branch, PICA.

Keywords

Vertebral artery · Posterior circulation
Anatomical variations · Imaging

B. Omotoso
Discipline of Clinical Anatomy, School of Laboratory Medicine and Medical Science, College of Health Sciences, University of KwaZulu-Natal, Durban, South Africa

R. Harrichandparsad
Department of Neurosurgery, School of Clinical Medicine, University of KwaZulu-Natal, Durban, South Africa
e-mail: harrichandparsad@ukzn.ac.za

I. G. Moodley
Jackpersad and Partners Inc.,, Durban, South Africa

N. Govender
Department of Basic Medical Sciences Faculty of Health Sciences, Durban University of Technology, Durban, South Africa
e-mail: nalinip@dut.ac.za

L. Lazarus (✉)
Discipline of Clinical Anatomy, School of Laboratory Medicine and Medical Science, College of Health Sciences, University of KwaZulu-Natal, Durban, South Africa
e-mail: Ramsaroopl@ukzn.ac.za

L. Shapiro (ed.), *How to use 3D Printing Innovations and Digital Storage to Democratize Anatomy Education*, Biomedical Visualization 6, https://doi.org/10.1007/978-3-031-68501-9_7

7.1 Introduction

The vertebrobasilar system of the brain is vulnerable to morphological deviations including hypoplasia, duplication, tortuosity, fenestration, and asymmetry of the vessels (Kim 2018b; Omotoso et al. 2021a, b; Tasdemir and Cihan 2023). These deviations precipitate the development of cerebrovascular diseases such as vertebrobasilar insufficiency, posterior circulation stroke, and aneurysms (Aghasadeghi and Dehghan 2022; Omotoso et al. 2022; Otu et al. 2023). More specifically, morphological and morphometrical variations of the vertebral artery (VA) segments (V1–V4) have been implicated in arterial injury, cerebrovascular accidents (strokes), aneurysm, and atherosclerosis (Guan et al. 2017; Decarvalho et al. 2019). Treatment modality options for clinically managing VA anomalies when accompanied by morphological and morphometric variations cannot be generalized. Effective management thus requires a case-by-case review of individual patients based on the location of the pathology in relation to the VA variation.

VA dominance and hypoplasia are correlated with basilar bending and posterior circulatory stroke; fenestrations with aneurysms, while tortuosity has been associated with atherosclerosis (Trivelato et al. 2016; Aghasadeghi and Dehghan 2022). However, these anomalies may occasionally coexist, with each segment of the VA having its associated variation. Other morphological variations include an origin of the posterior inferior cerebellar artery (PICA) from the V3 segment, persistent first intersegmental artery, atresia, and isolated VA (ref). Of note, differences in the diameter of the VA at a level superior to the first cervical vertebra may be implicated in the aetiology of vertebrobasilar insufficiency (Uzmansel et al. 2009). Moreover, a convoluted or tortuous VA with substantial medial or lateral displacement is considered dangerous when conducting dorsal and/or ventral procedures to the cervical spine (Sarmiento et al. 2021). Clinical consequences of a progressive medial deviation of the tortuous VA loops include vertebral body erosion, transverse foramen enlargement, and nerve root entrapment (Bruneau et al. 2006) as well as erode medially located bony structures such as the uncinate process (Kiresi et al. 2009). The subsequent erosion in vertebral bony structures due to loop formation and arterial tortuosity may be fatal during cervical instrumentation surgery, especially in older patients, as tortuosity increases with age (Curylo et al. 2000; Alicioglu et al. 2015).

Appropriate imaging of the VA and its variations is thus essential in identifying and managing potential anomalies. To date, commonly utilized imaging modalities include ultrasound, digital subtraction angiography (DSA), computed tomography angiography (CTA), magnetic resonance imaging (MRI), and magnetic resonance angiography (MRA) (Chen et al. 2004; Akgun et al. 2013; Uricchio et al. 2019; Salih et al. 2021). The choice of imaging modality is often dependent on the available resources (Salih et al. 2021). Albeit in the clinical setting where all three modalities are available, the preferred option in acute settings is CTA; MRA is more useful for screening and follow-up, whil DSA is clinically relevant during interventions for suspected injury or disease (Uricchio et al. 2019, Salih et al. 2021). In view of the unpredictable morphological and morphometric vascular variations (fenestration and aneurysm, hypoplasia and ischemic stroke, VA dissection), clinical experiences will improve the management of VA pathology associated with such variations.

This chapter, therefore, aims to describe the anatomical features of the extracranial (V1–V3) and intracranial (V4) components of the VA, emphasizing the clinical management of selected pathologies in the presence of anatomical variations (via a series of CT and DSA case reports) as well as the clinical utility of various imaging modalities during surgical intervention.

7.2 Embryology of the Vertebral Artery

Embryological development of the VAs occurs between the 32nd and 40th gestational days, through a series of longitudinal anastomoses linking the six (or seven) cervical segmental arteries, where each level of the cervical spine is served by a segmental artery (Luh et al. 1999). Other homologous longitudinal arteries formed between the segmental artery adjacent to the VAs include the common carotid, internal carotid, ascending pharyngeal, deep cervical, occipital, ascending cervical, and the anterior/posterior/lateral spinal arteries. The metameric nature of the VA combined with the complex anastomosis of primitive neural vessels forming a single vascular channel, may result in several anatomical variations including fenestrations, tortuosity, hypoplasia, the inconsistent PICA origin, and variations in the level of entering the transverse foramen (neuroangio.org; Accessed Date 2024 March 28). Noteworthy, recent reports from a group of established neurovascular surgeons indicated that some of the vessels associated with fenestration and variation in the level of entering the transverse foramen, may not be embryologically the same as "normal vert" or "true vert" (neuroangio.org; Accessed Date 2024, March 28). It is possible that these vessels are other adjacent longitudinal arteries that later reconstitute the VA depending on the segments at which the variation is located.

For instance, the V3 fenestration commonly described as a simultaneous presence of codominant lateral spinal artery and the "true vert," may later reconstitute the single VA at the V4 segment, such as in case 5 (Lasjaunias et al. 1985). Furthermore, extraosseous VAs are unlike the "true vert" but may be either the deep or ascending cervical artery and are evident in the level of entering the transverse foramen. This type of variation may vary from C5 to C3, as against the typical entry level at the transverse foramen of C6 vertebrae and can affect blood flow hemodynamics and influence arterial disease development (ref). Understanding the developmental anatomy and/or embryogenesis of the VA has the potential to improve the clinicians' perspectives regarding the incidence of variation as well as some anatomical variants associated with VA.

We provide a clinical description and management of five cases of VA morphologic variations and associated pathologies using computed tomography and digital subtraction angiography as the basis of clinical diagnosis. The patients were retrospectively identified from the database of Inkosi Albert Luthuli Central Hospital, Durban, South Africa. The Institutional Research Ethics Committee (University of KwaZulu-Natal Biomedical Research Ethics Committee-Ethical No: BREC/00004897/22) approved the study protocol. Patient informed consent was waived as this chapter utilized retrospective chart reviews. There was no patient contact, and no patient details were released from the images.

Case 1: Fenestration (Vertebra-Basilar Junction Aneurysm and Fenestration)

A 36-year-old female presented at a tertiary care referral hospital with sudden onset severe occipital headache associated with nausea and photophobia; and two witnessed seizures (new onset). The initial evaluation revealed no focal neurological deficits with a GCS of 15/15. A non-contrast brain CT shows a subarachnoid bleed anterior to the brainstem with a small amount of interventricular hemorrhage (IVH). A DSA revealed a fenestration at the vertebrobasilar junction with an aneurysm (at the proximal end) as the cause of the subarachnoid hemorrhage (Fig. 7.1a, b). *The bilateral VA is asymmetric with dominant right VA and hypoplastic left VA. Fenestration at the vertebrobasilar junction has been associated with an increased incidence of aneurysm. The aneurysm was treated with balloon-assisted coiling. Control runs after intervention illustrated complete aneurysm occlusion with preservation of the VAs and distal territory. The patient recovered well following the intervention.*

Fenestration of the vertebrobasilar system is the luminal division of an artery with a common origin into two separate and parallel channels along its course to reunite distally (Omotoso et al. 2022). This anatomical process is due to the incomplete fusion of primitive vessels during

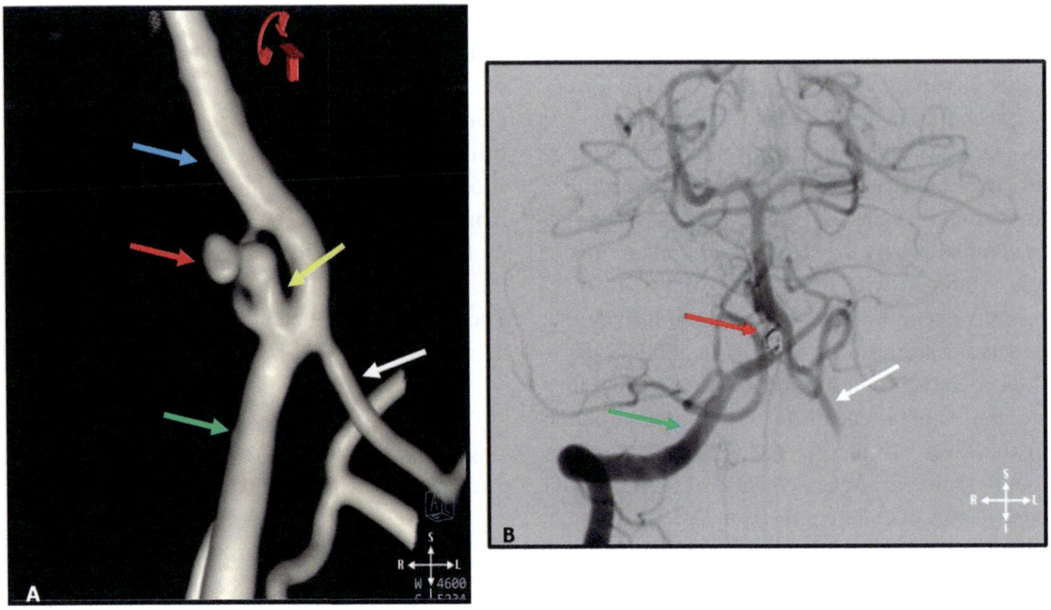

Fig. 7.1 (**a**–**b**) Case 1 illustrating (**a**) anterior view of 3D-DSA reconstructed vertebrobasilar junction showing the dominant right VA (green arrow), aneurysm (red arrow), and fenestration (yellow arrow); and hypoplastic left VA (white arrow). (**b**) posterior view of 2D-DSA of the vertebrobasilar junction and distal territory illustrating complete occlusion of the aneurysm (red arrow) with preservation of normal vertebral arteries (right VA—green arrow; left VA—white arrow) and distal territory. Key: DSA: digital subtraction angiography, VA vertebral artery

embryogenesis (Cooke et al. 2014). Noteworthy is that "fenestration" is used interchangeably with "duplication" in the literature (Ionete and Omojola 2006; Omotoso et al. 2022; Tsantili et al. 2023), however, both terms describe different morphological variations and require clarification. Fenestration involves a luminal division of an artery with a single origin whereas duplication typically involves double vascular origin, which later fuses in the course of the artery (Ionete and Omojola 2006). Fenestrations are commonly observed at the distal segments of the VA (V3, V4), including the vertebrobasilar junction and along the course of the basilar artery (Table 7.1).

The prevalence of basilar artery fenestration demonstrated in post-mortem studies, conventional angiography, MRA, and CTA ranges between 0.28 and 5.26% (Wollschlaeger et al. 1967; Sanders et al. 1993; Uchino et al. 2001; Omotoso et al. 2021). The size of the fenestrated segment may vary between 1 mm and 5 mm, and the shape can present as nearly circular, slit-like, or as long segment separations (Finlay and Canham 1994). Notably, the flow characteristics in the fenestrated segment are determined by the shape, size, length, and position of the fenestration (Tyfa et al. 2018). Furthermore, it was reported that nearly circular fenestration suffered a higher wall shear stress than the slit-like shape (Tyfa et al. 2018). There remains some controversy regarding the clinical relevance of fenestration, as some demonstrate fenestration as an incidental finding on imaging without any pathological significance (ref), and others report that fenestration could be linked to cerebrovascular diseases as well as increase the prevalence of aneurysms, especially at the vertebrobasilar junction (Omotoso et al. 2021; Ye et al. 2021).

In this case scenario, the proximal basilar artery fenestration which is equivalent to vertebrobasilar junction (Fig. 7.1a, b) and may be a potential cause for future aneurysm formation (Tyfa et al. 2018). More recently, hemodynamic differences were observed in the flow between the fenestrated and normal segments, with each limb of the fenestrated segment having a sig-

Table 7.1 Incidence of vertebrobasilar junction fenestration based on ethnicity and gender

Author (Year)	Country	Type of study	Number of cases	Sex Male/Female	Site of fenestration
Campos et al. (1987)	Canadian	DSA	21	7/14	Vertebrobasilar junction
Yoon et al. (2004)	South Korean	CTA	4	2/2	Vertebrobasilar junction
Uchino et al. (2012a)	Japanese	MRA	6	3/3	Vertebrobasilar junction
Consoli et al. (2013)	Italian	DSA	2	F	Vertebrobasilar junction
Kan et al. (2013)	American	DSA	1	F	Vertebrobasilar junction
Gupta et al. (2013)	Indian	CTA, DSA	4	3/1	Vertebrobasilar junction
Trivelato et al. (2016)	Brazilian	DSA	5	1/4	Vertebrobasilar junction
Zhu et al. (2016a)	Chinese	DSA	10	6/4	Vertebrobasilar junction
Karadağ et al. (2022)	–	CTA	7	–	Vertebrobasilar junction
Omotoso et al. (2021b)	South African	CTA	3	1/2	Vertebrobasilar junction

nificant hemodynamic role in the vertebrobasilar system (Dong et al. 2022). These hemodynamic changes may induce flow disturbance, especially at the proximal part of the fenestrated segment (Dong et al. 2022). Moreover, the hemodynamics of basilar artery fenestration using computational fluid dynamics confirms high wall shear stress and oscillatory shear index at the flow confluence of the fenestrated segment, especially at the medial wall (Dong et al. 2022). A consequence of higher wall shear stress and oscillatory shear index values includes the manifestation of multidirectional blood flow disorders, and subsequent endothelial dysfunction (Ku et al. 1985; Tyfa et al. 2018). Moreover, the increased risk of turbulence at the flow confluence of the fenestrated segment is indicative that this location may be a high-risk site for thrombus and unstable plaque formation, and consequent risk of ischemic stroke (Dong et al. 2022). Thus, the clinical consideration of all of the geometric factors associated with the size and location of fenestration is critical when planning surgical and endovascular procedures to treat multivessel brain vascular disorders.

Case 2: Asymmetry in Vertebral Artery Geometry

A 45-year-old female presented with symptomatic basilar invagination and Chiari 1 malformation without syringomyelia. The patient underwent posterior craniocervical junction decompression and fusion. Intraoperatively, it was discovered that C1 arches were thin and subsequently ineligible for instrumentation. Furthermore, the posterior arch of C1 was cartilaginous. Alternatively, C2 pars screws and C3 lateral mass screws were used. Post-operative CT illustrated that the right C2 pars screw was well positioned (Fig. 7.2a, b). *However, the left screw is slightly too lateral, with resultant encroachment of the vertebral foramen on the left. As a result, potential VA injury must be excluded with a CT angiogram. Post-operative evaluation shows that the patient is asymptomatic. This may be explained by a hypoplastic left VA, while the right VA is the dominant artery. This is one of the reasons why clinicians recommend routine evaluation of the VA with CTA during preoperative planning in patients undergoing posterior cervical surgery, especially when it involves the upper C-spine and the craniocervical junction.*

VA asymmetry is a congenital structural variation of the VA characterized by a difference in diameter between bilateral VAs of the same individual (Omotoso et al. 2021a). Of note when the asymmetry is apparent, the artery with the larger lumen is dominant, while the other is non-dominant and sometimes hypoplastic. Both left dominance, right dominance as well as codominance have been reported in population studies (Table 7.2). In the case of a dominant and hypoplastic VA, the dominant VA proceeds to form the basilar artery with little contribution from the hypoplastic VA, leading to basilar artery curvature or bending (Omotoso et al. 2021c; Anirudh et al. 2022; Meng et al. 2018). The frequency of VA dominance is higher in posterior circulation ischemic stroke patients (Sun et al. 2022). Moreover, the VA dominance and basilar artery bending may be associated with migraine pathophysiology including increased risk for posterior circulation infarction in migraineurs (Otu et al. 2023). It is thus imperative to identify and protect the dominant VA during surgical intervention in the vicinity of the VA. Damage to the dominant VA can clinically complicate the vascular circulation in the posterior circulation territory. Furthermore, iatrogenic injury to the VA can be life threatening if the dominant artery is involved in the case of asymmetry (Guan et al. 2017). Instrumentation procedures may require a case-by-case review of individual VAs to select a preferred intervention site so that the dominant VA can sustain blood flow in the vertebrobasilar system in case of injury to one of the VAs. In addition, the dominant VA must not be ligated when repairing VA injury, as it can result in permanent neurologic deficits (Decarvalho et al. 2019).

Case 3: Vertebral Artery Tortuosity

A 67-year-old female presented with a history of worsening headaches and confusion. A non-contrast CT brain scan shows a hyperdense lesion in the posterior third ventricle. An emergency ventriculoperitoneal shunt was inserted to address the hydrocephalus. An MR angiogram scan confirmed an intracranial aneurysm. The patient was planned for endovascular coiling, and standard access was performed on the right groin with a 6F sheath. The bilateral VA were of equal caliber and were very tortuous. The origin of the left VA was tortuous, with an unfavorable angle dipping inferiorly and then ascending superiorly (Fig. 7.3a–d). *The right VA was also very tortuous, with very poor stability of the guide catheter due to tortuosity of the aortic arch. However, the angle was favorable for access into the right VA from the right radial artery. Despite the multiple loop formation* (Fig. 7.3a–d), *sufficient proximal support of the guide catheter was obtained by accessing the right VA from the right radial artery. This allowed the coiling of the aneurysm with a microcatheter. The patient recovered well and was discharged in a stable condition 2 days after the procedure.*

The anatomical course of the VA is straight, except for the V3 segment, which demonstrates a serpentine course while passing through the transverse process of the atlas and axis (Omotoso 2021). This anatomical feature allows for movement of the head and neck without altering blood flow in the posterior circulation territory. However, besides the natural wide loop formed at the V3 segment, the VA can be tortuous along its course at the V1, V2, and V4 segments and may range from mild tortuosity and extend to a loop and coil formation (Omotoso 2021). The formation of VA tortuosity may be congenital or acquired and is an uncommon anatomic variation mostly reported at the V1, followed by the V2

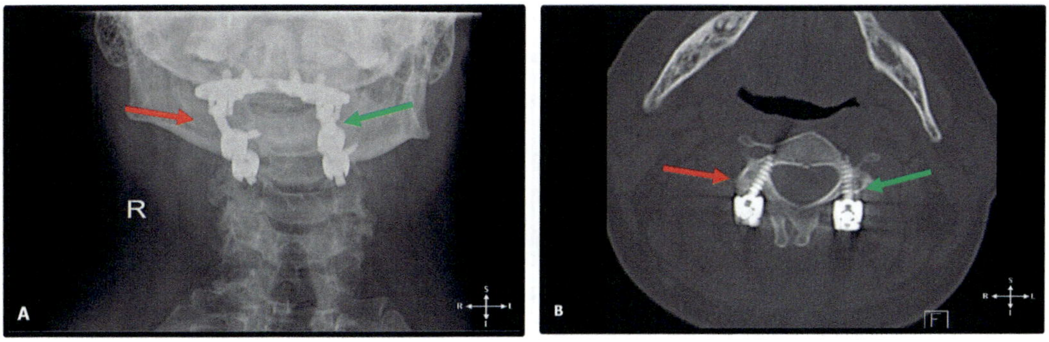

Fig. 7.2 (**a**–**b**) Case 2 illustrating CT posterior and inferior view of (**a**, **b**) the right C2 pars screw well positioned (red arrow) and (**a**, **b**) left C2 pars screw slightly too lateral with resultant encroachment of the transverse foramen of the left VA (green arrow). Key: CT: computed tomography, VA vertebral artery

Table 7.2 Vertebral artery asymmetry and pattern of dominance based on gender and ethnicity

Author (Year)	Sample size	Sex M/F	Study population	Type of study	VA Dominance Pattern	Total percentage
Ergun et al. (2016)	254	133/121	Turkish	DSA	Right	49.6%
Wang et al. (2016)	2370	1348/1022	Chinese	CT	Left	27.9%
Zhu et al. (2016b)	78	–	Chinese	MRA	Left	56.4%
Meng et al. (2018)	478	265/213	Chinese	MRI/MRA	Left	–
Abd El Gawad et al. (2019)	100	69/31	Egyptian	MDCTA	Co-dominance	89%
Vural and Çiçek (2019)	844	290/554	Turkish	Doppler ultrasonography	Left	–
Ngo et al. (2020)	108	–	Korean	MRA	Left	–
Omotoso et al. (2021a)	554	307/247	South African	MDCTA	Left	57.6
Şahin and Gökçe (2021)	215	146/69	Turkish	DSA	Left	47.9%
Tasdemir and Cihan (2023)	322	151/171	Turkish	MDCT	Left	39.8%

segment, and rarely reported at V4 (Table 7.3). The mechanism of tortuosity formation is imprecise (Kim et al. 2010; Han 2012), and thus difficult to assess the suggested congenital origin since vascular tortuosity increases with age (Curylo et al. 2000; Alicioglu et al. 2015; Elgueta et al. 2018). It was previously reported that vertebrobasilar artery tortuosity develops as a result of connective tissue disorders (Morris et al. 2011).

Loop formation can lead to significant and potentially dangerous medial or lateral artery displacement (Bruneau et al. 2006). For instance, failure to recognize a medially located VA may result in a life-threatening iatrogenic injury during decompression. VA loops may potentially increase the risk of damage during surgery, as well as cause neurovascular entrapment and vertebrobasilar insufficiency (Paksoy et al. 2003; Kim et al. 2010). Progressive medial deviation of loops of the tortuous VA may also cause vertebral body erosion, transverse foramen enlargement, and nerve root entrapment (Bruneau et al. 2006; Kiresi et al. 2009). Erosion in vertebral bony structures due to tortuosity can result in fatal complications during cervical instrumentation surgery, especially in older patients (Curylo et al. 2000; Alicioglu et al. 2015). In the case presented, bilateral tortuosity of the VA (Fig. 7.3a–d), is relatively rare and may influence the choice of endovascular approaches in managing basilar

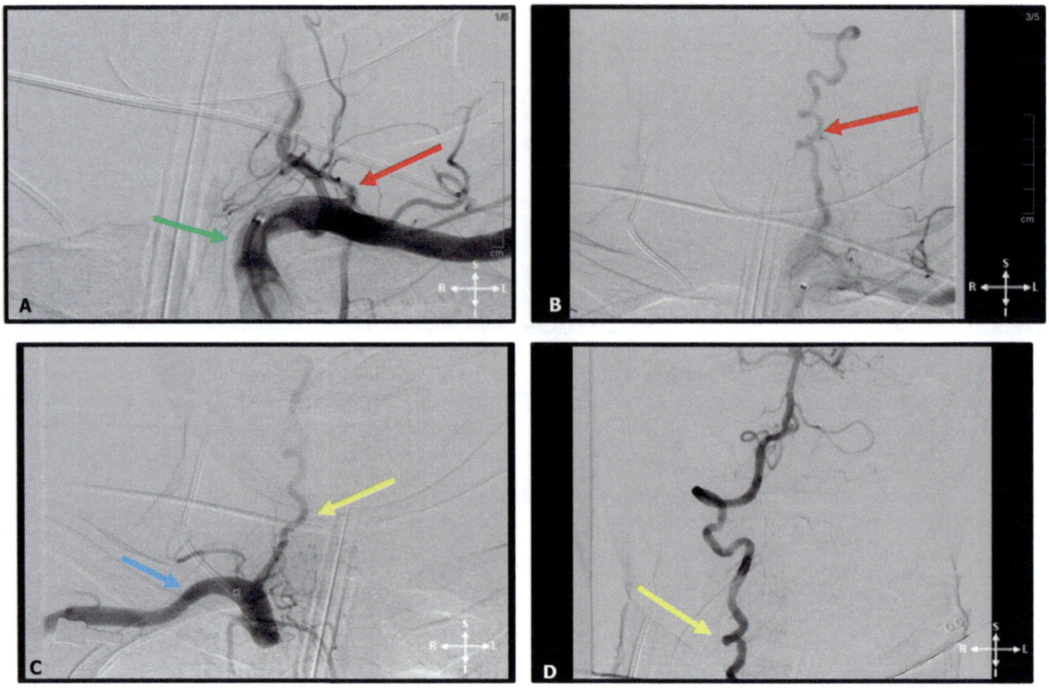

Fig. 7.3 (**a–d**) Case 3 illustrates DSA of proximal and distal end of the right and left VA, demonstrating (**a**) tortuous origin of the left VA (red arrow) arising from the subclavian artery (green arrow), (**b**) tortuous left VA along its course with multiple medial loop formation at the V2 segment (red arrow), (**c**) right VA (yellow arrow) arising from the subclavian artery (blue arrow), and (**d**) tortuous right VA with lateral loops at the V2 segment (yellow arrow). Key: DSA: digital subtraction angiography, VA vertebral artery

Table 7.3 Incidence of vertebral artery tortuosity based on ethnicity and gender

Author (year)	Country	Type of study	Sample size	Sex Male/ Female	Segment of VA Cervical level	Incidence (%)
Paksoy et al. (2003)	Turkish	MRI	173	4/9	V1	7.51
Bruneau et al. (2006)	France/ Belgium	MRI	250	–	V2	2
Ono et al. (2009)	Brazilian	MRI (Case report)	–	F	V2	–
Hong-tao et al. (2014)	Chinese	MRA (Case report)	2	1/1	V2, V4	–
Alicioglu et al. (2015)	Turkish	CTA	110	10/5	V2	13.6
Wakao et al. (2014)	Japanese	CTA	1054	4/6	V2	1
Yenigun et al. (2016)	Turkish	MRA	37	–	V1, V2	78.3, 21.6
Omotoso et al. (2021a)	South African	MDCTA	554		V1, V2	76.6,32.1

artery pathologies like aneurysms. Evaluating bilateral VA origins, the presence of dominant VA, the angle of origin from the subclavian artery, and the degree of tortuosity can improve clinical decision-making regarding which VA to access when targeting a lesion in the posterior circulation. Additionally, it may be preferable to use a radial/distal radial access rather than the standard femoral access if the anatomy is suitable.

Case 4: Posterior Inferior Cerebellar Artery (PICA) Aneurysm and Variable Origin from the Vertebral Artery

A 45-year-old male was found on the side of the road. The initial evaluation at the hospital revealed no external signs of trauma, and the patient was able to move all limbs with a GCS of 12/15 (E4M6V2). A non-contrast brain CT shows blood predominantly in the posterior fossa anterior to the brainstem, with intraventricular hemorrhage mainly in the fourth ventricle. An emergency external ventricular drain (EVD) was inserted on admission, and GCS improved to 14/15 (E4M6V4). CT angiogram revealed no obvious aneurysm but a spasm of bilateral VA, which is worse on the right. A formal digital subtraction angiogram was indicated. Right VA run reveals an aneurysm arising from the proximal segment of the posterior inferior cerebellar artery (PICA) (Fig. 7.4a). *This type of aneurysm is rare and represents 0.5% of all intracranial aneurysms. It is mostly dissecting in nature, and treatment often involves parent vessel occlusion or sacrifice with the risk of lateral medullary syndrome, also known as Wallenberg syndrome or posterior inferior cerebellar artery syndrome. The PICA origin from the right VA is at the oblique (C1) part of the V3 segment* (Fig. 7.4a). *PICA of extradural origin is less likely to supply brainstem perforators and is extremely unlikely to supply eloquent territories. In the present case, no perforators to the brainstem arise from the PICA on the right; instead, they arose directly from the intracranial VA (V4). This means that parent vessel occlusion of the right PICA will be well tolerated if needed. However, in this case, simple coiling with preservation of the PICA was performed. Left VA run is normal, aside from mild spasms secondary to the subarachnoid hemorrhage* (Fig. 7.4b). *The patient recovered well with no focal neurological deficits, and at 3 months follow-up, the patient improved to GCS 15/15.*

The PICA is the largest branch of the VA and usually arises from the intracranial part of the VA (V4 segment) (Omotoso et al. 2023). The PICA is of significant clinical importance and is involved in many diseases, including ischemic stroke, aneurysm, neurovascular compression syndrome (NVCS), arteriovenous malformation (AVM), and brain tumors (Lui et al. 2019). However, variable origins of the PICA may exist from different parts of the suboccipital segment (V3 segment, including the oblique (or C1), horizontal, and vertical parts) (Miao et al. 2020; Omotoso et al. 2023) (Table 7.4). The incidence of extracranial origin of PICA is between 0.4% and 20.8% (Miao et al. 2020, Omotoso et al. 2023) (Table 7.4). The extradurally located PICAs are also more susceptible to injury during posterior approaches to the lower brainstem and upper cervical spine (Miao et al. 2020). The trunk of PICA gives rise to perforating, choroidal, and cortical branches (Lui et al. 2019). The medullary perforators or perforating arteries to the lateral medulla typically originate from the PICA of intracranial origin. However, when PICA has an extradural origin, the medullary perforators will likely arise directly from the intracranial VAs (V4 segment) (Mercier et al. 2008). The possibility of the variable origins of PICA indicates that the perforators can either arise from the PICA or the V4 segment of the VA. In the latter case, sacrificing the V4 segment during intervention requires high-quality imaging to exclude the presence of perforators in the location of planned sacrifice and may be clinically useful for the treatment of PICA aneurysms. Even though PICA aneurysms are rare and challenging to manage, understanding the anatomy is essential for decision-making. Clinically the PICA is surgically exposed during various approaches and is vulnerable to damage. Thus, care must be taken to protect the medullary perforators during surgical intervention or endovascular procedure (Zhao and Tong 2022). Knowledge of the variable origin and preservation of the perforators that branch from the PICA or the VA are important during surgical interventions.

Iatrogenic injury to the medullary perforators can result in multiple Wallenberg's syndrome, also known as Lateral Medullary Syndrome (LMS) or posterior inferior cerebellar artery

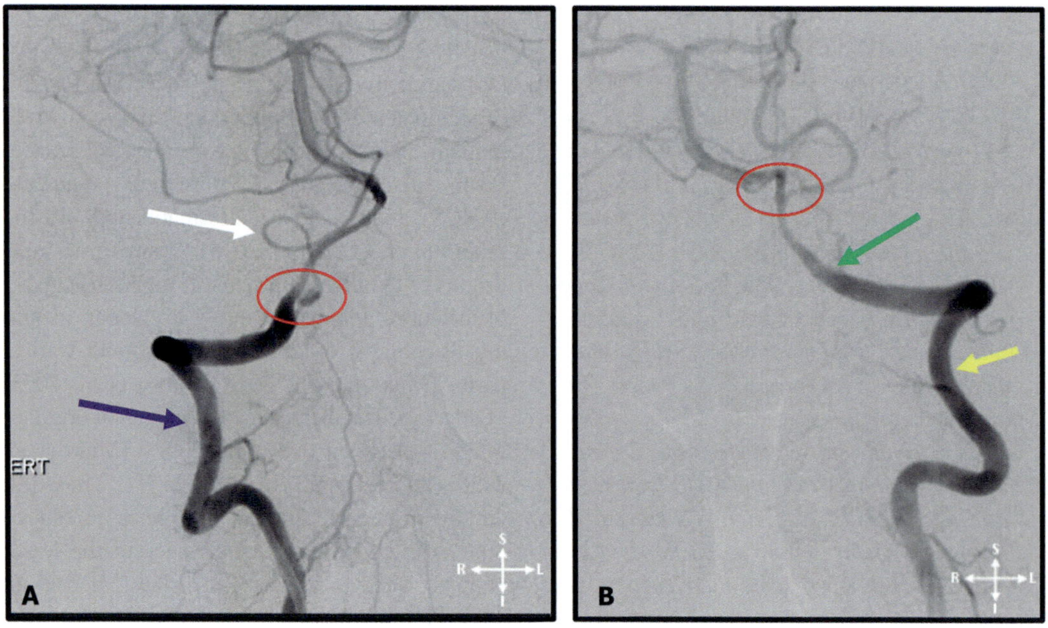

Fig. 7.4 (**a–b**) Case 4 illustrates the bilateral distal DSA VAs and the basilar arteries, demonstrating (**a**) an aneurysm arising from the proximal segment of the right PICA and PICA origin from the oblique part of V3 (red circle); the V3 segment of the right VA (purple arrow) and the PICA (white arrow). (**b**) demonstrates the V3 segment of the left VA (yellow arrow), the V4 segment of the left VA (green arrow), and the origin of the left PICA from the V4 segment (red circle). Key: DSA: digital subtraction angiography, PICA posterior inferior cerebellar artery, VA vertebral artery

syndrome (Benglis and Levi 2010). This syndrome is the most prevalent posterior circulation ischemic stroke and is the typical presentation of PICA territory infarcts (Miao et al. 2020). Almost 50% of the estimated 20% of ischemic strokes reported, appear in posterior circulation and manifest as LMS (Lui et al. 2019). This may be due to vascular damage to the lateral part of the medullary oblongata as a result of atherothrombotic occlusion of the VA or PICA and, less often, by medullary arteries (Saleem and Joe 2019; Lui et al. 2019) as well as by an isolated PICA dissection (Saleem and Joe 2019). Dissecting aneurysms and the extradural origin of the PICA are also considered risk factors of LMS (Komiya et al. 1988; Razak et al. 2011).

Case 5: Bilateral Persistent First Intersegmental Artery

Bilateral persistent first intersegmental artery is a rare variation of the VAs at the craniocervical junction (CCJ). In this case, it was found incidentally during DSA in a patient with right thalamic arteriovenous malformation (AVM). The left and right VA angiograms show persistent first intersegmental artery at C1/C2 bilaterally (Fig. 7.5a, b). *Prior knowledge and imaging of the VAs at the CCJ is essential for spine surgeons when planning posterior cervical decompression to be aware of potential extradural locations and rare variations of the VAs.*

Table 7.4 Prevalence of fenestration, variable PICA origin, and persistent first intersegmental artery at the suboccipital (V3) segment of the vertebral artery (VA)

Author (Year)	Sample size	Sex M/F	Study population	Type of study	Anatomical Variations (Number of patients)			Total percentage of variation (%)
					FIA	PICA	FEN	
Tokuda et al. (1985)	300	–	Japanese	DSA	2	2	3	2.3
Hong et al. (2008)	1013	446/567	South Korean	CTA	48	2	6	5.5
Yamaguchi et al. (2008)	140	69/71	Japanese	CTA	4	23	1	20
Duan et al. (2010)	98	66/32	Chinese	CTA	3	0	3	6
Uchino et al. (2012b)	2739	1615/1124	Japanese	MRA	87	30	25	5
O'Donnell et al. (2014)	975	591/384	American	CTA	1	4	1	0.6
Wakao et al. (2014)	387	269/118	Japanese	CTA	7	5	5	4.4
Fortuniak et al. (2016)	1800	968/832	Polish	CTA	0	11	3	0.8
Hong et al. (2016)	123	–	South Korean	CTA	3	1	0	3.6
Kim (2016)	546 314	–	South Korean	CTA	7 8	11 9	2 2	3.7 6
Vaněk et al. (2017)	511	328/183	Czech	DSA	2	21	1	4.7
Zhu et al. (2018)	678	–	Japanese	DSA	6	10	12	4
Isaji et al. (2018)	142	76/66	Japanese	CTA	4	27	1	22.5
Zhang et al. (2018)	200	105/95	Chinese	CTA	7	0	0	3.5
Xu et al. (2018)	100	61/39	Chinese	CTA	2	1	5	7
Arslan et al. (2019)	200	112/88	Turkish	CTA	1	2	0	1.5
Omotoso et al. (2021c)	554	307/247	South African	MDCTA	5	16	2	4.2

Key: FIA: persistent first intersegmental artery, FEN: fenestration, PICA: Posterior inferior cerebellar artery, MDCT—multidetector computed tomography, MRI—magnetic resonance imaging MRA—magnetic resonance angiography, DSA—digital subtraction angiography, CT—computed tomography

An abnormal intradural course of the V3 segment of the VA through the spinal canal between the vertebral body of C2 and C1 without passing through the transverse foramen of C1 has been previously defined with diverse terminologies such as an aberrant VA coursing intradurally at the C2 level (Kim 2016) and persistent first intersegmental artery (FIA) (Uchino et al. 2012a, b; Fortuniak et al. 2016). The latter is the most frequently used description. The prevalence of this variation ranges between 0.01% and 3.2% (Uchino et al. 2012a, b; O'Donnell et al. 2014; Arslan et al. 2019). In a recent systematic review, FIA was reported as the most common vascular variation at the suboccipital segment of the VA, with an overall incidence of 1.8% compared with fenestration and extradural PICA origin (Omotoso et al. 2023) (Table 7.4). This variation can be present without clinical symptoms and is sometimes symptomatic when it coexists with a disease condition and other vascular variations such as arteriosclerotic dilatation and tortuosity (Kim 2018a, 2018b).

In this case, bilateral persistent FIAs (Fig. 7.5), have been previously reported to

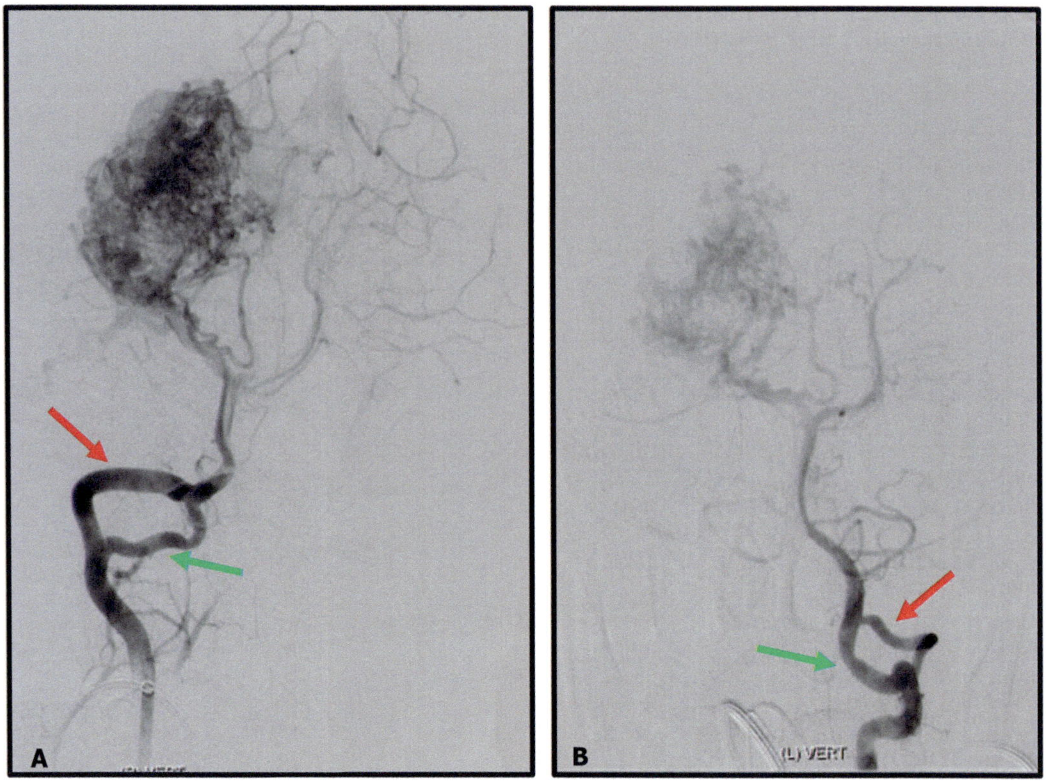

Fig. 7.5 (**a**–**b**) Case 5 illustrating DSA of the bilateral distal VAs (V3 andV4) and the basilar arteries, demonstrating (**a**, **b**) one limb of the fenestrated segment and FIA, the "true" VA with a typical course through the transverse foramen of the C1 (red arrows) and (**a**, **b**) the second limb premised as lateral spinal artery and segmental artery (green arrows). Key: DSA: digital subtraction angiography, PICA posterior inferior cerebellar artery, VA vertebral artery, FIA persistent first intersegmental artery

cause cervical cord myelopathy due to vascular compression when it coexists with other osseous anomalies at the craniovertebral junction (Takahashi et al. 2003). Persistent FIA has also been associated with cervical pain, occipital neuralgia, and accessory nerve palsy (Takahashi et al. 2003), which may be attributed to various embryological bases. Of note, when the first intersegmental artery persists without sufficient development of the typical VA branch during embryogenesis, the VA takes a variant course that enters the spinal canal between the C1 and C2 vertebral bodies (Uchino et al. 2012a, b). The simultaneous persistence of the FIA and the typical branch of the VA subsequently results in fenestration at the V3 segment of the artery (Uchino et al. 2012a, b). Earlier studies suggested that the intradural course of the distal VA at the C2 level is related to variations in size and connection of the lateral spinal artery (LSA) (Lasjaunias et al. 1985) or posterior spinal artery (PSA) (Siclari et al. 2007).

Understanding the embryogenesis of the lateral spinal artery is instrumental in improving the clinical evaluation of the various vascular variations encountered at the craniovertebral junction (neuroangio.org). For instance, the C1 origin of PICA has been described as a LSA segment. In this case of apparent VA fenestration at the C1 level (Fig. 7.5a, b), the more inferomedial vessel (green arrow) has been described as two separate arteries while the ascending region is the LSA, and the descending region is the segmental artery. The more lateral part (red arrow) is regarded as the "true" VA with a typical course through the transverse foramen of the C1 before proceeding intradurally (Fig. 7.5a, b). In the case of persistent FIA without fenestration (i.e., VA becomes intradural at C1), the "true" VA may be entirely missing; and the LSA acts like a VA and the unilateral persistent FIA can be easily overlooked (Uchino et al. 2012a, b). An awareness of this variant's anatomy and a careful review of images will assist in proper identification to prevent VA injury.

7.3 Conclusion

The anatomy of the vertebrobasilar tree is extensively variable and unpredictable. Each of the segments (V1–V4) has associated variations. It is important to highlight the clinical implications of morphologic variations such as fenestration, asymmetry, tortuosity, variable PICA origin, and FIA. This information is useful for diagnosing associated pathologies and planning surgical interventions in the vicinity of the VA to promote successful outcomes.

References

Abd El Gawad FA, Shaaban MH, Shuaib DM, Shallan HM (2019) Anatomical variations of the vertebral artery and its relation to the atlas vertebra-radiological and dry bone study. Eur J Anat 23:49–58

Aghasadeghi F, Dehghan A (2022) Evaluation of the association between the internal carotid artery and vertebral artery tortuosity and acute ischemic stroke using tortuosity index. Vascular 17085381221140163:424

Akgun V, Battal B, Bozkurt Y, Oz O, Hamcan S, Sari S, Akgun H (2013) Normal anatomical features and variations of the vertebrobasilar circulation and its branches: an analysis with 64-detector row CT and 3T MR angiographies. Sci World J 2013

Alicioglu B, Gulekon N, Akpinar S (2015) Age-related morphologic changes of the vertebral artery in the transverse process. Analysis by multidetector computed tomography angiography. Spine J 15:1981–1987

Anirudh EE, Omotoso BR, Lazarus L (2022) Radiological anatomy of the vertebrobasilar artery in a select South African population. Transl Res Anat 27:100205

Arslan D, Ozer MA, Govsa F, Kitis O (2019) Surgicoanatomical aspect in vascular variations of the V3 segment of vertebral artery as a risk factor for C1 instrumentation. J Clin Neurosci 68:243–249

Benglis D, Levi AD (2010) Neurologic findings of craniovertebral junction disease. Neurosurgery 66:A13–A21

Bruneau M, Cornelius JF, Marneffe V, Triffaux M, George B (2006) Anatomical variations of the V2 segment of the vertebral artery. J Operative Neurosurg 59:ONS-20-ONS-24

Campos J, Fox AJ, Vinuela F, Lylyk P, Ferguson GG, Drake CG, Peerless SJ (1987) Saccular aneurysms in basilar artery fenestration. Am J Neuroradiol 8:233–236

Chen CJ, Tseng YC, Lee TH, Hsu HL, See LC (2004) Multisection CT angiography compared with catheter angiography in diagnosing vertebral artery dissection. Am J Neuroradiol 25:769–774

Consoli A, Renieri L, Nappini S, Ricciardi F, Grazzini G, Scarpini G, Capaccioli L, Mangiafico S (2013) Endovascular treatment with 'kissing' flow diverter stents of two unruptured aneurysms at a fenestrated vertebrobasilar junction. J Neurointerventional Surg 5:e9–e9

Cooke DL, Stout CE, Kim WT, Kansagra AP, Yu JP, Gu A, Jewell NP, Hetts SW, Higashida RT, Dowd CF (2014) Cerebral arterial fenestrations. Interv Neuroradiol 20:261–274

Curylo LJ, Mason HC, Bohlman HH, Yoo JU (2000) Tortuous course of the vertebral artery and anterior cervical decompression: a cadaveric and clinical case study. Spine 25:2860–2864

Decarvalho SA, Abd-El-Barr MM, Groff MW (2019) Vascular complications in cervical spine surgery (anterior and posterior approach). Complications in neurosurgery. Elsevier

Dong J, Mei Y, Bai X, Tong X, Dmytriw AA, Yang B, Wang T, Patel AB, Yang X, Li M (2022) Hemodynamic differences between basilar artery fenestration and normal vertebrobasilar artery: a pilot study. Front Neurol 12:766174

Duan S, He H, Lv S, Chen L (2010) Three-dimensional CT study on the anatomy of vertebral artery at atlantoaxial and intracranial segment. Surg Radiol Anat 32:39–44

Elgueta MF, Jimenez JO, Wang NN, Lara AP, Chankowsky J, Charghi R, Tran DQ, Finlayson RJ (2018) Anatomical variations of the vertebral artery in the upper cervical spine: clinical relevance for procedures targeting the C1/C2 and C2/C3 joints. Reg Anesth Pain Med 43:367–371

Ergun O, Tatar IG, Birgi E, Hekimoglu B (2016) Evaluation of vertebral artery dominance, hypoplasia and variations in the origin: angiographic study in 254 patients. Folia Morphol (Warsz) 75:33–37

Finlay HM, Canham PB (1994) The layered fabric of cerebral artery fenestrations. Stroke 25:1799–1806

Fortuniak J, Bobeff E, Polguj M, Kośla K, Stefańczyk L, Jaskólski DJ (2016) Anatomical anomalies of the V3 segment of the vertebral artery in the polish population. Eur Spine J 25:4164–4170

Guan Q, Chen L, Long Y, Xiang Z (2017) Iatrogenic vertebral artery injury during anterior cervical spine surgery: a systematic review. J World Neurosurg 106:715–722

Gupta V, Ahuja CK, Khandelwal N, Kumar A, Gupta S (2013) Treatment of ruptured saccular aneurysms of the fenestrated vertebrobasilar junction with balloon remodeling technique: a short case series and review of the literature. Interv Neuroradiol 19:289–298

Han H-C (2012) Twisted blood vessels: symptoms, etiology and biomechanical mechanisms. J Vasc Res 49:185–197

Hong JT, Lee SW, Son BC, Sung JH, Yang SH, Kim IS, Park CK (2008) Analysis of anatomical variations of bone and vascular structures around the posterior atlantal arch using three-dimensional computed tomography angiography. J Neurosurg Spine 8:230–236

Hong JT, Kim IS, Kim JY, Lee HJ, Kwon JY, Kim MS, Sung JH (2016) Risk factor analysis and decision-making of surgical strategy for V3 segment anomaly: significance of preoperative CT angiography for posterior C1 instrumentation. Spine J 16:1055–1061

Hong-Tao Z, Shu-Ling Z, Dao-Pei Z (2014) Two case reports of bilateral vertebral artery tortuosity and spiral twisting in vascular vertigo. J BMC Neurol 14:14

Ionete C, Omojola MF (2006) MR angiographic demonstration of bilateral duplication of the extracranial vertebral artery: unusual course and review of the literature. Am J Neuroradiol 27:1304–1306

Isaji T, Yasuda M, Kawaguchi R, Aoyama M, Niwa A, Nakura T, Matsuo N, Takayasu M (2018) Posterior inferior cerebellar artery with an extradural origin from the V3 segment: higher incidence on the nondominant vertebral artery. J Neurosurg Spine 28:154–159

Kan P, Abla AA, Dumont TM, Snyder KV, Hopkins LN, Levy EI, Siddiqui AH (2013) Double-barrel stent-assisted coiling of a basilar artery fenestration aneurysm. J Neuroimaging 23:496–499

Karadağ A, Şenoğlu M, Turkis ÖF, Kinali B, Güvençer M, Pekçevik Y, Okromelidze L, Middlebrooks EH, Çamlar M, Tanriöver N (2022) Microsurgical perspective for uncommon variations associated with the vertebrobasilar junction: an anatomical and radiologic investigation. Cerrahpaşa Med J 46(2):97–100

Kim MS (2016) Developmental anomalies of the distal vertebral artery and posterior inferior cerebellar artery: diagnosis by CT angiography and literature review. Surg Radiol Anat 38:997–1006

Kim MS (2018a) C2 segmental-type vertebral artery diagnosed using computed tomographic angiography. J Korean Neurosurg Soc 61:194

Kim MS (2018b) Duplicated vertebral artery: literature review and clinical significance. J Korean Neurosurg Soc 61:28

Kim HS, Lee JH, Cheh G, Lee S-H (2010) Cervical radiculopathy caused by vertebral artery loop formation: a case report and review of the literature. J Korean Neurosurg Soc 48:465

Kiresi D, Gumus S, Cengiz SL, Cicekcibasi A (2009) The morphometric analysis of the V2 and V3 segments of the vertebral artery: normal values on MDCT. J Computerized Med Imaging Graphics 33:399–407

Komiya H, Saeki N, Iwadate Y, Sunami K, Yamaura A (1988) Posterior inferior cerebellar artery dissecting aneurysm presenting with Wallenberg's syndrome case report. Neurol Med Chir 28:404–408

Ku DN, Giddens DP, Zarins CK, Glagov S (1985) Pulsatile flow and atherosclerosis in the human carotid bifurcation. Positive correlation between plaque location and low oscillating shear stress. Arteriosclerosis 5:293–302

Lasjaunias P, Vallee B, Person H, ter Brugge K, Chiu M (1985) The lateral spinal artery of the upper cervical spinal cord: anatomy, normal variations, and angiographic aspects. J Neurosurg 63:235–241

Luh G, Dean B, Tomsick T, Wallace R (1999) The persistent fetal carotid-vertebrobasilar anastomoses. AJR Am. J. Roentgenol 172:1427–1432. https://doi.org/10.2214/ajr.172.5.10227532

Lui F, Tadi P, Anilkumar A (2019) Wallenberg syndrome. [updated 2023 June 25]. In: StatPearls [Internet]. StatPearls Publishing, Treasure Island (FL) 2023 Jan-. Available from: https://www.ncbi.nlm.nih.gov/books/NBK470174/. Accessed Date 2024 March 1

Meng X, Ding W, Wu X, DI, P. (2018) Clinical investigation and characterization of vertebrobasilar dolichoectasia and vertebral artery dominance. Discov Med 25:151–158

Mercier P, Brassier G, Fournier H, Picquet J, Papon X, Lasjaunias P (2008) Vascular microanatomy of the pontomedullary junction, posterior inferior cerebellar arteries, and the lateral spinal arteries. Interv Neuroradiol 14:49–58

Miao H-L, Zhang D-Y, Wang T, Jiao X-T, Jiao L-Q (2020) Clinical importance of the posterior inferior cerebellar artery: a review of the literature. Int J Med Sci 17:3005–3019

Morris SA, Orbach DB, Geva T, Singh MN, Gauvreau K, Lacro RVJC (2011) Increased vertebral artery tortuosity index is associated with adverse outcomes in children and young adults with connective tissue disorders. J Circu 124:388–396

Ngo MT, Kwak HS, Chung GH (2020) Change in basilar artery length and bending according to aging and vertebral artery dominance: a longitudinal study. Sci Rep 10:1–8

O'Donnell CM, Child ZA, Nguyen Q, Anderson PA, Lee MJ (2014) Vertebral artery anomalies at the craniovertebral junction in the US population. Spine 39:E1053–E1057

Omotoso BR (2021) An anatomical exploration of the extracranial (V1-V3) and intracranial (V4) components of the vertebral arteries in a select KwaZulu-Natal population. (Doctoral Dissertation) https://researchspace.ukzn.ac.za/items/85a8cb63-f508-4240-b1db-2fc5c0eaae7c. Accessed Date 2024 Apr 30

Omotoso BR, Harrichandparsad R, Satyapal KS, Moodley IG, Lazarus L (2021) Radiological anatomy of the intracranial vertebral artery in a select South African cohort of patients. Scientific reports 11(1):12138. https://doi.org/10.1038/s41598-021-91744-9

Omotoso B, Harrichandparsad R, Moodley I, Satyapal K, Lazarus L (2021a) An anatomical investigation of the proximal vertebral arteries (V1, V2) in a select South African population. Surg Radiol Anat 43:1–13

Omotoso BR, Harrichandparsad R, Satyapal KS, Lazarus L (2021b) Radiological anatomy of the suboccipital segment of the vertebral artery in a select South African population. Eur J Anat 25:553–562

Omotoso BR, Harrichandparsad R, Satyapal KS, Moodley IG, Lazarus L (2021c) Radiological anatomy of the intracranial vertebral artery in a select South African cohort of patients. Sci Rep 11:1–9

Omotoso BR, Harrichandparsad R, Moodley IG, Satyapal KS, Lazarus L (2022) Fenestration of the vertebrobasilar junction detected with multidetector computed tomography angiography. Folia Morphol (Warsz) 81:510–514

Omotoso BR, Harrichandparsad R, Lazarus L (2023) Prevalence of anatomical variations at the suboccipital (V3) segment of the vertebral artery: a systematic review. Neuroradiology 65:1677–1684

Ono SE, Kawasaki CS, Coelho LODM, Lucaski FF, Carvalho Neto A (2009) Widening of intervertebral foramen by tortuous vertebral artery. Arq Neuropsiquiatr 67:115–116

Otu E, Şen S, Örmeci T, Yüzbaşioğlu N (2023) Association between vertebral artery dominance and basilar artery curvature in migraineurs: an anatomical magnetic resonance imaging study. Neuroradiol J 19714009231224444:314

Paksoy Y, Levendoglu FD, Ögün CÖ, Üstün ME, Ögün TC (2003) Vertebral artery loop formation: a frequent cause of cervicobrachial pain. Spine J 28:1183–1188

Razak A, Clark D, Farooq MU, Kassab MY (2011) Wallenberg's syndrome with extradural–extracranial origin of the posterior inferior cerebellar artery. Neurol Sci 32:711–713

Şahin H, Gökçe M (2021) An association between vertebral artery dominance and steno-occlusion of the vertebral artery segments: a retrospective angiographic study. J Clin Neurosci 92:203–206

Saleem F, Joe MD (2019) Lateral medullary syndrome. StatPearls PublishingStatPearls Publishing LLC, Treasure Island (FL)

Salih M, Moore JM, Ogilvy CS (2021) Computed tomography angiography versus digital subtraction angiography as a primary diagnostic tool in nontraumatic subarachnoid hemorrhage: cost-effectiveness analysis study. World Neurosurg 152:e398–e407

Sanders WP, Sorek PA, Mehta BA (1993) Fenestration of intracranial arteries with special attention to associated aneurysms and other anomalies. Am J Neuroradiol 14:675–680

Sarmiento JM, Cohen JD, Babadjouni RM, Quintero-Consuegra MD, Gonzalez NR, Perry TG (2021) Evaluation of tortuous vertebral arteries before cervical spine surgery: illustrative case. J Neurosurg 1

Siclari F, Burger I, Fasel J, Gailloud P (2007) Developmental anatomy of the distal vertebral artery in relationship to variants of the posterior and lateral spinal arterial systems. Am J Neuroradiol 28:1185–1190

Sun Y, Shi Y-M, Xu P (2022) The clinical research progress of vertebral artery dominance and posterior circulation ischemic stroke. Cerebrovasc Dis 51:553–556

Takahashi T, Tominaga T, Hassan T, Yoshimoto T (2003) Cervical cord compression with myelopathy caused by bilateral persistence of the first intersegmental arteries: case report. Neurosurgery 53:234–237

Tasdemir R, Cihan O (2023) Multidetector computed tomography evaluation of origin, V2 segment variations and morphology of vertebral artery. Folia Morphol (Warsz) 82:274–281

Tokuda K, Miyasaka K, Abe H, Abe S, Takei H, Sugimoto S, Tsuru M (1985) Anomalous atlantoaxial portions of vertebral and posterior inferior cerebellar arteries. Neuroradiology 27:410–413

Trivelato F, Abud D, Nakiri G, de Castro Afonso L, Ulhoa A, Manzato L, Rezende M (2016) Basilar artery fenestration aneurysms: endovascular treatment strategies based on 3D morphology. Clin Neuroradiol 26:73–79

Tsantili A, Karampelias V, Samolis A, Chrysikos D, Antonopoulos I, Spanidis Y, Protogerou V, Troupis T (2023) Anatomical variations of human vertebral and basilar arteries: a current review of the literature. Morphologie 107:169–175

Tyfa Z, Obidowski D, Reorowicz P, Stefańczyk L, Fortuniak J, Jóźwik K (2018) Numerical simulations of the pulsatile blood flow in the different types of arterial fenestrations: comparable analysis of multiple vascular geometries. Biocybernetics Biomed Eng 38:228–242

Uchino A, Kato A, Takase Y, Kudo S (2001) Basilar artery fenestrations detected by MR angiography. Radiat Med 19:71–74

Uchino A, Saito N, Okada Y, Kozawa E, Nishi N, Mizukoshi W, Inoue K, Nakajima R, Takahashi M (2012a) Fenestrations of the intracranial vertebrobasilar system diagnosed by MR angiography. Neuroradiology 54:445–450

Uchino A, Saito N, Watadani T, Okada Y, Kozawa E, Nishi N, Mizukoshi W, Inoue K, Nakajima R, Takahashi M (2012b) Vertebral artery variations at the C1–2 level diagnosed by magnetic resonance angiography. Neuroradiology 54:19–23

Uricchio M, Gupta S, Jakowenko N, Levito M, Vu N, Doucette J, Liew A, Papatheodorou S, Khawaja AM, Aglio LS (2019) Computed tomography angiography versus digital subtraction angiography for post clipping aneurysm obliteration detection: a meta-analysis. Stroke 50:381–388

Uzmansel D, Kurtoglu Z, Bagdatoglu C, Kara A, Yildiz A (2009) Unilateral variant vertebral artery with an aneurysm of the basilar tip: a case report. Int J Anatomical Variations 2:96–98

Vaněk P, Bradáč O, De Lacy P, Konopková R, Lacman J, Beneš V (2017) Vertebral artery and osseous anomalies characteristic at the craniocervical junction diagnosed by CT and 3D CT angiography in normal Czech population: analysis of 511 consecutive patients. Neurosurg Rev 40:369–376

Vertebral Artery. https://neuroangio.org/anatomy-and-variants/vertebral-artery/. Accessed Date 2024 March 28

Vural A, Çiçek EED (2019) Is the asymmetry between the vertebral arteries related to cerebral dominance? Turk J Med Sci 49:1721–1726

Wakao N, Takeuchi M, Nishimura M, Riew KD, Kamiya M, Hirasawa A, Kawanami K, Imagama S, Sato K, Takayasu M (2014) Vertebral artery variations and osseous anomaly at the C1-2 level diagnosed by 3D CT angiography in normal subjects. Neuroradiology 56:843–849

Wang L, Zhang J, Xin S (2016) Morphologic features of the aortic arch and its branches in the adult Chinese population. J Vasc Surg 64:1602–1608. e1

Wollschlaeger G, Wollschlaeger PB, Lucas FV, Lopez VF (1967) Experience and result with postmortem cerebral angiography performed as routine procedure of the autopsy. Am J Roentgenol 101:68–87

Xu S, Ruan S, Song X, Yu J, Xu J, Gong R (2018) Evaluation of vertebral artery anomaly in basilar invagination and prevention of vascular injury during surgical intervention: CTA features and analysis. Eur Spine J 27:1286–1294

Yamaguchi S, Eguchi K, Kiura Y, Takeda M, Kurisu K (2008) Posterolateral protrusion of the vertebral artery over the posterior arch of the atlas: quantitative anatomical study using three-dimensional computed tomography angiography. J Neurosurg Spine 9:167–174

Ye D, Huang J, Wang S, Sheng S, Liu M (2021) Cerebral arterial fenestration associated with stroke and other cerebrovascular diseases. Neuro Rep 32:1279–1286

Yenigun A, Ustun M, Tugrul S, Dogan R, Ozturan OJTJOL, Otology (2016) Classification of vertebral artery loop formation and association with cervicogenic dizziness. J Laryngol Otol 130:1115–1119

Yoon SM, Chun YI, Kwon Y, Kwun BDJSN (2004) Vertebrobasilar junction aneurysms associated with fenestration: experience of five cases treated with Guglielmi detachable coils. Surg Neurol 61:248–254

Zhang H, Chai W, Wang S, Wang Y, Li H (2018) Persistent first intersegmental artery (PFIA) visualized by three-dimensional computed tomography angiography in Chinese population. Int J Surg 52:233–236

Zhao H, Tong X (2022) Bypass surgery management of complex proximal and distal posterior inferior cerebellar artery aneurysms. J Craniofacial Surg 33:2049–2054

Zhu D-Y, Fang Y-B, Wu Y-N, Li Q, Duan G-L, Liu J-M, Xu Y, Hong B, Zhao W-Y, Huang Q-H (2016a) Treatment of fenestrated vertebrobasilar junction-related aneurysms with endovascular techniques. J Clin Neurosci 28:112–116

Zhu W, Wang Y-F, Dong X-F, Feng H-X, Zhao H-Q, Liu C-F (2016b) Study on the correlation of vertebral artery dominance, basilar artery curvature and posterior circulation infarction. Acta Neurol Belg 116:287–293

Zhu S-W, Yang Y, Liu Y-G, Cao J-W, Li F (2018) Anatomical features and clinical significance of radiculomuscular artery variants involving the suboccipital segment of vertebral artery. Clin Neuroradiol 28:75–80

Polycystic Ovarian Syndrome: Current Insights

8

Sagaren Naidu, Sapna Ramdin, Thajasvarie Naicker, Jagidesa Moodley, Girija Naidoo, and Nalini Govender

Abstract

Polycystic ovary syndrome (PCOS) is a common endocrine disorder that affects women in the reproductive age group from adolescence to menopause. It is characterized by anovulatory menstrual irregularities and infertility, hyperandrogenism, and polycystic ovaries on ultrasound morphology. It is also associated with significant cardiometabolic risks, including type 2 diabetes, hypertension, dyslipidemia, and obesity. A review of the literature indicates that current management interventions for PCOS are largely symptom based and focus on the improvement of hyperandrogenism, menstrual cycle regularity, and fertility. As the precise cause of PCOS remains unknown, treatment of this disorder targets the amelioration of hormonal imbalances and metabolic dysfunction that characterizes PCOS.

Therefore, lifestyle modifications and hormonal therapy are the most widely used forms of treatment. Metformin is also considered in treatment regimens because of its usefulness in improving insulin resistance, reducing androgen levels and body mass index. The treatment of PCOS should include regular check-ups to effectively manage the symptoms associated with the disorder. The etiology of PCOS is an enigma. However, understanding the pathophysiology allows for a more efficient and earlier diagnosis of women who are vulnerable to developing PCOS. This would greatly aid the management of affected women with PCOS and enhance their quality of life. This chapter provides a narrative review of the multifaceted features associated with PCOS development, the clinical manifestations as well as the management of affected women in the various stages of their reproductive life.

Keywords

PCOS · Pathophysiology · Clinical manifestations · Management · Neuroendocrine Hyperandrogenism

S. Naidu (✉)
Durban Fertility Clinic, Chelmsford Medical Centre 3, St Augustine Hospital, Durban, South Africa
e-mail: sagie@durbanfertilityclinic.co.za

S. Ramdin · G. Naidoo · N. Govender (✉)
Department of Basic Medical Sciences, Durban University of Technology, Durban, South Africa
e-mail: nalinip@dut.ac.za

T. Naicker
Discipline of Optics and Imaging, Doris Duke Medical Research Institute, College of Health Sciences, University of KwaZulu-Natal, Durban, KwaZulu-Natal, South Africa

J. Moodley
Women's Health and HIV Research Group, College of Health Sciences, University of KwaZulu-Natal, Durban, South Africa

L. Shapiro (ed.), *How to use 3D Printing Innovations and Digital Storage to Democratize Anatomy Education*, Biomedical Visualization 6, https://doi.org/10.1007/978-3-031-68501-9_8

8.1 Introduction

Polycystic ovary syndrome (PCOS), is a complex neuroendocrine, metabolic, and reproductive disorder with psychological features affecting 8–13% of women of childbearing age (Ollila et al. 2023; March et al. 2010; Azziz et al. 2006; Teede et al. 2018; Diamanti-Kandarakis et al. 2006; Bozdag et al. 2016). Up to 70% of affected women remain undiagnosed (March et al. 2010), due to the PCOS diagnostic criteria used (Deswal et al. 2020). The most widely accepted guidelines for a clinical diagnosis of PCOS require the presence of a minimum of two out of three principal features, viz. physical or biochemical features of hyperandrogenism, ovulatory dysfunction, and polycystic ovarian morphology (Teede et al. 2018). A range of comorbidities are also associated with the disorder, including insulin resistance, obesity, type 2 diabetes, and cardiometabolic conditions thus contributing to a range of heterogeneous phenotypes. This prevalence is derived from a modest amount of research in high-income countries. Presentation varies by ethnicity and in high-risk populations such as indigenous women where prevalence and complications may be higher (Yasmin et al. 2022). However, recent studies in countries like India (Bharathi et al. 2017; Deswal et al. 2019) and China (Yang et al. 2022) indicate a similar prevalence. A large proportion of some subpopulations remain undiagnosed and unreported as a result of topographical settings (Wolf et al. 2018), race and/or ethnicity, phenotype, genetic and environmental factors combined with the varied criteria used to define PCOS (Kauffman et al. 2002; Ali 2015).

There are several constraints and controversies associated with the diagnosis and management of PCOS emanating from variation of definitions of components of the diagnostic criteria, significant variability in the heterogenous phenotypes with or without obesity, ethnicity, and in clinical features throughout the reproductive life (Teede et al. 2018). These features lead to the diverse diagnosis and management of affected women among various healthcare providers globally (Saini et al. 2016). These difficulties are compounded by an inability to discern the diverse clinical features of PCOS (Brakta et al. 2017), resulting in delayed diagnosis, management, and poor satisfaction of women afflicted by PCOS (Gibson-Helm et al. 2017). Additionally, the wide range of diagnostic guidelines creates challenges when ascertaining its prevalence.

This reproductive phenomenon referred to as "amenorrhea associated with polycystic ovaries" was first described by Stein and Leventhal in 1935, in a case series of seven women found at laparotomy (Stein and Leventhal 1935). It was the first report to link ovarian pathology with clinical hyperandrogenism and oligomenorrhea. This clinical triad formed the basis for PCOS diagnosis characterized by amenorrhea associated with hirsutism and bilateral polycystic ovaries and remains the basis for the diagnosis of PCOS (Stein and Leventhal 1935). During the intervening decades, many aspects of the syndrome remained enigmatic regardless of a plethora of research conducted on PCOS. Since then, a myriad of other symptoms has also been correlated with PCOS such as acne, hirsutism, obesity, and infertility. Collectively these clinical symptoms were called Stein Leventhal Syndrome until the introduction of ultrasound in the late 1960 and early 1970 that described ovarian morphology. A polycystic ovarian morphology was described as having 12 or more follicles of 2–9 mm under ultrasound examination (Balen et al. 2009). Furthermore, not all women with polycystic ovaries have PCOS and not all women with PCOS have polycystic ovaries. There are a broad spectrum of symptoms that range from mild to severe, and some women may only have a few clinical symptoms while others have more. This conundrum makes it difficult for clinicians to make an exact diagnosis of PCOS thus accounting for the discrepancy in prevalence rates. Also, the controversy concerning the diagnosis and treatment contributes to the overall complexities of the syndrome. In this chapter, we provide a narrative review of the multifaceted features associated with PCOS development, the clinical manifestations as well as the management of PCOS is described.

8.2 Methodology

A literature search was done using PubMed and the Google search engines. Keywords including "polycystic ovarian syndrome," "pathogenesis," "risk factors," "genes," and "angiogenesis" were used to find the relevant literature. All review and original reports describing the etiology of PCOS as well as its association with angiogenesis were considered in developing this narrative review.

8.3 Diagnostic Criteria and PCOS Phenotypes

8.3.1 Diagnostic Criteria

Globally, PCOS is diagnosed based on the Rotterdam criteria (2003), which mandates the presence of a minimum of two criteria demonstrated in Fig. 8.1. Numerous attempts were made to classify PCOS and standardize the diagnosis. The diagnostic criteria have been defined three times, by the National Institute of Health (NIH) in 1990, the European Society of Human Reproduction and Embryology (ESHRE), and the American Society for Reproductive Medicine (ASRM) in 2003, known as the Rotterdam criteria and by the Androgen Excess and PCOS Society (AE-PCOS) in 2006 (Azziz et al. 2009; Carmina 2004; Garad et al. 2011; Zawadski and Dunaif 1992). The Rotterdam Criteria has been endorsed by the International consensus group on PCOS in 2018 (Teede et al. 2018). However, the diagnostic criteria differ in terms of clinical, biological, and image-based outcomes (Okoroh et al. 2012). For example, diagnosis is confirmed via the NIH 1990 guidelines only if both hyperandrogenism and chronic anovulation are present, whereas for the AE-PCOS Society 2006, both hyperandrogenism and ovarian dysfunction are necessary for the diagnosis. In contrast, the Rotterdam 2003 and the NIH 2012/International PCOS Guidelines 2018, confirm diagnosis when at least 2 of the clinical factors (hyperandrogenism, oligo- and/or anovulation, and polycystic ovaries) are evident. In 1990, the NIH established the first set of guidelines whilst the Rotterdam 2003 guidlines aimed to augment the NIH diagnostic delineation in contast to the evidence based delineation offered by the AE-PCOS Societyin 2006.

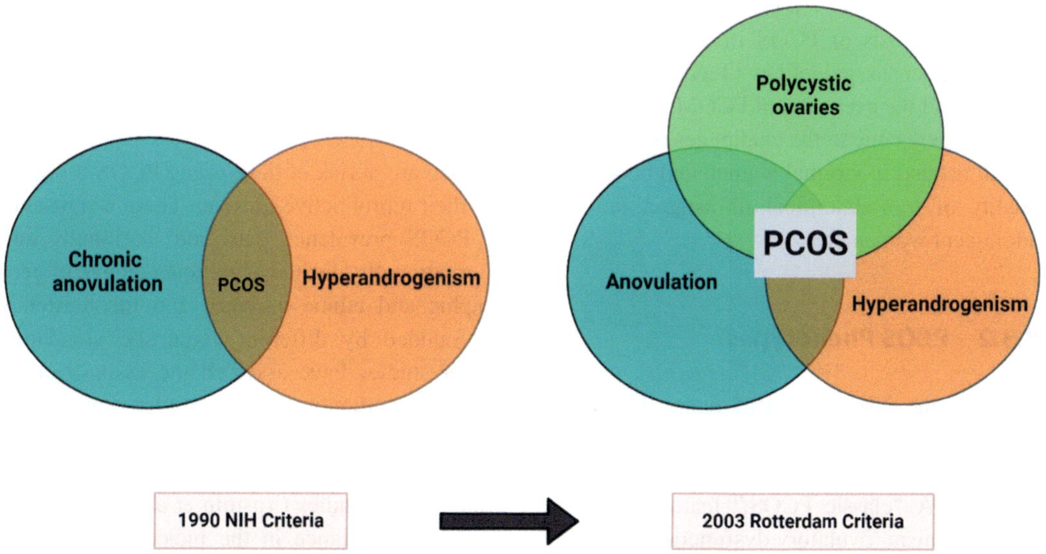

Fig. 8.1 Criteria used for the diagnosis of PCOS, illustrating the shift from the 1990 NIH consensus to the 2003 Rotterdam consensus Adapted from Broekmans et al. (2006). This figure was generated using BioRender (https://biorender.com/)

In 2012, a more topical definition emanated from the NIH Evidence-based PCOS Methodology Workshop Panel, which recommended a comprehensive revision to the Rotterdam 2003 criteria that enhances the sub-phenotype identification (Johnson et al. 2012). Moreover, the various diagnostic groups explicitly discount women with inherent pathologies such as thyroid abnormalities and congenital adrenal hyperplasia that cause hyperandrogenism and/or menstrual abnormalities (Azziz et al. 2009; Draft Summary and Recommendations of International Evidenced-Based Guideline for the Assessment and Management of Polycystic Ovary Syndrome 2018). The Rotterdam criteria were approved by the 2018 International PCOS Guidelines and cautiously advocated for phenotyping of women. Notably, a minimum of 2 years post-menarche, together with hyperandrogenism and ovulatory abnormalities are prerequisites for accurate diagnosis (Draft Summary and Recommendations of International Evidenced-Based Guideline for the Assessment and Management of Polycystic Ovary Syndrome 2018). Hence, the diagnosis of this heterogeneous syndrome continues to be based on the Rotterdam criteria, of hyperandrogenism, anovulatory menstrual irregularities, and/or with polycystic ovaries on ultrasound (Ndefo et al. 2013). However, the NHI criteria are still relevant for the diagnosis of PCOS in adolescents. The recent consensus guidelines (Teede et al. 2023) recommend the exclusion of PCOM as a criterion in adolescents due to the challenges of transvaginal ultrasound in virginal women and poor image quality of transabdominal ultrasound in obese adolescent women.

8.3.2 PCOS Phenotypes

In 2012, the NIH consensus panel decided on four phenotypes:

(a) Type A "classic PCOS," features hyperandrogenism, ovulatory dysfunction, and polycystic ovarian morphology.
(b) Type B only hyperandrogenism and ovulatory dysfunction.
(c) Type C includes hyperandrogenism and polycystic ovaries.
(d) Type D included non-hyperandrogenic PCOS with polycystic ovary morphology and ovulatory dysfunction.

However, these classifications also failed to consider the variations between lean and obese PCOS phenotypes (Guastella et al. 2010; Legro et al. 2013). Obese PCOS patients often manifest with insulin resistance and dyslipidemia, whereas lean PCOS patients exhibit mild and/or no insulin resistance accompanied by a normal lipid profile (Azziz et al. 2016; Teede et al. 2018). Thus, the etiology of PCOS still remains an unexplained condition despite its association with genetic and familial factors, poor dietary control, metabolic conditions, psychological stress, and environmental factors (Bulsara et al. 2021). Lifestyle factors such as physical inactivity and unhealthy diets containing higher carbohydrates and fat serve to increase the risk of insulin resistance and obesity, predisposing women to PCOS (Sirmans and Pate 2014; Pramodh 2020).

In this narrative review, we therefore describe the multifaceted features associated with PCOS development, the clinical manifestations as well as the management of PCOS.

8.4 Prevalence, Ethnicity, and Racial Disparities

Women are at risk of developing PCOS throughout their reproductive lifespan. There is a paucity of PCOS prevalence data both nationally and internationally, due to a significant degree of geographic and ethnic variance and inconsistency confounded by different diagnostic standards. Fewer studies have assessed the association of race and/or ethnicity of this complex multisystem disorder. Limitations of these studies include the small sample size, selection bias, and the lack of comparative studies (Yasmin et al. 2022). Large degrees of variance in the incidence of PCOS observed among different ethnic groups may be attributed to discrepancies in sample size and diagnostic criteria used (Ganie et al. 2019; Sendur and Yildiz 2021). In addition to variations in diagnostic criteria, geographic location, and lifestyle factors such as diet and physical activity also influence the prevalence of PCOS (Siddiqui et al. 2022). In 2018, international evidence-based PCOS guidelines recommended that differences in ethnic and racial background should be taken into account in the presentation of PCOS (Teede et al. 2018).

The prevalence of PCOS in women from different ethnicities, including Asian, Hispanic, African American, Greek Islander, and indigenous Australians has been summarized (Wolf et al. 2018). However, the review did not include any information on women of Middle Eastern and African ancestry. Several individual studies have been conducted to estimate the prevalence of PCOS among the general women population of individual countries in the Middle East such as Oman (Al Khaduri et al. 2014), Qatar (Sharif et al. 2016), Saudi Arabia (Alraddadi et al. 2018), and United Arab Emirates (Saidunnisa et al. 2016). A higher frequency of PCOS and associated complications have been observed in indigenous women of various geographic locations when compared to other communities (Sendur and Yildiz 2021; Davis et al. 2002; Zhao and Qiao 2013). In another systematic review analysis, a 6–10% global PCOS prevalence was reported among adolescent women according to different diagnostic criteria, NIH (6%), Rotterdam (10%), and the AE-PCOS Society (10%) (Bozdag et al. 2016). Additionally, in the United States, a prevalence range of 10.3–47.5% was reported (Okoroh et al. 2012).

A 2019 meta-analysis conducted across 204 countries indicated that approximately 66 million women were confirmed PCOS cases while 2.1 million were incident cases (Safiri et al. 2022). Furthermore in a more recent report, an estimated 19,766.07 PCOS cases were documented in central sub-Saharan Africa (Gao et al. 2023) whereas in Africa as a continent, the prevalence ranges between 16 and 32% (Oriji and Onwuegbulam 2019). The high prevalence rates may be attributed to increased societal understanding and consciousness of the disorder. Of note, the high prevalence may emanate from the varied diagnostic criteria as well as the use of earlier definitions that were limited to morphological confirmation of polycystic ovaries (Ali et al. 2022). In a recent study conducted in Cape Town, South Africa, the data indicates a probable PCOS incidence of 12.2% ($n = 29$) in women of mixed ancestry (Pringle et al. 2022). In 154 Caucasian women from Spain, the NIH diagnostic criteria confirmed a prevalence rate of 6.5% (Asunción et al. 2000). In a retrospective evaluation of women ($n = 728$) at a single maternity hospital, from Adelaide, Australia, the prevalence rates according to each criterion were: NIH (8.7 ± 2.0%), Rotterdam (11.9 ± 2.4%), and AES (10.2 ± 2.2%) (March et al. 2010). A notable increase was observed in the prevalence using the Rotterdam (17.8 ± 2.8%) and AES (12.0 ± 2.4%) criteria, following the inclusion of data from women who did not undergo an ultrasound examination of the ovaries (March et al. 2010).

Racial variations emanate from the expression of 5-α reductase, which assists in the conversion of testosterone to dihydrotestosterone in PCOS pathogenesis (Glintborg et al. 2010; Zhao and Qiao 2013). East Asian women have a lower concentration of 5-α reductase, which may be the reason for their lower level of hirsutism (Zhao and Qiao 2013). In contrast, women of Middle Eastern, Mediterranean, Indian, and South Asian origin are more susceptible to developing PCOS and/or hirsutism compared to East Asian or Caucasian women (Glintborg et al. 2010; Fauser et al. 2012; Zhao and Qiao 2013; Huang and Yong 2016; Chan et al. 2017). Incidence data based on the NIH criteria from a systematic review, highlights the lowest rates in Chinese women (5.6%), followed by Caucasians (5.5%), Middle Eastern (6.1%), and women of African ancestry (7.4%) (Ding et al. 2017). Moreover, using the NIH criteria, the prevalence of PCOS in Mexican women was 6.0% (95% CI: 1.9–10.1%), however, this increased to 6.6% (95% CI: 2.3–10.9%) when assessed using the Rotterdam criteria (Moran et al. 2010a). In contrast, the incidence of PCOS was 13% in Mexican-American women living in Los Angeles (Chan et al. 2017).

High prevalence rates were reported among Indian (Deswal et al. 2019) and indigenous

Australian women (Davis et al. 2002). In rural Haryana, India, 4.21% (n = 94) of women were diagnosed with PCOS out of a sample population of 2400 (Deswal et al. 2019). Additionally, 29% (n = 27) of these women resided in rural regions compared to 71% (n = 67) from urban areas, indicative that Westernized diets and lifestyle preferences may predispose PCOS development (Deswal et al. 2019). In a study conducted among indigenous Australian women, the 26% who were diagnosed with PCOS, also presented with high incidences of obesity, hirsutism, and type 2 diabetes (Davis et al. 2002). Conversely, in a population of 1310 women with PCOS, a 29.1% prevalence was reported in South Asians versus 70.9% in Caucasian women (Mani et al. 2015). The lower prevalence rates observed in South Asian women may be attributed to younger age, reduced systolic and diastolic blood pressures, and lower BMI and/or obesity compared to Caucasian women (Mani et al. 2015). However, hyperandrogenic features and acanthosis nigricans were higher in South Asian than in Caucasian women (Mani et al. 2015). Additionally, genome-wide association studies conducted in women with PCOS have been able to ascertain the presence of 19 specific genetic susceptibility *loci* (Hiam et al. 2019). Additionally, 11 of these specific genetic susceptibility *loci* were observed in European and Han Chinese women, suggestive of a shared genetic basis of PCOS between different racial groups (Hiam et al. 2019; Chen et al. 2011; Goodarzi et al. 2012).

Racial and ethnic diversity are also linked with poor metabolic outcomes (Glintborg et al. 2010; Fauser et al. 2012; Chan et al. 2017; Sendur and Yildiz 2021). Data from a recent systematic review confirmed a higher susceptibility to the development of metabolic syndrome in South Asian, Indian, and Norwegian women with PCOS (Sendur and Yildiz 2021). Of note, African American women appear to be more vulnerable to the onset of hypertension than Caucasian women in the United States (Fauser et al. 2012; Sendur and Yildiz 2021; Wijeyaratne et al. 2002). Notably, Hispanic women have a higher hirsutism predisposition (93.8% vs. 86.8%), abnormal free androgen index (75.8% vs. 56.5%), abnormal homeostasis model assessment (52.3% vs. 38.4%), hyperglycemia (14.8% vs. 6.5%), and lower sex hormone-binding globulin (SHBG) levels compared with non-Hispanic Caucasians (Engmann et al. 2017). Significantly higher rates of metabolic syndrome and hypertriglyceridemia were also reported in Hispanic women compared to non-Hispanic African American women (Engmann et al. 2017).

In summary,

- PCOS is relatively mild in Caucasian women.
- Caucasian women have a higher body mass index in North America and Australia.
- Mediterranean, Hispanic, and Middle Eastern women have more severe hirsutism.
- South East Asian (Indian) and indigenous Australians have increased central obesity, insulin resistance, acanthosis nigricans, type 2 diabetes, and metabolic risks.
- East Asians, especially in China, have milder hirsutism and a lower body mass index.
- African women have a higher BMI and metabolic syndrome.
- African American adolescents and young adults with PCOS have an increased metabolic risk than their Caucasian counterparts.

Thus, understanding the differences in clinical presentation can help with the diagnosis.

8.5 Clinical Features

It is important to appreciate that PCOS is referred to as a syndrome, reflecting different clinical manifestations. The key clinical features include irregular menstrual cycles, hyperandrogenism, and polycystic ovaries on pelvic ultrasonography, infertility as a consequence of anovulation, obesity, and insulin resistance (Table 8.1) (Sirmans and Pate 2014; Walters et al. 2018; Witchel et al. 2019). The etiology remains unknown and may involve a combination of genetic factors, poor dietary control, and metabolic conditions (Bulsara et al. 2021). The clinical complications vary in pregnant and non-pregnant women, manifesting from adolescence to menopause. We provide

Table 8.1 Clinical manifestations and/or symptoms

Reproductive	Metabolic	Psychological
Hyperandrogenism (Teede et al. 2010; Zhang et al. 2019) Ovulatory dysfunction (Ndefo et al. 2013; Shahid et al. 2022) Menstrual dysfunction (Ndefo et al. 2013) Infertility (Ashraf et al. 2019; Shahid et al. 2022)	Insulin resistance (Chen and Pang 2021; Zeng et al. 2020) Dyslipidemia (Chen and Pang 2021) Obesity (Joham et al. 2022; Tong et al. 2022) Type 2 diabetes (Zeng et al. 2020; Kakoly et al. 2019)	Anxiety (Sirmans and Pate 2014) Depression (Teede et al. 2010) Reduced quality of life (Teede et al. 2010)

an overview of menstrual disorders, hyperandrogenism, polycystic ovarian morphology, ovulatory dysfunction, insulin resistance, age categories, and ultrasound morphology since they comprise the major pathological mechanisms.

8.5.1 Menstrual Disorders

Menstrual disorders are the most common clinical manifestation of PCOS and affected women present with irregular bleeding due to chronic anovulation (Kaunitz 2019). Although menstrual abnormalities are common during the reproductive years in these women, the incidence of spontaneous ovulation occurs in at least one-third of cases (Laven et al. 2002). Fairley and Taylor (2003) reported that 70% of women in their cohort presented with irregular menstruation due to chronic anovulation (Fairley and Taylor 2003), inclusive of secondary amenorrhea and oligomenorrhea and menometrorrhagia (American College of Obstetricians and Gynecologists 2018). Oligomenorrhea is defined as menstrual cycles that are greater than 35 days long or 4–9 menstrual cycles a year (Riaz and Parekh 2022), whereas secondary amenorrhea refers to the absence of menses for at least 3 months when pregnancy is excluded (Casas and Chuang 2020). Metrorrhagia is characterized by heavy, irregular menstruation due to breakthrough bleeding from endometrial hyperplasia as a consequence of excessive estrogen stimulation from aromatization of high androgen levels (Singh and Puckett 2023). This menstrual abnormality may vary from light to heavy flow and consequent anaemia. Women presenting with amenorrhea or oligomenorrhea have about a 90% chance of being diagnosed with PCOS and about 95% of affected women will present with amenorrhea or oligomenorrhea (Fauser et al. 2012). Furthermore, the affected women presenting with amenorrhea usually have a more severe form of hyperandrogenism and a more severe PCOS phenotype (Fauser et al. 2012). Of note, the menstrual cycles in women with PCOS normalize as they approach menopause, creating diagnostic challenges in this group (Fauser et al. 2012; Elting et al. 2000).

8.5.2 Hyperandrogenemia

8.5.2.1 Clinical

Excess androgens are a key feature of PCOS and contribute to the clinical phenotype of these patients. Although excessive circulating androgens will contribute to the ovulatory and menstrual dysfunction of these women, the most common signs include hirsutism, acne, and alopecia (female pattern hair loss) (Lizneva et al. 2016). The modified Ferriman-Gallway score is most commonly used to assess hirsutism, as a marker for hyperandrogenism, even when ethnic differences are considered (Yildiz et al. 2010).

Hirsutism is confirmed in almost 70% of women with PCOS, suggestive of the clinical importance of biochemical verification of excess androgen in women suspected of PCOS (Fauser et al. 2012). On the other hand, isolated acne and alopecia are not commonly associated with hyperandrogenemia and are not good markers of androgen excess. Biochemical evaluation for excess androgens, viz., measuring total and free testosterone, dehydroepiandrosterone sulfate

(DHEAS), and androstenedione, is essential, especially among those who present with hirsutism that is uncertain or even nonexistent (Lizneva et al. 2016).

The differential diagnosis in women affected by hyperandrogenism includes 21-hydroxylase-deficient non-classic congenital adrenal hyperplasia, classic congenital adrenal hyperplasia, hyperandrogenic insulin-resistant acanthosis nigricans (HAIRAN) syndrome, androgen-secreting neoplasms, and idiopathic hirsutism (Azziz et al. 2004). There may be rare instances where hyperandrogenism may result from Cushing's disease, hyperprolactinemia, hypothyroidism, and complications arising from prescribed prescription (Legro et al. 2013).

In women of reproductive age, hyperandrogenemia is a consequence of PCOS because of excess androgen (Fruzzetti et al. 1995), hyperinsulinemia (Landay et al. 2009), and excess synthesis of adrenal androgens (Köşüş et al. 2012). The ACTH concentrations are also considerably elevated in PCOS (Siddiqui et al. 2022), implicating an imbalance in adrenal cortisol steroidogenesis with amplification of androgen synthesis (Siddiqui et al. 2022). Hyperandrogenism is reported to resolve to some extent during perimenopause in women with PCOS (Winters et al. 2000) due to ovarian and adrenal aging and consequent reduction in androgen synthesis (Brown et al. 2011).

8.5.2.2 Biochemical

In females, the adrenal gland and ovarian theca cells are the major sites of testosterone production. Elevated levels of total or free Testosterone are considered as significant biochemical markers of hyperandrogenism (Lizneva et al. 2016). In a study conducted in women with PCOS, free testosterone levels were increased in nearly 70% of serum specimens (Azziz et al. 2009). Moreover, DHEAS, an androgen metabolite and pro-androgen, may be considered as an alternative diagnostic biomarker. Azziz and co-workers observed an increase in DHEAS in almost 25% of their PCOS cohort, albeit, age-related cut-off values must be considered due to the indirect association between age and DHEAS levels. It is important to note that biochemical measurements of androgen are not a replacement for the clinical assessment of hyperandrogenism but are most valuable in the absence of overt signs of hyperandrogenism. Recently, the measurement of total and free testosterone has been recommended to ascertain biochemical hyperandrogenemia (Teede et al. 2023). The estimation of free testosterone is determined by calculating the free androgen index. In the absence of raised testosterone and free androgen index, quantifying androstenedione and DHEAS should be considered. However, DHEAS reduces with progressive age and therefore has a poor specificity. The biochemical assessment of hyperandrogenism is valuable in the absence of clinical features of hyperandrogenism.

8.5.3 Ultrasound Polycystic Ovarian Morphology

Ultrasound evaluation and detection of polycystic ovarian morphology (PCOM) is an important component of PCOS diagnosis, providing a comprehensive morphology of the ovaries (Dewailly et al. 2014). These morphologic features include an increase in ovary size, the presence of multiple small follicles of similar size, an increased ovarian stromal volume and blood flow, increased echogenicity as well as the peripheral distribution of follicles (Rubin et al. 2008; Reichert 2012).

However, limited data is available regarding the ovarian morphology in normal adolescents and those diagnosed with PCOS due to the restrictions associated with transvaginal ultrasound imaging (Zhu et al. 2016). The effectiveness of transabdominal ultrasound imaging is limited by image quality particularly in obese girls (Zhu et al. 2016). This may limit the clinical differentiation between polycystic ovaries and multifollicular ovaries, since multifollicular ovaries are common during puberty (Adams et al. 1985; Foong et al. 2004). Multifollicular ovaries are normal or slightly larger in size, inclusive of about 8–12 follicles, and usually result from a dysregulated gonadotropin control (Fig. 8.2). Although magnetic resonance imaging may be clinically useful as an alternative diagnostic tool for PCOM in adolescence, it remains a costly option (Zhu et al. 2016). Albeit follicle counts

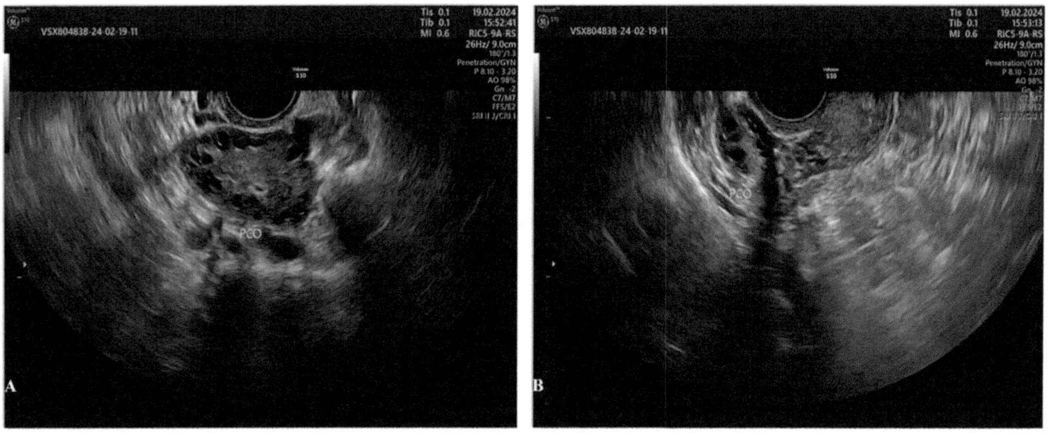

Fig. 8.2 (**a–b**) Polycystic ovarian morphology on ultrasound demonstrating small follicles on the periphery of ovary with thickened stroma; enlarged ovaries that are usually more spherical in shape; multiple small follicles of similar size arranged around the periphery, giving the appearance of a "string-of-pearls; the increased and hyperechoic stroma occupying the center of the ovaries; and higher intraovarian stromal blood flow

greater than 25 per ovary were clinically endorsed based on technological improvements in 2D and 3D ultrasound (Dewailly et al. 2014). There are no definitive criteria for polycystic morphology on ultrasound in adolescent girls and it is therefore not recommended in adolescents.

8.5.4 Antimullerian Hormone Levels

PCOM in adults may be defined by measuring Antimullerian Hormone (AMH) levels, however, this hormone demonstrates a weak specificity and cannot be used as a single test for PCOS diagnosis. The levels of AMH peak between 20 and 25 years in the general population and therefore its use among adolescents is not advised. Moreover, variations in AMH levels are evident across the menstrual cycle, among those using combined oral contraceptives, and also lower in patients with a higher BMI (Teede et al. 2023).

8.5.5 Insulin Resistance and Biochemical Assessment

8.5.5.1 Clinical Assessment

Insulin resistance (IR) occurs when target cells are unresponsive to insulin, resulting in compensatory hyperinsulinemia (Teede et al. 2010), increased androgen secretion (Diamanti-Kandarakis and Dunaif 2012) and consequent abdominal fat deposition in women with PCOS (Escobar-Morreale and San Millán 2007; Escobar-Morreale 2018). The clinical consequences of IR and the hallmark of PCOS are inclusive of acanthosis nigricans, obesity, and hyperandrogenemia (Higgins et al. 2008; Phiske 2014). Patients with acanthosis nigricans present with hyperpigmentation and thickening of skin, with a velvety appearance that is uniformly disseminated on the neck, underarm, groin, antecubital, and popliteal fossae (Phiske 2014).

8.5.5.2 Biochemical Assessment

The biochemical diagnosis of IR remains controversial due to variations in laboratory standards; and can be calculated as follows:

8.5.5.2.1 Homeostasis Model Assessment (HOMA)-Insulin Resistance

Homeostasis model assessment (HOMA) pioneered by Matthews and co-workers in 1985, is a robust clinical and epidemiological tool, which provides equations for estimating IR and β-cell function from simultaneous fasting measures of insulin and glucose levels (Matthews et al. 1985). Diagnosis of IR is confirmed when the result is >2.71, based on the following calculation: HOMA-IR = Fasting glucose (mmol) × Fasting insulin (uU/mL)/22.5.

8.5.5.2.2 Fasting Insulin Level

Quantifying insulin levels is believed to be one of the most useful ways to evaluate IR (Thomas et al. 2000), however, its clinical significance is limited due to the lack of standardization and increased caution of false positives (Singh and Saxena 2010).

8.5.5.2.3 Glucose/Insulin Ratio

The glucose/insulin ratio is a noteworthy indicator of IR as it confers a high sensitivity and specificity of insulin (Vuguin et al. 2001; Silfen et al. 2001; Legro et al. 1998). Ratios of <4.5 and <7 are considered abnormal in adults and in prepubertal children, respectively (James et al. 2011).

8.5.5.2.4 Quantitative Insulin Sensitivity Check Index (QUICKI)

The QUICKI is a reliable and accurate indicator of insulin sensitivity, demonstrating exceptional positive predictive power (Chen et al. 2003, 2005). A QUICKI <0.357 is associated with a greater susceptibility to IR and manifestations of metabolic anomalies (Hřebíček et al. 2002). It is calculated as: QUICKI = 1/(log[Insulin μU/mL] + log[Glucose mg/dL]) (Duncan et al. 1995).

8.5.5.2.5 Glucose Insulin Product

The glucose insulin product is considered an accurate indicator of whole-body insulin sensitivity and a more accurate indicator of insulin sensitivity (Phiske 2014). If the plasma level of glucose is elevated and the insulin response is amplified, the IR may be more serious (Singh and Saxena 2010).

8.5.5.2.6 Log (Homeostasis Model Assessment-Insulin Resistance): Log (HOMA-IR)

The Log (HOMA-IR) is a valuable tool in evaluating IR and it may be more applicable to use HOMA in instances where insulin resistance or sensitivity is not of primary interest (Singh and Saxena 2010). Alternate IR biomarker measurements include adiponectin (Yamamoto et al. 2004), soluble CD36 (Handberg et al. 2006), tumor necrosis factor alpha (Zinman et al. 1999), insulin growth factor binding protein-1 (Motaghedi et al. 2007), hemoglobin A1C (HbA1C) (Osei et al. 2003), ferritin (Jehn et al. 2004), high sensitivity C-reactive protein (Meng et al. 2007), C3 complement (Weyer et al. 2000), HbA1C, weight, fasting glucose and lipoprotein profile, blood pressure, and an alanine transferase test for fatty liver (James et al. 2011). Albeit women with lean PCOS may not show any clinical signs of IR.

Recent recommendations from the consensus group advise that insulin resistance is a clinical diagnosis and do not recommend biochemical testing (Teede et al. 2023). However, oral glucose tolerance tests should be done for every patient with insulin resistance prior to any fertility treatment.

8.5.6 Cardiometabolic Risks

8.5.6.1 Obesity

The impact of hyperinsulinemia on steroidogenesis and reproduction as well as IR are integral mechanisms mediating the effects of weight gain and obesity in PCOS (Barber et al. 2019). Weight gain may also occur due to high carbohydrate diets, sedentary lifestyles, and mental health challenges. Additionally, the development of PCOS can contribute to weight gain and vice versa. Visceral fat deposition is a key role-player in the metabolic dysfunction associated with obesity in PCOS (Zhuang et al. 2009). This validates the need for effective weight loss in PCOS women who are overweight (BMI: 25.0–29.9 kg/m^2) and/or obese (BMI = >30 kg/m^2). Even a modest weight loss of 5–10% is believed to be significant in managing PCOS, positively influencing the metabolic, reproductive, and hyperandrogenic aspects of PCOS (Barber et al. 2006; Azziz et al. 2001). However, adherence to lifestyle changes for maintaining weight loss is difficult and thus unsuccessful (Mann et al. 2007). Several strategies including novel insulin-sensitizing agents, appetite suppressants, and bariatric surgery exist in addition to lifestyle changes, that have the potential to maintain weight loss and reduce obesity (Saboor Aftab et al. 2014; Zac-Varghese et al. 2011).

8.5.6.2 Metabolic Syndrome

Hyperglycemia, dyslipidemia, central obesity, and hypertension are inclusive of the metabolic syndrome (MetS) (Alberti et al. 2009). In PCOS, MetS is mainly mediated through IR via different mechanisms (Sangaraju et al. 2022). For example, hypertension is a consequence of endothelial dysfunction and lack of nitric oxide bioavailability, dyslipidemia results from IR due to increased triglyceride production and secretion of non-ester fatty acid secretion, whereas hyperglycemia is due to compensatory hyperinsulinemia and pancreatic *B*-cell depletion (Essah et al. 2007; Lim et al. 2019).

8.5.6.3 Type 2 Diabetes

There is compelling evidence that impaired glucose tolerance, gestational diabetes, and type 2 diabetes (T2DM) are associated with hyperinsulinemia in PCOS (Boomsma et al. 2006; Moran et al. 2010b; De Sousa 2016). An earlier meta-analysis revealed that women with PCOS are more vulnerable to impaired glucose tolerance (OR:2.48;95% CI 1.63–3.77) and T2DM (OR: 4.43; 95% CI 4.06–4.82), in comparison to women without PCOS (Moran et al. 2010b). Furthermore, a similar trend was observed when analyzed only for BMI-matched studies, i.e., higher odds ratios were observed in PCOS women for impaired glucose tolerance (OR: 2.54; 95% CI 1.44–4.47) and T2DM (OR: 4.00; 95% CI 1.97–8.10) when compared to women without PCOS. Women with PCOS also have higher levels of advanced glycosylation end-products, irrespective of the presence of T2DM and IR, indicative of tissue damage induced by hyperglycemia (Diamanti-Kandarakis et al. 2008). Management should therefore integrate calorie-restricted diets and lifestyle improvements.

8.5.6.4 Dyslipidemia

Women with PCOS are also more susceptible to dyslipidemia (Lo et al. 2006), as demonstrated by increased serum levels of very low-density lipoprotein (VLDL), low-density lipoprotein (LDL), triglycerides, high-density lipoprotein (HDL), and cholesterol (Srikanthan et al. 2006). This dysregulation may arise largely from IR, as the downregulation of lipolysis by insulin is reduced, resulting in an upregulation of adipolysis (Wild et al. 2010). These lipid abnormalities are linked to insulin resistance and have been shown to predict the onset of ischemic heart disease due to atherosclerosis (Wild et al. 2011). Dyslipidemia in PCOS women is believed to be the consequence of obesity, IR, and hyperandrogenism (Sangaraju et al. 2022). Elevations in the hepatic production of apolipoprotein B-containing VLDL are linked to IR and hypertriglyceridemia (Sangaraju et al. 2022). Excess testosterone levels in PCOS patients can also cause androgen receptor-mediated IR and dyslipidemia by reducing HDL (high-density lipoprotein).

Since dyslipidemia commonly manifests in PCOS patients and is a confirmed risk factor for cardiovascular disease (CVD), the vulnerability of women with PCOS to CVD is thus augmented as a consequence of hypertension and IR (Hardy et al. 2013). The subclinical features of CVD such as calcification of the coronary arteries (Christian et al. 2003), and greater left ventricular mass (Orio et al. 2004) and thickness of the carotid intima-media (Talbott et al. 2000), is also increased in women with PCOS. A recent systematic review based on 23 studies, confirmed a higher risk of being diagnosed with T2DM, hypertension, and dyslipidemia, in women with PCOS in comparison to women without PCOS (Wekker et al. 2020).

8.5.7 Mental Health

Body image and self-worth are predictors of anxiety and depression (Simbar et al. 2020). In a study conducted on obese women with PCOS, a significant negative association was reported between mental well-being and obesity (Barber et al. 2019). This association was further exacerbated by features such as hirsutism, menstrual irregularities, and fertility problems (Barber et al. 2019). Depression and anxiety are also strongly correlated with altered body image stress, eating disorders like bulimia and anorexia nervosa, and sexual dysfunction (Alur-Gupta et al. 2024).

Patients with depression are less likely to follow medical treatment recommendations or adhere to lifestyle intervention programs which may be psychological barriers to treatment adherence. Management with cognitive behavioral therapy and inclusion of mental healthcare practitioners are integral to the multidisciplinary team for those affected with PCOS (Cowan et al. 2023).

8.5.8 Clinical Features Stratified by Age

The clinical complications vary in severity from adolescence until menopause.

8.5.8.1 In Adolescence

The development of PCOS in adolescence ranges from genetic predisposition to in utero and postnatal exposure to androgens. However, these theories are speculative, and more research is needed. Special criteria for the PCOS diagnosis in adolescents are not yet established, and currently, the Rotterdam diagnostic criteria of adults are used (Meczekalski et al. 2023). The normal endocrine variant of adolescence, characterized by transient post-pubertal anovulation and multicystic ovaries, is not easily distinguishable from the persistent anovulation and the polycystic ovarian morphology related to PCOS (Witchel et al. 2019).

Anovulatory menstrual bleeding and hyperandrogenism (clinical and/or biochemical) are key features that characterize PCOS during adolescence (Dumesic et al. 2015; Witchel et al. 2020). In addition, oligomenorrhea must be present for at least 2 years after menarche. Young women may present with obesity, hyperandrogenemia, and signs of hyperinsulinemia and diagnosis may be hindered due to the similarities shared by the syndrome and pubertal development (Teede et al. 2018; Witchel et al. 2016; Ibáñez et al. 2017). Ovarian imaging is also problematic in this age group, since the transvaginal ultrasound approach is inapplicable and abdominal ultrasound imaging quality is challenging in obese girls (Zhu et al. 2016). Since the risk of PCOS related to premature pubarche in adolescents is almost 15–20% (Rosenfield 2007), a clear distinction between polycystic and multicystic ovaries is necessary (Livadas and Diamanti-Kandarakis 2013).

Attempts have been made to provide a definition of PCOS in adolescents (Sultan and Paris 2006), based on the clinical consideration of integrating four of the following criteria:

1. Clinical hyperandrogenemia is characterized by persistent and serious-degree acne and hirsutism.
2. Biochemical hyperandrogenemia: levels of serum testosterone >50 ng/dl and LH/FSH ratio >2.
3. Insulin resistance and hyperinsulinemia: acanthosis nigricans, visceral adiposity, and impaired glucose tolerance.
4. Oligomenorrhea persisting 2 years post menarche.
5. Polycystic ovarian morphology on ultrasound.

8.5.8.2 Reproductive Years

8.5.8.2.1 Non-pregnant Adult Women

Adult women with PCOS may present with reproductive, metabolic, and/or psychological symptoms that vary in severity and complexity (Table 8.1) (Pramodh 2020; Katulski et al. 2015). Dermatologic manifestations including acne, hirsutism, androgenic alopecia (Christ and Cedars 2023), and skin pigmentation due to acanthosis nigricans (Keen et al. 2017) are common in these women. Additionally, they are more vulnerable to anxiety and depression, as well as eating disorders and consequent weight gain, obesity, and insulin resistance (Pirotta et al. 2020), which is greater in obese PCOS women (Purwar and Nagpure 2022). These women also present with chronic anovulatory menstrual cycles (Rababa'h et al. 2022), which is a significant cause of infertility (Teede et al. 2010). Women with PCOS are also predisposed to anemia caused by heavy irregular breakthrough bleeding (Singh and Puckett 2023).

Insulin resistance together with hyperinsulinemia, also elevates the risk of CVD development (Osibogun et al. 2020). Older PCOS women have a greater risk of cardiometabolic anomalies such as type II diabetes, dyslipidemia, hypertension, and obesity as well as endometrial abnormalities like infertility (Liu et al. 2021b), cancer (Pk et al. 2022), and hyperplasia (Barber et al. 2019; Marchesan et al. 2021; Ding et al. 2018; Hoeger et al. 2021). These risks are intensified in obese women with diabetes, hypertension, and low parity (Zhang et al. 2020). Furthermore, a familial and/or genetic history of dyslipidemia exacerbates the risk of developing PCOS, and is particularly increased in first-degree family members (Sirmans and Pate 2014).

8.5.8.2.2 Pregnant Adult Women

Infertility is due to high levels of circulating androgens and luteinizing hormones that cause chronic anovulation. Subfertility can be described by the effects of obesity, metabolic, and endocrine abnormalities, on ovarian function, oocyte quality, and endometrial receptivity (Haqq et al. 2014). Consequences of IR and ovarian hyperandrogenemia include premature granulosa cell luteinization, disrupted folliculogenesis (Rosenfield and Ehrmann 2016), and a compromising oocyte quality leading to early miscarriages (Fauser et al. 2012).

Pregnant women with PCOS are vulnerable to maternal complications including gestational diabetes, hypoglycemia, preeclampsia, and gestational hypertension (Katulski et al. 2015; Alchami et al. 2015; Vanky and Løvvik 2020; Zhou et al. 2021). Gestational diabetes emanates from IR and glucose intolerance caused by pancreatic β-cell dysfunction (Choudhury and Rajeswari 2021) whereas BMI prior to pregnancy and maternal age, may be implicated in gestational hypertension (Zhou et al. 2021). The onset and progression of these complications increase the risk of miscarriage, late-onset unexplained intrauterine deaths, and adverse birth outcomes including congenital abnormalities, macrosomia growth restriction/reduced birth weight (Kollmann et al. 2015), cesarean births, increased perinatal morbidity and mortality, and neonatal complications such as hypoglycemia, jaundice, and respiratory distress (Doherty et al. 2015; Yu et al. 2016; Schneider et al. 2019). This risk may be worsened in women with an obese phenotype (Aversa et al. 2020).

8.5.8.2.3 Premenopausal, Menopausal, and Postmenopausal Women

The transition from the normal reproductive period to menopause is associated with anovulatory menstrual cycles and ovaries that are smaller in volume with fewer follicles, making diagnosis difficult (Tanbo and Fedorcsak 2021; Alsamarai et al. 2009). However, there should be a high index of suspicion in patients with obesity, IR, and signs of androgen excess with late-onset acne and hirsutism. Excess levels of insulin together with positive oral glucose tolerance tests are reported in both pre- and postmenopausal women with PCOS, thus increasing their vulnerability to increased IR (Puurunen et al. 2011). Likewise, PCOS is the most common cause of androgen excess among postmenopausal women (Elhassan et al. 2018). Surrogate PCOS indicators such as increased BMI, waist circumference and waist-to-hip ratios, increased IR indexes, increased fasting glucose concentrations, decreased HDL cholesterol, increased triglycerides, and LDL are consistent with an increased risk of cardiometabolic complications in the perimenopausal and menopausal women with PCOS (Millán-de-Meer et al. 2023).

A systematic review and meta-analysis of peri- and postmenopausal women confirms that most of the clinical and biochemical characteristics of premenopausal women with PCOS, as well as many of its cardiometabolic associations, persist during their late-reproductive years and after menopause (Escobar-Morreale 2018). The risk of endometrial cancer is higher due to unopposed estrogen stimulation of the endometrium as a result of aromatization of testosterone to estrone (Wan et al. 2016). In postmenopausal women, PCOS is categorized by androgen excess with cardiometabolic features such as weight gain and obesity (Millán-de-Meer et al. 2023). Interestingly, a steady improvement in IR, androgens, and lipids was observed in lean PCOS women from early to late-reproductive age, indicative that weight management may reduce cardiometabolic risk (Livadas et al. 2020).

8.6 Pathophysiology

PCOS affects the hypothalamic–pituitary–ovarian (HPO) axis and its pathophysiology is multifactorial and polygenic (Szeliga et al. 2022) (Fig. 8.3). Gonadotrophin biosynthesis and secretion are influenced by hypothalamic, paracrine, and endocrine factors, with a considerable overlap between all three (Ulloa-Aguirre et al. 2017). Neuroendocrine abnormalities include an increased gonadotrophin-releasing hormone pulse frequency with consequent hypersecretion of luteinizing hormone (LH) to impact ovarian androgen synthesis, folliculogenesis, and oocyte development (Baskind and Balen 2016). This disturbed ovarian–pituitary and hypothalamic feedback accentuates the gonadotrophin abnormalities. The presence of enlarged polycystic ovaries suggests that the ovary is the primary site of endocrine abnormality, particularly hyperandrogenism (Rosenfield 2015). Notably, the *Kiss1* gene has been implicated in regulating gonadotropin discharge and ovulation and the consequent hypothalamic–pituitary dysregulation implies a dysfunctional Kiss 1 system (Witchel and Tena-Sempere 2013).

8.6.1 Ovarian Steroidogenesis and Dysregulation

Ovarian hormone biosynthesis is mediated through the HPO axis as well as other paracrine and environmental factors (Rosenfield and Ehrmann 2016). Granulosa cells are involved in this biosynthesis and are described by the two-cell gonadotrophin theory, in which steroidogenesis occurs through a complex interaction between the granulosa and theca cells (Drummond 2006). A balance between ovarian androgen secretion and estrogen synthesis is necessary to maintain optimal ovarian functioning, particularly ovulation (Rosenfield and Ehrmann 2016). Even though these pathways are dependent on LH and FSH concentrations, intrafollicular modulators are essential for follicular development (Rosenfield and Ehrmann 2016).

The main source of androgen production is the ovary under the influence of LH, and the adrenal gland under the influence of adrenocorticotrophin (ACTH) in equal parts (Yesiladali et al. 2022). Androgens promote the growth of primary follicles to the stage of pre-antral and small antral follicles, and, in the presence of excess androgens, this process is accelerated compared with the normal ovary (Homburg 2009). Small amounts of androgens are central to follicular maturation and ovulation (Sen et al. 2014). Consequently in women with PCOS, the ovarian dysregulation of androgen synthesis results in LH hyper-responsiveness, theca cell stimulation, causing hyperandrogenemia (Magoffin 2007). Excess androgen incites a proliferation of small antral follicles and consequent follicular maturation arrest and polycystic ovarian morphology (Rosenfield and Ehrmann 2016).

8.6.2 Hyperandrogenism

Hyperandrogenism occurs during puberty due to the hypersecretion of ovarian and adrenal androgens, viz., dehydroepiandrosterone (DHEA), DHEAS, testosterone, dihydrotestosterone, and androstenedione (Burger 2002; Siddiqui et al. 2022). Injury to the primary theca cell results in hyperandrogenemia and consequent hypothalamic–pituitary–ovarian axis dysregulation (Rosenfield 2015). Hypersecretion of testosterone, androstenedione, DHEA, and DHEAS in PCOS induces premature ovarian follicular development and consequent anovulation and oligo-amenorrhea (Siddiqui et al. 2022; Andrade et al. 2016).

Fig. 8.3 Pathophysiology and clinical manifestations of PCOS. Adapted from Moran and Norman (2004), Singh et al. (2023). This figure was generated using BioRender (https://biorender.com/). Antimullerian hormone (AMH); Androgen receptor (AR); Steroid 21-hydroxylase (CYP21); Cytochrome P450 17α-hydroxylase/17,20-lyase (CYP17); Cytochrome P450 aromatase (CYP19); Cholesterol Side-Chain Cleavage Enzyme Gene (CYP11a); Calpain10 (CAPN10); Follicular-stimulating hormone (FSH); Follicular-stimulating hormone receptor (FSHR); Fat mass obesity (FTO); Insulin (INS); Insulin receptor (INSR); Luteinizing hormone (LH); Sex hormone-binding globulin (SHBG)

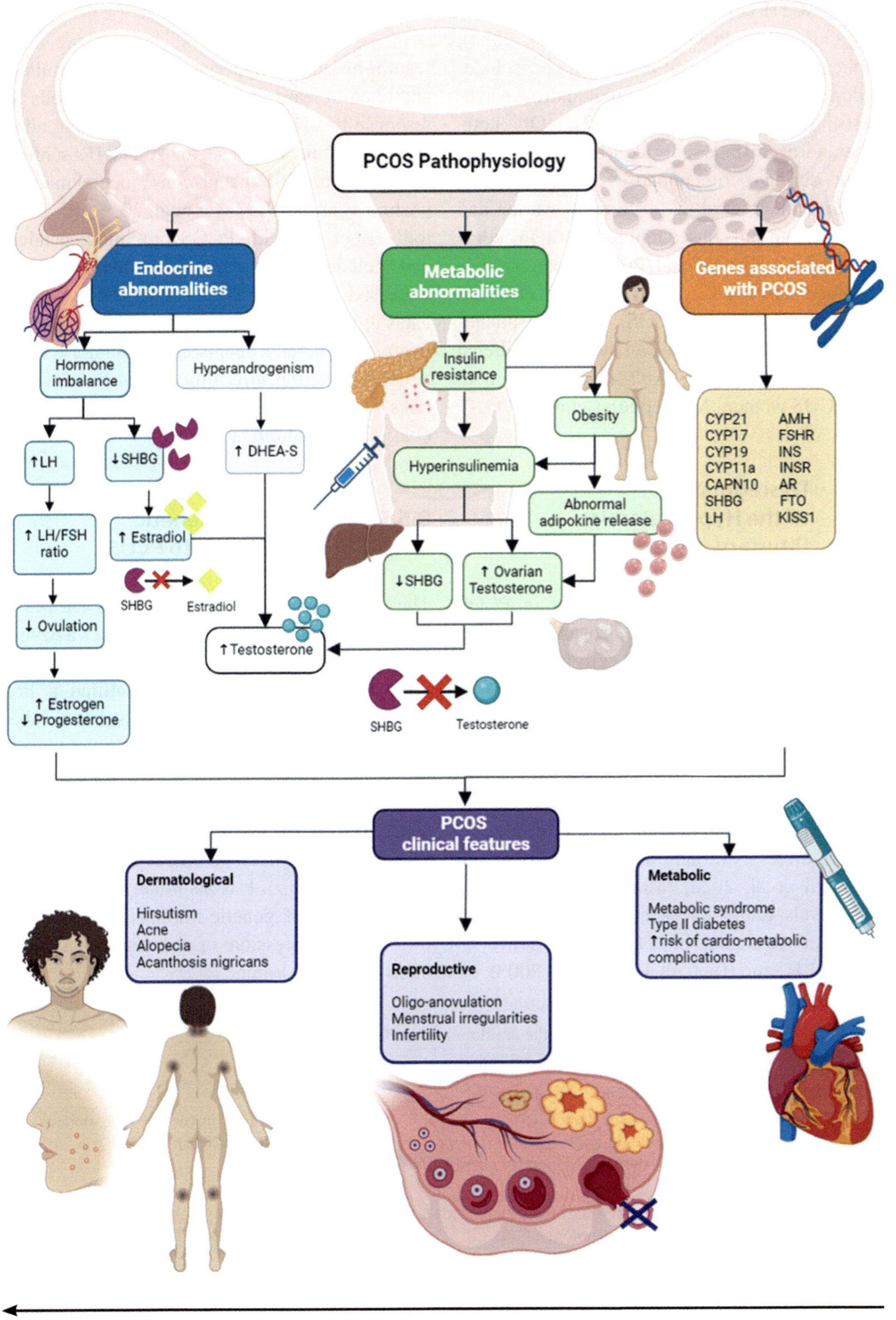

PCOS Pathophysiology
Endocrine abnormalities
Metabolic abnormalities
Genes associated with PCOS
Hormone imbalance
Hyperandrogenism
↑LH
↓SHBG
↑ DHEA-S
↑ LH/FSH ratio
↑ Estradiol
SHBG
Estradiol
↓ Ovulation
↑ Testosterone
↑ Estrogen
↓ Progesterone
Insulin resistance
Obesity
Hyperinsulinemia
Abnormal adipokine release
↓SHBG
↑ Ovarian Testosterone
SHBG
Testosterone
CYP21
CYP17
CYP19
CYP11a
CAPN10
SHBG
LH
AMH
FSHR
INS
INSR
AR
FTO
KISS1
PCOS clinical features
Dermatological
Hirsutism
Acne
Alopecia
Acanthosis nigricans
Reproductive
Oligo-anovulation
Menstrual irregularities
Infertility
Metabolic
Metabolic syndrome
Type II diabetes
↑risk of cardio-metabolic complications

The ACTH concentrations are also considerably elevated in PCOS (Siddiqui et al. 2022), implicating an imbalance in adrenal cortisol steroidogenesis with amplification of androgen synthesis (Siddiqui et al. 2022). Of note, hyperandrogenism is reported to resolve to some extent during perimenopause in women with PCOS (Winters et al. 2000) due to ovarian and adrenal aging and consequent reduction in androgen synthesis (Brown et al. 2011). Noteworthy, a sustained androgen hypersecretion is observed during menopause in PCOS, while postmenopausal women demonstrate an increase in adrenal and ovarian androgen levels in comparison to non-PCOS women (Markopoulos et al. 2011).

8.6.3 Dysregulation of the Hypothalamic–Pituitary–Ovarian Axis

Excess levels of LH are implicated in PCOS pathogenesis as it increases androgen production and secretion by ovarian theca cells (Ashraf et al. 2019). Changes in the pulse frequencies of gonadotropin-releasing hormone (GnRH) upregulate LH secretion and downregulate follicle-stimulating hormone (FSH) secretion (Rosenfield and Ehrmann 2016; Dumesic et al. 2015). The increase in pulse frequency of GnRH elevates LH which drives ovarian androgen production (Siddiqui et al. 2022). Moreover, the reduced FSH levels inhibit the adequate conversion of androgen to estradiol. In PCOS there is a reversal of FSH:LH ratio (Richard and Ricardo 2003).

8.6.4 Hyperinsulinemia

Insulin-resistant hyperinsulinemia stimulates androgen release from ovarian theca cells in response to LH, affecting a corresponding adrenal androgenic response to ACTH (Rosenfield and Ehrmann 2016) and consequent hirsutism (Ashraf et al. 2019; Mayer et al. 2015). Risk factors associated with PCOS are interdependent and include genetic and environmental factors, increased androgen levels, and obesity which leads to the clinical manifestations of hyperinsulinemia, hyperandrogenism, anovulatory menstrual dysfunction, infertility, and cardiometabolic risks (Ajmal et al. 2019; Balen 2004; Bulsara et al. 2021; Siddiqui et al. 2022).

8.6.5 Familial and Genetic Predisposition to PCOS

The epidemiology of PCOS suggests the possibility of a genetic link as it demonstrates a pattern of familial inheritance (Kahsar-Miller et al. 2001). Genetic studies have identified a link between PCOS and disordered insulin metabolism, and indicate that the syndrome might be the presentation of a complex genetic trait disorder (Table 8.2). The phenotype of women with polycystic ovaries and of the PCOS itself can be very variable, it is difficult to elucidate the genotype (Khan et al. 2019). It is also likely that different combinations of genetic variants will result in differential expression of the separate components of the syndrome.

Table 8.2 Candidate Genes associated with the development of PCOS

Gene	Physiological role	Link to PCOS
Genes involved in ovarian and adrenal steroidogenesis		
CYP21	During steroidogenesis, it encodes a catalytic enzyme responsible for converting 17-hydroxyprogesterone to 11-deoxycortisol (Liu et al. 2021a)	Heterozygote CYP21 mutations are associated with PCOS (Tal et al. 2015)
CYP17	Encodes the enzyme (P450c17α) which catalyzes the conversion of pregnenolone to 17-hydroxypregnenolone and progesterone to 17-hydroxyprogesterone (Artini et al. 2006)	Increased CYP17 expression was observed in ovarian theca cells of PCOS women (Ferrara et al. 2003) Defective transcription of CYP17 in theca cells may contribute to increased secretion of ovarian androgens in PCOS (Ferrara et al. 2003)

(continued)

Table 8.2 (continued)

Gene	Physiological role	Link to PCOS
CYP19	Encodes the enzyme (aromatase p450) which is required for estrogen formation (Scotti et al. 2014)	Aromatase p450 activity is dysregulated in PCOS (Tal et al. 2013a) Women with PCOS display decreased aromatase activity (Tal et al. 2013a)
CYP11a	Encodes an enzyme that is necessary for the conversion of cholesterol to pregnenolone (Tal et al. 2013b)	CYP11a gene variation is associated with hyperandrogenism noted in PCOS (Raja-Khan et al. 2010) Association factor of PCOS (Raja-Khan et al. 2010, Wang et al. 2018)
Genes exerting a steroid hormone effect		
AR	Located on chromosome X and encodes a 90-kb long tridomain protein (Llurba et al. 2015)	AR gene contains a CAG repeat polymorphism involved in PCOS development (Chen et al. 2022)
SHBG	Located on chromosome 17 (p12-p13); encodes human SHBG (Geldenhuys et al. 2018)	SHBG gene polymorphism increases PCOS risk (Rana et al. 2020)
Genes involved in gonadotropin action and regulation		
LH and its receptor	Encodes for LH and its receptor (Wickenheisser et al. 2000)	Gene variants of LH and its receptor are linked with PCOS susceptibility (Di Pietro et al. 2018) LHβ gene polymorphisms related to PCOS development (Di Pietro et al. 2018)
AMH	Encodes a protein associated with infertility (Ma et al. 2018)	Certain AMH gene variants are possible indicators of PCOS (Artini et al. 2009) AMH gene mutations may cause hyperandrogenism associated with PCOS (Artini et al. 2009)
FSHR	Encodes a protein G-coupled receptor (Tal et al. 2014)	Polymorphism is linked to ovarian function (Donma et al. 2019) Increased FSHR gene frequency was observed in PCOS (Donma et al. 2019)
KISS1	Encodes the hypothalamic peptide Kisspeptin, which is involved in the mediation of the HPG axis and regulation of LH and FSH release (Oride et al. 2015)	PCOS women exhibit elevated kisspeptin levels suggesting that an upregulation in KISS1 activity stimulates HPG axis, resulting in irregular menstrual cycles and excessive androgen release (Tang et al. 2019)
Genes involved in insulin action and secretion		
INS	Sandwich gene between tyrosine hydroxylase and IGF-II (Shang et al. 2012)	VNTR polymorphism regulates INS transcriptional rate and is linked to PCOS (Wickenheisser et al. 2000; Dinsdale and Crespi 2021)
INSR	Encodes turmeric protein composed of two alpha and two beta chains (Dinsdale et al. 2021)	D19S884 is linked to PCOS and is located in the same chromosome region as the INSR gene (chromosome 19p13.2) (Ono et al. 2014)
CAPN10	Encodes the Calpain-10 protein and is involved in insulin secretion and action (Schüring et al. 2012)	Insulin resistance is linked to PCOS and CAPN10 polymorphisms leads to PCOS and is considered a PCOS candidate gene (Ashraf et al. 2019)
Gene involved in obesity and type 2 diabetes		
FTO	Encodes the enzyme alpha-ketoglutarate (Pangath et al. 2022)	Increased single nucleotide polymorphism (rs9939609) in PCOS-affected women (Holoch et al. 2014)

Androgen receptor (AR); Sex hormone-binding globulin (SHBG); Luteinizing hormone (LH); Antimullerian hormone (AMH); Follicular-stimulating hormone (FSH); Follicular-stimulating hormone receptor (FSHR); Hypothalamic–pituitary–gonadal axis (HPG); Insulin (INS); Insulin receptor (INSR); Calpain10 (CAPN10); Fat mass obesity (FTO)

8.7 Clinical Investigations

Various diagnostic tests are employed to assess PCOS, viz.:

- Polycystic Ovarian Morphology
- Transvaginal ultrasound (TVU) is characterized by 12–25 follicles per ovary. However, TVU is inapplicable in virginal adolescents and thus a pelvic ultrasound with a full bladder is advised. Ultrasound imaging is also unsuitable in obese patients and magnetic resonance imaging is recommended albeit being uneconomical.
- Endocrine tests are used to measure FSH/LH ratios and Prolactin levels.
- Androgen excess: Total and Free Testosterone index, Androstenedione, DHEA, and SHBG.
- Hyperinsulinemia: Fasting Insulin level, the HOMA Index, and QUICKI.
- Type 2 Diabetes: OGTT, Fasting Blood Glucose, and HbA1C tests.
- Dyslipidemia is confirmed by conducting a fasting lipogram.
- Liver and thyroid function tests.
- Weight assessment, viz., Body Mass Index (BMI) and Waist: Hip Ratios.

8.8 Management of PCOS

The management of PCOS must be individualized to the specific needs of each patient depending on their age and reproductive phase (Fig. 8.4). The goals of therapy should be on improving the clinical manifestations of hyperandrogenemia and hyperinsulinemia, controlling the menstrual cycle and ovulation inductors for infertility, and precluding cardiometabolic complications (Singh et al. 2023). Irregular menses, hirsutism, and infertility remain the most stressful symptoms experienced by women with PCOS (Kitzinger and Willmott 2002). Numerous alternative therapeutic options are recommended for managing PCOS, with lifestyle and dietary changes forming the basis. Pharmacological and non-pharmacological options may also reduce the symptoms associated with menstrual aberrations, androgen excess, and infertility, while various options exist for controlling cardiometabolic comorbidities (Singh et al. 2023) (Fig. 8.5). A multidisciplinary management approach inclusive of a gynecologist, dietician, cardiologist, and medical endocrinologist is also advised. The spectrum of metabolic irregularities experienced by women with PCOS cannot be exclusively remedied by a single therapy. Metabolic function may only be improved if lifestyle and dietary modifications are amalgamated with treatment that targets symptoms.

8.8.1 Oral Contraceptives

The first line of management includes oral contraceptives (OCs), which are aimed at regulating menstruation and hyperandrogenic features such as acne and hirsutism (Shah and Patil 2018). Estrogen (ethinyl estradiol) and progesterone, which induces the antiandrogenic effects of cyproterone acetate and drosperinone, are key ingredients in the combined OC. The negative feedback action exerted by the OC subsequently downregulates ovarian androgen synthesis and reduces the clinical features of hyperandrogenism (American College of Obstetricians and Gynecologists 2010). OCs also stimulate hepatic SHBG production, reduces the circulating concentration of free androgens, prevents the peripheral conversion of testosterone to dihydrotestosterone (DHT), and the affinity of DHT to the androgen receptor as well as reduce adrenal androgen secretion (Rotterdam and ESHRE 2004). The physiological impact of OCs on the hypothalamic–pituitary axis and on ovarian steroidogenesis, subsequently reduces hyperandrogenism. In view of the various benefits, OCs are clinically useful in managing PCOS-associated androgenic alopecia, hirsutism, acne, and menstrual abnormalities (Zimmerman et al. 2014). Moreover, the antiandrogenic compounds found in third-generation combination OCs are reported to enhance the metabolic, lipid, and adipokine profiles in PCOS (Mendoza et al. 2014).

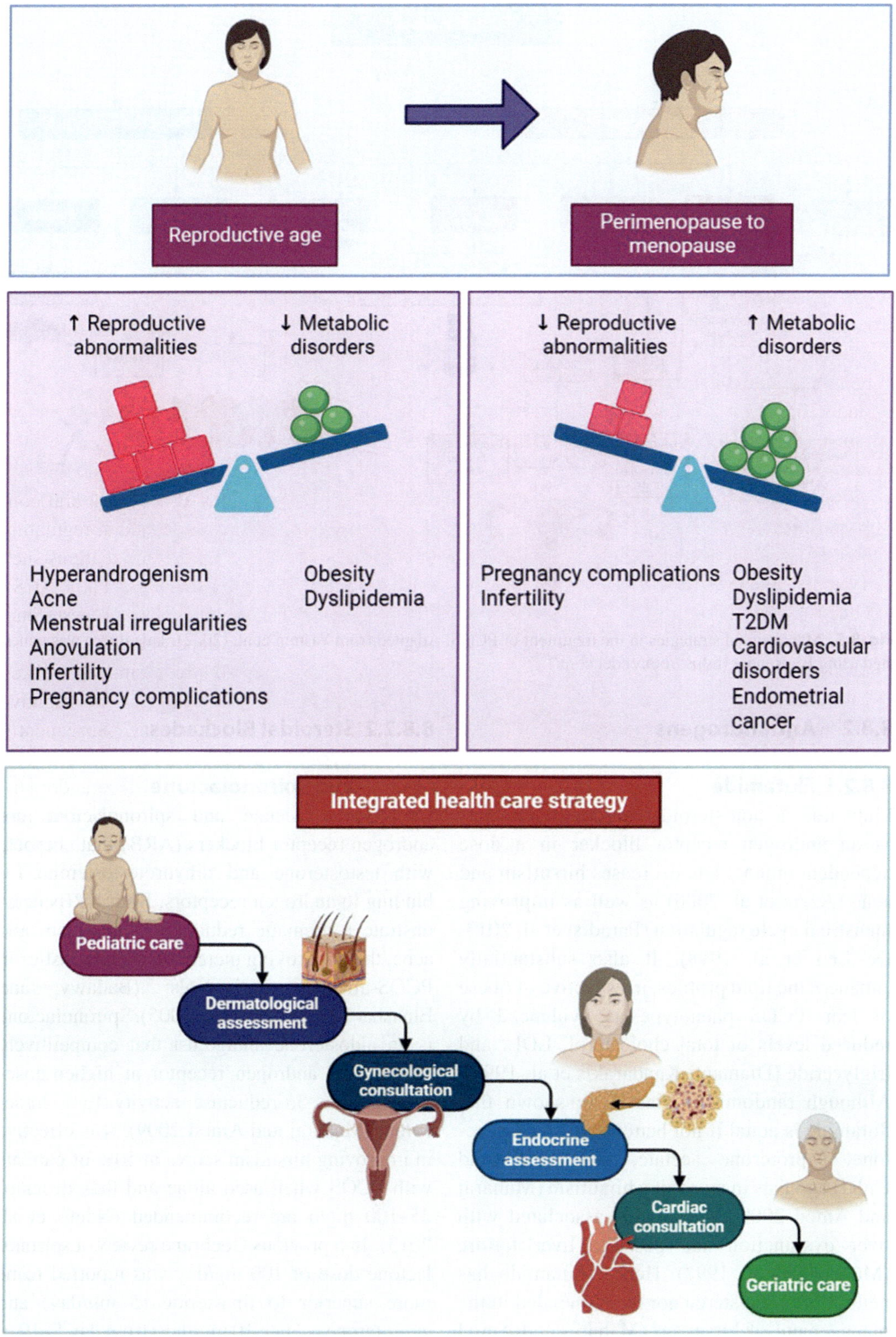

Fig. 8.4 Clinical features and integrated health care. (**a**) Changes in clinical features in women of reproductive age until menopause. (**b**) Recommended healthcare strategies. Adapted from Fauser et al. (2012). This figure was generated using BioRender (https://biorender.com/)

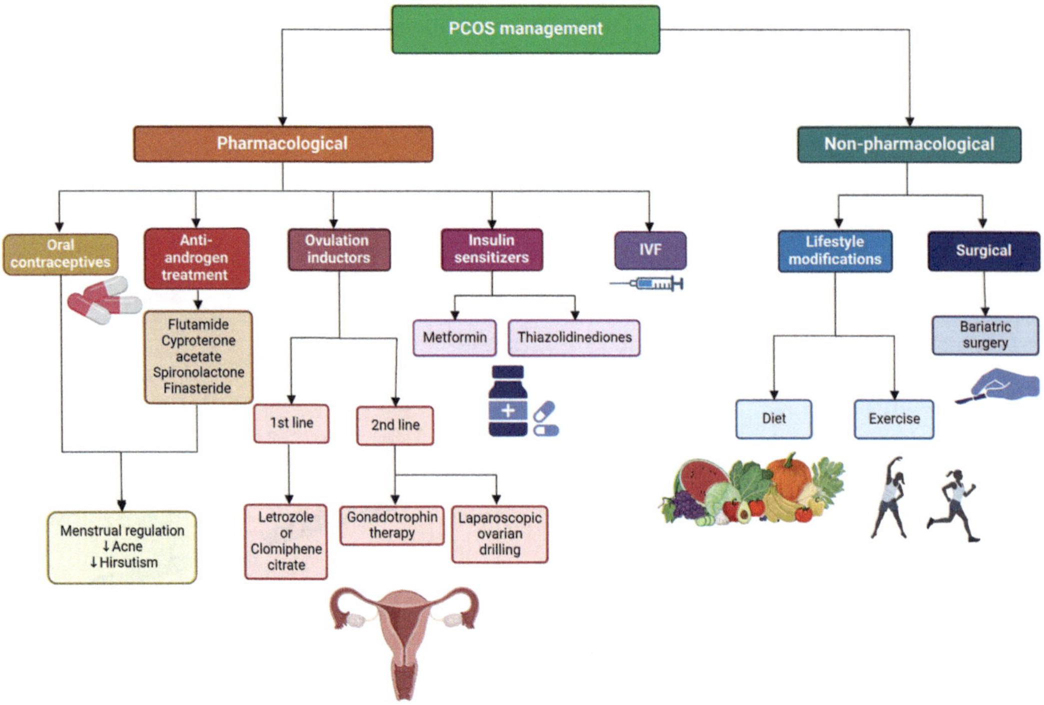

Fig. 8.5 Management strategies in the treatment of PCOS. Adapted from Yasmin et al. (2022). This figure was generated using BioRender (https://biorender.com/)

8.8.2 Antiandrogens

8.8.2.1 Flutamide

Flutamide, a non-steroidal antiandrogen, acts as an androgen receptor blocker in a dose dependent manner and decreases hirsutism and acne (Azziz et al. 2000) as well as improving menstrual cycle regulation (Paradisi et al. 2013; De Leo et al. 1998). It also substantially enhances the lipid profiles, irrespective of obese or lean PCOS phenotype as evidenced by reduced levels of total cholesterol, LDL, and triglyceride (Diamanti-Kandarakis et al. 1998). Although randomized trials have shown that flutamide is equal if not better than spironolactone, cyproterone acetate, finasteride, and GnRH agonists in managing hirsutism (Maharaj and Amod 2009), the drug is associated with liver dysfunction and possible liver failure (Müderris et al. 1997). Hence, flutamide has neither been registered nor recommended in the management of hirsutism (Maharaj and Amod 2009).

8.8.2.2 Steroidal Blockades: Cyproterone Acetate and Spironolactone

Cyproterone acetate and spironolactone are androgen receptor blockers (ARB) that compete with testosterone and dihydrotestosterone for binding to androgen receptors. Both ARBs demonstrate a dramatic reduction in hirsutism and acne, thus improving metabolic characteristics in PCOS-affected individuals (Badawy and Elnashar 2011; Zulian et al. 2005). Spironolactone is an aldosterone antagonist that competitively inhibits the androgen receptor at higher doses and inhibits 5α-reductase activity to a lesser extent (Maharaj and Amod 2009). It is effective in improving hirsutism scores in 50% of patients with PCOS when used alone and thus doses of 25–100 mg/d are recommended (Ndefo et al. 2013). In a previous Cochrane review, a spironolactone dose of 100 mg/day was reported to be more superior to finasteride (5 mg/day) and cyproterone acetate 10 mg/day (from day 5–19 of the menstrual cycle) (Farquhar et al. 2003).

8.8.2.3 Finasteride

The 5-alpha reductase inhibitor Finasteride, when administered at a dosage between 2.5 and 7.5 mg/day, prevents the conversion of testosterone into dihydrotestosterone (Azziz et al. 2000). Albeit the effectiveness of finasteride in managing hirsutism is comparable to, or somewhat less effective than spironolactone and flutamide (Maharaj and Amod 2009).

8.8.3 Cosmetic Measures

Cosmetic treatment combined with pharmacotherapy is commonly utilized in managing hirsutism. Commonly used cosmetic methods include laser electrolysis in conjunction with topical eflornithine cream (Vaniqua®) (Shapiro and Lui 2001), which seems to be the most efficient method; followed by hair removal options such as bleaching, waxing, and shaving. Of note, minoxidil is effective in managing androgenic alopecia (Gupta et al. 2024).

Oral contraceptives and antiandrogenic drugs are also effective in combination or alone to reduce hyperandrogenism and improve symptoms in patients with PCOS (Maharaj and Amod 2009). Furthermore, the response rates and effectiveness increase by 75% with the synergistic use of these drugs. A combination of OCs and androgen blockades is equally effective in the treatment of hirsutism when compared to flutamide; and OCs in combination with CPA also showed a significant improvement in hirsutism (Maharaj and Amod 2009). There is a paucity of data regarding the use of CPA alone when compared to the combination of OCs and CPA. However, side effects of cyproterone acetate and spironolactone include a decrease in libido and impaired liver function (Maharaj and Amod 2009). These drugs should be considered as first-line therapy in adolescents and younger women where fertility is not a priority. Short courses of OCs and AR blocking agents should be considered in perimenopausal women after assessing risk factors like smoking and thromboembolism (Hale et al. 2014).

8.8.4 Insulin-Sensitizing Agents

Management strategies that focus on insulin resistance should incorporate pharmaceutical and lifestyle modifications. Insulin-sensitizing agents enhance insulin responsiveness in target cells by lowering insulin secretion and regulating glucose tolerance (Grundy 2002).

8.8.4.1 Metformin

The biguanide metformin decreases liver gluconeogenesis, by glucose absorption, and enhances the responsiveness of peripheral tissues toward insulin (Grundy 2002; Rashid et al. 2022). Metformin improves menstrual cyclicity and ovulation, as well as reduces circulating androgen concentrations (Rashid et al. 2022). More recently, a significant reduction in BMI was observed in PCOS women treated with metformin as well as those managed by lifestyle modifications, however, a downregulation in testosterone was only observed in the metformin-treated group (Curi et al. 2012).

Findings from a randomized controlled trial reported a notable diminution in BMI in both obese and severely obese PCOS women treated with metformin devoid of any lifestyle modifications (Harborne et al. 2005). Metformin significantly improves dyslipidemia either by directly influencing the hepatic metabolism of free fatty acids or indirectly by ameliorating hyperinsulinemia (Sin et al. 2011). It is clinically useful in precluding and managing type 2 diabetes among those with impaired glucose tolerance (Garber et al. 1997). The recommended daily dose of metformin in patients without impaired glucose tolerance and type 2 diabetes, ranges between 500 mg and 2000 mg (Richardson et al. 2021).

8.8.4.2 Thiazolidinediones

The thiazolidinediones, pioglitazone, and Rosiglitazone amplify insulin sensitivity which reduces insulin and androgen levels in the blood (Marshall and Dunaif 2012). In women with PCOS, pioglitazone decreases insulin resistance, hyperandrogenism, and ovulatory disorders as well as fasting serum insulin and free androgen levels and simultaneously upregulates SHBG levels (Madnani et al. 2013; Brettenthaler et al. 2004). A meta-analysis revealed that metformin when used together with TZD significantly enhances insulin resistance and lipid metabolism as well as reduces total testosterone concentrations compared to using metformin alone (Xu et al. 2017).

8.8.5 Fertility Management

Infertility issues in PCOS women should involve lifestyle modification for weight loss and the use of ovulation inductors (Glendining and Campbell 2023). Additionally, laparoscopic ovarian drilling should be considered in Clomiphene resistance and in vitro fertilization (IVF) in selected cases.

8.8.5.1 Ovulation Induction Agents

Ovulation induction agents are recommended for managing patients with anovulation.

8.8.5.1.1 Letrozole

Letrozole was initially utilized in clomiphene resistance and is now regarded as a first-line treatment for ovulation induction. Letrozole inhibits aromatization of androgens to estrogens, resulting in estrogen deficiency and consequent elevated FSH release for follicular development (Elizur and Tulandi 2008). A meta-analysis done in PCOS women confirms that letrozole clinically improves pregnancy and live birth rates per patient compared to clomiphene citrate (Roque et al. 2015). This may be attributed to the absence of anti-estrogenic effects on the endometrium and cervical mucus that clomiphene citrate has.

8.8.5.1.2 Clomiphene Citrate

Clomiphene citrate (CC), a selective estrogen receptor modulator (SERM), amplifies FSH release in response to follicular maturation (Sawant and Bhide 2019). It exerts its effects on the HPO axis via downregulation of estrogen receptors and the consequent negative feedback of estrogen on the pituitary gonadotropin release. This subsequently enhances ovarian follicular activity in response to an amplified gonadotropin secretion that stimulates follicular maturation (Kettel et al. 1993). Noteworthy, CC elevates the GnRH pulse amplitude but does not affect the pulse frequency in anovulatory PCOS women (Kettel et al. 1993; Practice Committee of the American Society for Reproductive Medicine 2013). The management is initiated with a preliminary 5-day oral daily dose of 50 mg CC during days 2–5 of the preovulatory phase following the commencement of progestin-induced or spontaneous menses (Practice Committee of the American Society for Reproductive Medicine 2013). This is followed by 50 mg increments in subsequent cycles until ovulation is induced or the highest dosage of 150 mg is reached for 3–6 menstrual cycles (Practice Committee of the American Society for Reproductive Medicine 2013).

Among PCOS women who are managed with CC, an expected 22% will conceive each cycle following CC treatment, while 75%–80% will ovulate (Practice Committee of the American Society for Reproductive Medicine 2013). Since the side effects of CC are negligible, it is considered safe to use and the antiestrogenic effects may be neutralized with estrogen supplements. Albeit periodic ultrasounds are warranted to screen the uterus and ovaries to reduce the predisposition to antiestrogenic endometrial consequences and multifollicular growth (Sawant and Bhide 2019). In patients managed with CC, the rates of CC-induced twin and triplet pregnancies are 5–7% and 1.1%, respectively (Sawant and Bhide 2019). Of note, letrozole and CC are often more effective when used in combination with metformin.

8.8.5.1.3 Gonadotrophins

The gonadotropins are considered second-line ovulation induction agents since the first-line oral ovulation inductors have been ineffective, particularly in women who experience adverse antiestrogenic endometrial effects and are impervious to oral therapy (Sawant and Bhide 2019). Prediction of response and inducing monofollicular development to reduce multiple pregnancies is difficult in PCOS. Low FSH dosages administered during the early preovulatory phase together with careful monitoring with ultrasound are essential to avoid ovarian hyperstimulation syndrome (OHSS) (Balen et al. 2016).

8.8.5.2 Surgery: Laparoscopic Ovarian Drilling

Laparoscopic ovarian drilling (LOD) is a surgical option for improving ovarian function. (Gjönnaess 1984). Laparoscopic ovarian cautery as well as varying types of lasers have been used for drilling and about 10 holes per ovary are usually punctured in the ovarian surface and stroma (Collée et al. 2021). Of note, the LOD and ovarian wedge resection procedures are similar and can result in ovarian androgen-producing tissue damage and subsequent reduction in peripheral conversion of androgens to estrogens (Bordewijk et al. 2020).

Increased FSH levels together with reduced serum androgen and LH levels following LOD, result in an estrogenic environment that is conducive to follicular recruitment, maturation, and ovulation (Armar et al. 1990; Greenblatt and Casper 1987; Lebbi et al. 2015). Ovarian drilling may also ameliorate hirsutism and acne, regulate menstrual cycles thus enhancing ovulation and pregnancy (Debras et al. 2019; Lemieux et al. 1999). However, LOD can predispose obese women to potential perioperative and surgical difficulties as well as postoperative adhesions and premature ovarian failure (Sawant and Bhide 2019).

There is a lack of consensus regarding the technique and number of cautery holes required per ovary. However, there is widespread support for the technique proposed by Amer et al. (2003). The technique involves a 4-point cauterization of the ovarian capsule, using a monopolar electrosurgical needle, each for 4 s at 40 watts and a depth of 4 mm. Thus, LOD should only be done as second-line treatment for ovulation induction in clomiphene-resistant cases of PCOS as an alternative to gonadotropin therapy.

8.8.5.3 Assisted Reproductive Technology

8.8.5.3.1 In vitro Fertilization (IVF)

This technique is offered to PCOS women with anovulation in the absence of other causes of infertility. It is a recognized third-line treatment alternative when first- or second-line ovulation induction has failed. Although IVF is extremely successful and can increase pregnancy rates by up to 50% after just one round of treatment in younger women with a healthy ovarian reserve, its clinical use remains problematic (Sawant and Bhide 2019). Challenges include an increased risk of OHSS, multiple births, and exorbitant financial costs especially for women living in poorly resourced countries.

Higher levels of antimullerian hormone are present in PCOS women, increasing their risk of controlled ovarian hyperstimulation (COH) and OHSS development, requiring personalized stimulation approaches (Pundir et al. 2012).

In PCOS women undergoing IVF, the GnRH antagonist protocol is more superior to the GnRH agonist protocol, confirming significant reductions in the risk and incidence of OHSS as well as enabling adequate control over the adverse effects triggered by gonadotropins (Teede et al. 2018; Pundir et al. 2012). Furthermore, elective single embryo transfer (eSET) procedures, especially in the young have resulted in a decrease in the incidence of multiple births (Youssef et al. 2011). IVF use has been substantially improved following the introduction of the GnRH antagonist protocol, GnRH agonist trigger, elective freeze-all policies, and eSET.

8.8.5.3.2 In Vitro Maturation

This treatment approach involves the harvesting of immature oocytes from unstimulated or mildly stimulated ovaries (Das and Son 2023). For women with PCOS, In Vitro Maturation (IVM) may remove or minimize the predisposition of OHSS following COH (Huang et al. 2010). However, the technique can be extremely labor intensive and time consuming as well as requiring specialized facilities with experienced and knowledgeable staff to carry out the procedure.

8.8.5.4 Lifestyle Modification

In women with PCOS, dietary and behavioral modifications as well as physical exercise support the recovery of metabolic and reproductive functions (Gu et al. 2022). Intentional lifestyle changes implemented in obese PCOS women may improve mental health and reduce stress levels, thus improving the quality of life. In obese women, ovulation is dysregulated due to lower levels of SHBG and higher levels of free testosterone in blood and target tissue thus increasing the risk of insulin resistance, metabolic syndrome, and type 2 diabetes (Li et al. 2017). Metformin combined with lifestyle modifications is reported to improve the overall quality of life in these patients (Kim et al. 2020). Moderate aerobic exercise and a low-caloric diet are recommended. Even a minimal loss of weight (5–10%) can improve metabolic dysfunction, reproductive abnormalities, and the psychological symptoms associated with PCOS (Moran et al. 2017). More than 40% of patients fail to lose weight with lifestyle modification due to insulin resistance, depression, and anxiety.

Surgical methods are also an option for managing weight loss in PCOS. More than 40% of patients fail to lose weight with lifestyle modification due to insulin resistance, depression, and anxiety. In cases of severely obese patients, bariatric surgery is a recommended option should other treatment options be unsuccessful (Malik and Traub 2012). Improvements in surgical techniques and less invasive procedures have made bariatric surgery safer. Nonetheless, bariatric surgery does impose risks like surgical and dietary complications; and it is recommended that pregnancy should only be considered 12 months post-surgery (Hoeger et al. 2021).

8.8.6 Management Based on Age Categories

8.8.6.1 In Adolescents

The appropriate therapeutic approach to adolescents with PCOS remains a major concern for clinicians (Mastorakos et al. 2006). There are no established therapeutic rules regarding the management of anovulation, hyperandrogenemia, and insulin resistance (Mastorakos et al. 2006). Management strategies are largely symptom-based and should ideally commence with educating adolescents about the syndrome and the implementation of lifestyle changes (Teede et al. 2018). Essential lifestyle changes include weight management and the integration of behavioral changes such as routine exercise and diet modification (Dokras et al. 2016; Lanzo et al. 2015; Moran et al. 2009; Pasquali and Gambineri 2009). Dietary adjustments and routine exercise reduce the impact of menstrual irregularities, hyperandrogenism, insulin resistance, and body composition in affected individuals (Harris-Glocker et al. 2010). Their mental health state such as depression, anxiety, body image perception, and disordered eating, also require monitoring (Harris-Glocker et al. 2010).

Drug administration constitutes the second step in the therapeutic approach of PCOS and it does not substitute the role of lifestyle modification and weight loss. To date, there are no Food and Drug Administration (FDA) or European Medicines Agency (EMA) approved pharmacological interventions available for PCOS treatment (Evans and Hoeger 2020). Drug therapy involves the use of antiandrogenic oral contraceptives (OCPs) for menstrual regulation and the treatment of acne and hirsutism. The combined oral contraceptives (COC) containing progesterone that are antiandrogenic (cyproterone acetate and drospirenone) and antiandrogens are widely used in managing hyperandrogenism (Hachey et al. 2020) as well as regulating the menstrual cycle (Bredella et al. 2013; Escobar-Morreale 2018; Horn and Geraci 2013).

Additionally, insulin-sensitizing agents like metformin are commonly used in combination with COCs to improve IR, hyperinsulinemia, and androgen concentration in PCOS adolescents (Bredella et al. 2013, Escobar-Morreale 2018, Horn and Geraci 2013). Since insulin resistance enhances anovulation and hyperandrogenemia in PCOS women, treatment with insulin sensitizers, mainly metformin, may offer a holistic therapeutic approach to PCOS (Pasquali 2018). The use of metformin in conjunction with dietary modifications and exercise may also suppress androgen levels, reduce body mass index and subcutaneous adipose tissue, lower hepatic glucose production, and aid in regulating the menstrual cycle (Escobar-Morreale 2018; Witchel et al. 2016). Additionally, treatment with flutamide, an antiandrogen, may have beneficial effects on ovulatory function and metabolic aberrations in adolescents with functional ovarian hyperandrogenism, particularly those with a history of premature pubarche (Ibáñez and de Zegher 2006). Therapeutic benefit may be maximized when flutamide is combined with metformin treatment, especially in those with a history of precocious puberty (Ibáñez and de Zegher 2006; Ibánez et al. 2008). Ovulation induction drugs like letrozole, clomiphene citrate, and gonadotrophins and assisted reproductive techniques are prescribed for the management of infertility. Appetite suppressants, insulin-sensitizing drugs, and low carbohydrate diets are advised for weight loss.

8.8.6.2 Management During Reproductive Years

Lifestyle modifications play an integral role in managing PCOS in women of reproductive age. Physical activity and weight loss are recommended for overweight and/or obese women (Artini et al. 2018). A 5–10% reduction in weight can effectively restore the menstrual cycle and ovulation (Balen et al. 2006; Kiddy et al. 1992) as well as improve the pregnancy outcome rates following the use of ovulation-inducing agents (Crosignani et al. 2003). Since IR is implicated in PCOS, metformin is widely prescribed to improve ovulation in PCOS women (Morley et al. 2017). Managing hyperandrogenism is of utmost importance when taking care of PCOS women. In patients who do not wish to fall pregnant, OCPs are commonly recommended (Williams et al. 2016; Pasquali 2018), while estrogen–progestin combination pills aid in managing both hormonal and hyperandrogenism (Vrbikova and Cibula 2005). OCPs improve hirsutism, acne, and regulate menstrual cycles whereas cosmetic treatments such as depilation, epilation, or laser hair removal can be employed instead of OCPs for managing mild hirsutism (Pasquali and Gambineri 2014).

Additionally, antiandrogens such as spironolactone and finasteride, can improve hirsutism and reduce excessive hair growth in women, as spironolactone works as an androgen receptor blocker and finasteride as a 5-alpha-reductase inhibitor that blocks endogenous steroids (Pasquali 2018). However, antiandrogen treatment should be administered in conjunction with contraceptives due to their teratogenic effects and potential to feminize male fetuses (Pasquali 2018; Fauser et al. 2012). Letrozole in combination with Metformin is preferred over Clomiphene citrate as a first-line oral ovulation induction agent (Teede et al. 2018). Parenteral gonadotrophins are used as second-line therapy. Laparoscopic ovarian drilling should be considered as second-line therapy in the case of clomiphene resistance (Fernandez et al. 2011). IVF is considered in cases of other causes of infertility or where ovulation induction has failed (Costello et al. 2019). However, careful consideration should be given to the use of antagonist protocols to avoid OHSS.

8.8.6.3 Management in Perimenopause, Menopause, and Postmenopause

The participation of a multidisciplinary team comprising a medical endocrinologist, cardiologist, and dietician should also be a primary concern. Treatment goals should include lifestyle changes directed at weight loss and management of metabolic abnormalities to reduce cardiometabolic complications. Obesity and hyperinsulinism can be treated with Metformin and other emerging weight-loss drugs (Millán-de-Meer et al. 2023). Short courses of OCs containing progesterone along with antiandrogenic compounds (cyproterone acetate and drospirenone) and spironolactone can be utilized in the treatment of hyperandrogenism and irregular menstruation (Pasquali and Gambineri 2014). During the phase of menopause, hormone replacement treatment (HRT) with antiandrogenic properties and weight-promoting progesterone should be taken into consideration following the screening of HRT-related risk factors (Langer et al. 2021). Regular screening should be performed for endometrial hyperplasia and cancer.

8.8.7 Novel and Emerging Treatments

8.8.7.1 Statins

Affected women with PCOS commonly display signs of dyslipidemia or are at risk of abnormal lipid metabolism as described above in the pathophysiology of PCOS. Statins, such as pravastatin, atorvastatin, simvastatin, fluvastatin, and rosuvastatin, have been shown to reduce the synthesis of steroids, which improves dyslipidemia and features of androgen excess (Chen et al. 2021).

8.8.7.2 Insulin-Sensitizing Agents

While Metformin is the most widely used insulin-sensitizing agent, it has adverse gastrointestinal side effects like nausea, vomiting, and diarrhea. This has motivated the search for alternative insulin-sensitizing agents for managing insulin resistance in PCOS, including sodium glucose co-transporter inhibitors, and incretin mimetics.

8.8.7.2.1 Sodium Glucose Co-transporters (SGLT1 and SGLT2)

The sodium glucose co-transporters viz., SGLT1 and SGLT2 inhibitors are a recent class of drugs used in the management of type 2 diabetes (Sinha and Ghosal 2022). Intestinal glucose uptake is mediated by SGLT1 whereas in the kidneys, glucose is regulated by SGLT2 (Azizogli et al. 2023). Both co-transporters are also clinically useful in weight loss and also enhances insulin sensitivity, indicative of their potential as alternatives to insulin-sensitizing drugs (Lempesis et al. 2023). Preliminary clinical trial findings highlight the clinical utility of SGLT2 inhibitors as novel therapeutics, however, their use in PCOS treatments is not yet approved (Sinha and Ghosal 2022).

8.8.7.2.2 Incretin Mimetics

Incretins are hormones synthesized in the gut in response to food, and are reported to increase weight loss and improve insulin resistance (Cena et al. 2020). These drugs stimulate glucose-dependent insulin release, and include glucose-dependent insulinotropic peptide (GIP) and glucagon-like peptide 1 (GLP-1) (Baggio and Drucker 2007).

8.8.7.2.3 Inositol

Myo-inositol (MI) and D-chiro-inositol (DCI) are stereoisomers of inositol and are hexahydroxycyclohexanes which share the same molecular formula as glucose (Kalra et al. 2016). They are integral as secondary messengers of insulin, and lower levels exacerbate insulin resistance (Croze and Soulage 2013). The use of myoinositol and D-chiro-inositol (MDI) treatment to restore deficient MI and DCI levels, may mitigate metabolic, ovulatory, and hyperandrogenic aspects of PCOS (Kalra et al. 2016).

8.8.7.3 Calcium and Vitamin D Supplements

Physiologically, vitamin D is important in reproduction as it modifies AMH signaling which influences ovarian follicular development and luteinization, as well as mediates FSH sensitivity and progesterone synthesis (Singh et al. 2023). Reduced 25-hydroxyvitamin D (25(OH)D) levels may compound menstrual anomalies, obesity, insulin resistance, hyperandrogenism, and infertility. Vitamin D therapy can also reduce unusual elevations in serum AMH levels which subsequently improves folliculogenesis in vitamin-D-deficient PCOS patients (Irani et al. 2014). Metformin administered in conjunction with a calcium/vitamin D supplement restores menstrual consistency, follicular maturation, insulin sensitivity, and dyslipidemia (Shojaeian et al. 2019).

8.9 Conclusion

This narrative review highlights variations in the prevalence of PCOS due to a global non-standardized diagnostic criterion. It recommends the use of the Rotterdam diagnostic criteria and that defined phenotypes are recognized. Of note, the hypothalamic–pituitary axis is dysregulated in PCOS and thus results in androgen excess, hyperinsulinism, and polycystic ovarian morphology. Notably, this chapter outlines different causative mechanisms such as familial inheritance that may predispose a dysregulation of the ovarian-pituitary and hypothalamic feedback. Clinical care is tailored to the presentation in the reproductive phase of diagnosis and therapy directed to the presenting symptoms. A multidisciplinary team approach of care to management is vital to prevent long-term cardiometabolic complications. The cornerstone of therapy is based on lifestyle modification of weight loss and exercise combined with pharmacological therapy. Due to the multifactorial nature of this disorder, healthcare institutions, and governmental departments of health should make provisions to raise PCOS awareness relating to its symptoms and complications, particularly in low-income communities and areas with limited access to healthcare.

References

Adams J, Polson D, Abdulwahid N, Morris D, Franks S, Mason H, Tucker M, Price J, Jacobs H (1985) Multifollicular ovaries: clinical and endocrine features and response to pulsatile gonadotropin releasing hormone. Lancet 326:1375–1379

Ajmal N, Khan SZ, Shaikh R (2019) Polycystic ovary syndrome (PCOS) and genetic predisposition: a review article. Eur J Obstet Gynecol Reprod Biol: X 3:100060

Al Khaduri M, Al Farsi Y, Al Najjar TAA, Gowri V (2014) Hospital-based prevalence of polycystic ovarian syndrome among Omani women. Middle East Fertility Soc J 19:135–138

Alberti KG, Eckel RH, Grundy SM, Zimmet PZ, Cleeman JI, Donato KA, Fruchart J-C, James WPT, Loria CM, Smith JR (2009) Harmonizing the metabolic syndrome: a joint interim statement of the international diabetes federation task force on epidemiology and prevention; national heart, lung, and blood institute; American heart association; world heart federation; international atherosclerosis society; and international association for the study of obesity. Circulation 120:1640–1645

Alchami A, O'Donovan O, Davies M (2015) PCOS: diagnosis and management of related infertility. Obstet Gynaecol Reproductive Med 25:279–282

Ali AT (2015) Polycystic ovary syndrome and metabolic syndrome. Ceska gynekologie 80:279–289

Ali AT, Al-Ani O, Al-Ani F, Guidozzi F (2022) Polycystic ovary syndrome and metabolic disorders: a review of the literature. Afr J Reprod Health 26:89–99

Alraddadi SM, Borzangi KA, Almuher EM, Albaik AF, Aljawad LA, Shaqroon AA, Alhazmi J (2018) Prevalence of polycystic ovarian syndrome with associated risk factors and impact on well-being among women in reproductive aged (18–45) years in Al-Madinah 2017. World J Pharm Res 7:231–247

Alsamarai S, Adams JM, Murphy MK, Post MD, Hayden DL, Hall JE, Welt CK (2009) Criteria for polycystic ovarian morphology in polycystic ovary syndrome as a function of age. J Clin Endocrinol Metab 94:4961–4970

Alur-Gupta S, Dokras A, Cooney LG (2024) Management of PCOS must include assessment and treatment of mental health symptoms. Fertil Steril 121:384

Amer SAK, Li TC, Cooke ID (2003) A prospective dose-finding study of the amount of thermal energy required for laparoscopic ovarian diathermy. Hum Reprod 18(8):1693–1698

American College of Obstetricians and Gynecologists (2010) ACOG practice bulletin no. 110: noncontraceptive uses of hormonal contraceptives. Obstet Gynecol 115:206–218

American College of Obstetricians and Gynecologists (2018) ACOG practice bulletin no. 194: polycystic ovary syndrome. Obstet Gynecol 131:e157–e171

Andrade VHLD, Mata A, Borges RS, Costa-Silva DR, Martins LM, Ferreira PMP, Cunha-Nunes LC, Silva BBD (2016) Current aspects of polycystic ovary syndrome: a literature review. Rev Assoc Med Bras 62:867–871

Armar NA, Mcgarrigle HH, Honour J, Holownia P, Jacobs HS, Lachelin GC (1990) Laparoscopic ovarian diathermy in the management of anovulatory infertility in women with polycystic ovaries: endocrine changes and clinical outcome. Fertil Steril 53:45–49

Artini PG, Monti M, Matteucci C, Valentino V, Cristello F, Genazzani AR (2006) Vascular endothelial growth factor and basic fibroblast growth factor in polycystic ovary syndrome during controlled ovarian hyperstimulation. Gynecol Endocrinol 22:465–470

Artini PG, Ruggiero M, Parisen Toldin MR, Monteleone P, Monti M, Cela V, Genazzani AR (2009) Vascular endothelial growth factor and its soluble receptor in patients with polycystic ovary syndrome undergoing IVF. Hum Fertil 12:40–44

Artini PG, Obino MER, Sergiampietri C, Pinelli S, Papini F, Casarosa E, Cela V (2018) PCOS and pregnancy: a review of available therapies to improve the outcome of pregnancy in women with polycystic ovary syndrome. Exp Rev Endocrinol Metab 13:87–98

Ashraf S, Nabi M, Rashid F, Amin S (2019) Hyperandrogenism in polycystic ovarian syndrome and role of CYP gene variants: a review. Egyptian J Med Human Genet 20:1–10

Asunción M, Calvo RM, San Millán JL, Sancho J, Avila S, Escobar-Morreale HCF (2000) A prospective study of the prevalence of the polycystic ovary syndrome in unselected Caucasian women from Spain. J Clin Endocrinol Metabol 85:2434–2438

Aversa A, La Vignera S, Rago R, Gambineri A, Nappi RE, Calogero AE, Ferlin A (2020) Fundamental concepts and novel aspects of polycystic ovarian syndrome: expert consensus resolutions. Front Endocrinol 11:516

Azizogli AR, Vitti MR, Mishra R, Osorno L, Heffernan C, Kumar VA (2023) Comparison of SGLT1, SGLT2, and dual inhibitor biological activity in treating type 2 diabetes mellitus. Adv Therapeut 6:2300143

Azziz R, Carmina E, Sawaya ME (2000) Idiopathic hirsutism. Endocr Rev 21:347–362

Azziz R, Ehrmann D, Legro RS, Whitcomb RW, Hanley R, Fereshetian AG, O'keefe M, Ghazzi MN, Group

P T S (2001) Troglitazone improves ovulation and hirsutism in the polycystic ovary syndrome: a multicenter, double blind, placebo-controlled trial. J Clin Endocrinol Metabol 86:1626–1632
Azziz R, Sanchez L, Knochenhauer E, Moran C, Lazenby J, Stephens K, Taylor K, Boots L (2004) Androgen excess in women: experience with over 1000 consecutive patients. J Clin Endocrinol Metabol 89:453–462
Azziz R, Carmina E, Dewailly D, Diamanti-Kandarakis E, Escobar-Morreale HF, Futterweit W, Janssen OE, Legro RS, Norman RJ, Taylor AE (2006) Criteria for defining polycystic ovary syndrome as a predominantly hyperandrogenic syndrome: an androgen excess society guideline. J Clin Endocrinol Metabol 91:4237–4245
Azziz R, Carmina E, Dewailly D, Diamanti-Kandarakis E, Escobar-Morreale HF, Futterweit W, Janssen OE, Legro RS, Norman RJ, Taylor AE (2009) Task force on the phenotype of the polycystic ovary syndrome of the androgen excess and PCOS society. The androgen excess and PCOS society criteria for the polycystic ovary syndrome: the complete task force report. Fertil Steril 91:456–488
Azziz R, Carmina E, Chen Z, Dunaif A, Laven JS, Legro RS, Lizneva D, Natterson-Horowtiz B, Teede HJ, Yildiz BO (2016) Polycystic ovary syndrome. Nat Rev Dis Primers 2:16057
Badawy A, Elnashar A (2011) Treatment options for polycystic ovary syndrome. Int J Women's Health 3:25–35
Baggio LL, Drucker DJ (2007) Biology of incretins: GLP-1 and GIP. Gastroenterology 132:2131–2157
Balen A (2004) The pathophysiology of polycystic ovary syndrome: trying to understand PCOS and its endocrinology. Best Pract Res Clin Obstet Gynaecol 18:685–706
Balen AH, Dresner M, Scott EM, Drife JO (2006) Should obese women with polycystic ovary syndrome receive treatment for infertility? vol 332. Br Med J Publish Group, p 434
Balen A, Homburg R, Franks S (2009) Defining polycystic ovary syndrome. British medical journal publishing. Group 338:a2968
Balen AH, Morley LC, Misso M, Franks S, Legro RS, Wijeyaratne CN, Stener-Victorin E, Fauser BC, Norman RJ, Teede H (2016) The management of anovulatory infertility in women with polycystic ovary syndrome: an analysis of the evidence to support the development of global WHO guidance. Hum Reprod Update 22:687–708
Barber TM, Mccarthy M, Wass J, Franks S (2006) Obesity and polycystic ovary syndrome. Clin Endocrinol 65:137–145
Barber TM, Hanson P, Weickert MO, Franks S (2019) Obesity and polycystic ovary syndrome: implications for pathogenesis and novel management strategies. Clin Med Insights: Reproductive Health 13:1179558119874042
Baskind NE, Balen AH (2016) Hypothalamic–pituitary, ovarian and adrenal contributions to polycystic ovary syndrome. Best Pract Res Clin Obstet Gynaecol 37:80–97
Bharathi RV, Swetha S, Neerajaa J, Madhavica JV, Janani DM, Rekha S, Ramya S, Usha B (2017) An epidemiological survey: effect of predisposing factors for PCOS in Indian urban and rural population. Middle East Fertility Soc J 22:313–316
Boomsma C, Eijkemans M, Hughes E, Visser G, Fauser B, Macklon N (2006) A meta-analysis of pregnancy outcomes in women with polycystic ovary syndrome. Hum Reprod Update 12:673–683
Bordewijk EM, Ng KYB, Rakic L, Mol BWJ, Brown J, Crawford TJ, Van Wely M (2020) Laparoscopic ovarian drilling for ovulation induction in women with anovulatory polycystic ovary syndrome. Cochrane Database Syst Rev 2:CD001122
Bozdag G, Mumusoglu S, Zengin D, Karabulut E, Yildiz BO (2016) The prevalence and phenotypic features of polycystic ovary syndrome: a systematic review and meta-analysis. Hum Reprod 31:2841–2855
Brakta S, Lizneva D, Mykhalchenko K, Imam A, Walker W, Diamond MP, Azziz R (2017) Perspectives on polycystic ovary syndrome: is polycystic ovary syndrome research underfunded? J Clin Endocrinol Metabol 102:4421–4427
Bredella MA, Mcmanus S, Misra M (2013) Impact of metformin monotherapy versus metformin with oestrogen-progesterone on lipids in adolescent girls with polycystic ovarian syndrome. Clin Endocrinol 79:199–203
Brettenthaler N, de Geyter C, Huber PR, Keller U (2004) Effect of the insulin sensitizer pioglitazone on insulin resistance, hyperandrogenism, and ovulatory dysfunction in women with polycystic ovary syndrome. J Clin Endocrinol Metabol 89:3835–3840
Broekmans F, Knauff E, Valkenburg O, Laven J, Eijkemans M, Fauser B (2006) PCOS according to the Rotterdam consensus criteria: change in prevalence among WHO-II anovulation and association with metabolic factors. BJOG Int J Obstet Gynaecol 113:1210–1217
Brown ZA, Louwers YV, Fong SL, Valkenburg O, Birnie E, de Jong FH, Fauser BC, Laven JS (2011) The phenotype of polycystic ovary syndrome ameliorates with aging. Fertil Steril 96:1259–1265
Bulsara JP, Patel P, Soni A, Acharya S (2021) A review on brief insight into polycystic ovarian syndrome. Endocrine Metabolic Sci 100085
Burger HG (2002) Androgen production in women. Fertil Steril 77(Suppl 4):S3–S5
Carmina E (2004) Diagnosis of polycystic ovary syndrome: from NIH criteria to ESHRE-ASRM guidelines. Minerva Ginecol 56:1–6
Casas RS, Chuang CH (2020) Menstruation and secondary amenorrhea. Sex-and gender-based women's health: a practical guide for primary care, pp 71–81
Cena H, Chiovato L, Nappi RE (2020) Obesity, polycystic ovary syndrome, and infertility: a new avenue for GLP-1 receptor agonists. J Clin Endocrinol Metabol 105:e2695–e2709

Chan JL, Kar S, Vanky E, Morin-Papunen L, Piltonen T, Puurunen J, Tapanainen JS, Maciel GAR, Hayashida SAY, Soares JR (2017) Racial and ethnic differences in the prevalence of metabolic syndrome and its components of metabolic syndrome in women with polycystic ovary syndrome: a regional cross-sectional study. Am J Obstet Gynecol 217:189. e1–189. e8

Chen W, Pang Y (2021) Metabolic syndrome and PCOS: pathogenesis and the role of metabolites. Metabolites 11:869–869

Chen H, Sullivan G, Yue LQ, Katz A, Quon MJ (2003) QUICKI is a useful index of insulin sensitivity in subjects with hypertension. Am J Physiol-Endocrinol Metab 284:E804–E812

Chen H, Sullivan G, Quon MJ (2005) Assessing the predictive accuracy of QUICKI as a surrogate index for insulin sensitivity using a calibration model. Diabetes 54:1914–1925

Chen Z-J, Zhao H, He L, Shi Y, Qin Y, Shi Y, Li Z, You L, Zhao J, Liu J (2011) Genome-wide association study identifies susceptibility loci for polycystic ovary syndrome on chromosome 2p16. 3, 2p21 and 9q33. 3. Nat Genet 43:55–59

Chen J, Huang C, Zhang T, Gong W, Deng X, Liu H, Liu J, Guo Y (2021) The effects of statins on hyperandrogenism in women with polycystic ovary syndrome: a systematic review and meta-analysis of randomized controlled trials. Reprod Biol Endocrinol 19:1–11

Chen AY, Seifer DB, Tal R (2022) The role of Angiogenic factor dysregulation in the pathogenesis of polycystic ovarian syndrome. Polycystic Ovary Syndrome: Curr Emerg Concepts 449–487

Choudhury AA, Rajeswari VD (2021) Gestational diabetes mellitus-A metabolic and reproductive disorder. Biomed Pharmacother 143:112183

Christ JP, Cedars MI (2023) Current guidelines for diagnosing PCOS. Diagnostics 13:1113

Christian RC, Dumesic DA, Behrenbeck T, Oberg AL, Sheedy PF, Fitzpatrick LA (2003) Prevalence and predictors of coronary artery calcification in women with polycystic ovary syndrome. J Clin Endocrinol Metabol 88:2562–2568

Collée J, Mawet M, Tebache L, Nisolle M, Brichant, G (2021) Polycystic ovarian syndrome and infertility: overview and insights of the putative treatments. Gynecological Endocrinology, 37, 869–874.

Costello MF, Garad RM, Hart R, Homer H, Johnson L, Jordan C, Mocanu E, Qiao J, Rombauts L, Teede HJ, Vanky E, Venetis CA, Ledger WL (2019) A review of second- and third-line infertility treatments and supporting evidence in women with polycystic ovary syndrome. Med Sci (Basel) 7

Cowan S, Lim S, Alycia C, Pirotta S, Thomson R, Gibson-Helm M, Blackmore R, Naderpoor N, Bennett C, Ee C, Rao V, Mousa A, Alesi S, Moran L (2023) Lifestyle management in polycystic ovary syndrome—beyond diet and physical activity. BMC Endocr Disord 23:14

Crosignani PG, Colombo M, Vegetti W, Somigliana E, Gessati A, Ragni G (2003) Overweight and obese anovulatory patients with polycystic ovaries: parallel improvements in anthropometric indices, ovarian physiology and fertility rate induced by diet. Hum Reprod 18:1928–1932

Croze ML, Soulage CO (2013) Potential role and therapeutic interests of myo-inositol in metabolic diseases. Biochimie 95:1811–1827

Curi DD, Fonseca AM, Marcondes JAM, Almeida JAM, Bagnoli VR, Soares JR, Baracat EC (2012) Metformin versus lifestyle changes in treating women with polycystic ovary syndrome. Gynecol Endocrinol 28:182–185

Das M, Son WY (2023) In vitro maturation (IVM) of human immature oocytes: is it still relevant? Reprod Biol Endocrinol 21:110

Davis SR, Knight S, White V, Claridge C, Davis BJ, Bell R (2002) Preliminary indication of a high prevalence of polycystic ovary syndrome in indigenous Australian women. Gynecol Endocrinol 16:443–446

De Leo V, Lanzetta D, D'antona D, La Marca A, Morgante G (1998) Hormonal effects of flutamide in young women with polycystic ovary syndrome. J Clin Endocrinol Metabol 83:99–102

De Sousa S (2016) Metabolic syndrome, diet and exercise. Best practice & research. Clin Obstet Gynaecol 37:140–151

Debras E, Fernandez H, Neveu M, Deffieux X, Capmas P (2019) Ovarian drilling in polycystic ovary syndrome: long term pregnancy rate. Eur J Obstet Gynecol Reprod Biol: X 4:100093

Deswal R, Nanda S, Ghalaut VS, Roy PS, Dang AS (2019) Cross-sectional study of the prevalence of polycystic ovary syndrome in rural and urban populations. Int J Gynecol Obstet 146:370–379

Deswal R, Narwal V, Dang A, Pundir CS (2020) The prevalence of polycystic ovary syndrome: a brief systematic review. J Human Reprod Sci 13:261

Dewailly D, Lujan ME, Carmina E, Cedars MI, Laven J, Norman RJ, Escobar-Morreale HF (2014) Definition and significance of polycystic ovarian morphology: a task force report from the androgen excess and polycystic ovary syndrome society. Hum Reprod Update 20:334–352

Di Pietro M, Pascuali N, Parborell F, Abramovich D (2018) Ovarian angiogenesis in polycystic ovary syndrome. Reproduction 155:R199–R209

Diamanti-Kandarakis E, Dunaif A (2012) Insulin resistance and the polycystic ovary syndrome revisited: an update on mechanisms and implications. Endocr Rev 33:981–1030

Diamanti-Kandarakis E, Mitrakou A, Raptis S, Tolis G, Duleba AJ (1998) The effect of a pure antiandrogen receptor blocker, flutamide, on the lipid profile in the polycystic ovary syndrome. J Clin Endocrinol Metabol 83:2699–2705

Diamanti-Kandarakis E, Kandarakis H, Legro RS (2006) The role of genes and environment in the etiology of PCOS. Endocrine 30:19–26

Diamanti-Kandarakis E, Katsikis I, Piperi C, Kandaraki E, Piouka A, Papavassiliou AG, Panidis D (2008) Increased serum advanced glycation end-products is a distinct finding in lean women with polycystic ovary syndrome (PCOS). Clin Endocrinol 69:634–641

Ding T, Hardiman PJ, Petersen I, Wang F-F, Qu F, Baio G (2017) The prevalence of polycystic ovary syndrome in reproductive-aged women of different ethnicity: a systematic review and meta-analysis. Oncotarget 8:96351

Ding D-C, Chen W, Wang J-H, Lin S-Z (2018) Association between polycystic ovarian syndrome and endometrial, ovarian, and breast cancer: a population-based cohort study in Taiwan. Medicine 97:e12608

Dinsdale NL, Crespi BJ (2021) Endometriosis and polycystic ovary syndrome are diametric disorders. Evol Appl 14:1693–1715

Dinsdale N, Nepomnaschy P, Crespi B (2021) The evolutionary biology of endometriosis. Evol Med Public Health 9:174–191

Doherty DA, Newnham JP, Bower C, Hart R (2015) Implications of polycystic ovary syndrome for pregnancy and for the health of offspring. Obstet Gynecol 125:1397–1406

Dokras A, Sarwer DB, Allison KC, Milman L, Kris-Etherton PM, Kunselman AR, Stetter CM, Williams NI, Gnatuk CL, Estes SJ (2016) Weight loss and lowering androgens predict improvements in health-related quality of life in women with PCOS. J Clin Endocrinol Metabol 101:2966–2974

Donma O, Ozturk EZ, Ekmekci OB, Ocal P, Topcu B, Ekmekci H, Gungor ZB, Sonmez H, Donma MM (2019) Molecular links between angiogenesis and inflammation in polycystic ovary syndrome. Medicina 50:163–167

Draft Summary and Recommendations of International Evidenced-Based Guideline for the Assessment and Management of Polycystic Ovary Syndrome (2018) Draft Summary and Recommendations of International Evidenced-Based Guideline for the Assessment and Management of Polycystic Ovary Syndrome [Online]. http://pcos-cre.edu.au/wp-content/uploads/2018/02/PCOS-guideline-summary-and-recommendations_PC-DRAFT_FINAL-V2-050218-1.pdf. Accessed 7 Aug 2023

Drummond AE (2006) The role of steroids in follicular growth. Reprod Biol Endocrinol 4:16

Dumesic DA, Meldrum DR, Katz-Jaffe MG, Krisher RL, Schoolcraft WB (2015) Oocyte environment: follicular fluid and cumulus cells are critical for oocyte health. Fertil Steril 103:303–316

Duncan M, Singh B, Wise PH, Carter G, Alaghband-Zadeh J (1995) A simple measure of insulin resistance. Lancet 346:120–121

Elhassan YS, Idkowiak J, Smith K, Asia M, Gleeson H, Webster R, Arlt W, O'Reilly MW (2018) Causes, patterns, and severity of androgen excess in 1205 consecutively recruited women. J Clin Endocrinol Metabol 103:1214–1223

Elizur SE, Tulandi T (2008) Drugs in infertility and fetal safety. Fertil Steril 89:1595–1602

Elting MW, Korsen TJ, Rekers-Mombarg LT, Schoemaker J (2000) Women with polycystic ovary syndrome gain regular menstrual cycles when ageing. Hum Reprod 15:24–28

Engmann L, Jin S, Sun F, Legro RS, Polotsky AJ, Hansen KR, Coutifaris C, Diamond MP, Eisenberg E, Zhang H, Santoro N (2017) Racial and ethnic differences in the polycystic ovary syndrome metabolic phenotype. Am J Obstet Gynecol 216:493.e1–493.e13

Escobar-Morreale HF (2018) Polycystic ovary syndrome: definition, aetiology, diagnosis and treatment. Nat Rev Endocrinol 14:270–284

Escobar-Morreale HF, San Millán JL (2007) Abdominal adiposity and the polycystic ovary syndrome. Trends Endocrinol Metab 18:266–272

Essah PA, Wickham EP, Nestler JE (2007) The metabolic syndrome in polycystic ovary syndrome. Clin Obstet Gynecol 50:205–225

Evans A, Hoeger KM (2020) Polycystic ovary syndrome in adolescence: toward a better diagnosis and treatment. Curr Opin Endocrine Metab Res 12:105–111

Fairley DH, Taylor A (2003) Anovulation. Bmj 327:546–549

Farquhar C, Lee O, Toomath R, Jepson R (2003) Spironolactone versus placebo or in combination with steroids for hirsutism and/or acne. Cochrane Database Syst Rev Cd000194

Fauser BC, Tarlatzis BC, Rebar RW, Legro RS, Balen AH, Lobo R, Carmina E, Chang J, Yildiz BO, Laven JS (2012) Consensus on women's health aspects of polycystic ovary syndrome (PCOS): the Amsterdam ESHRE/ASRM-sponsored 3rd PCOS consensus workshop group. Fertil Steril 97:28–38. e25

Fernandez H, Morin-Surruca M, Torre A, Faivre E, Deffieux X, Gervaise A (2011) Ovarian drilling for surgical treatment of polycystic ovarian syndrome: a comprehensive review. Reprod Biomed Online 22:556–568

Ferrara N, Frantz G, Lecouter J, Dillard-Telm L, Pham T, Draksharapu A, Giordano T, Peale F (2003) Differential expression of the angiogenic factor genes vascular endothelial growth factor (VEGF) and endocrine gland-derived VEGF in normal and polycystic human ovaries. Am J Pathol 162:1881–1893

Foong S, Session D, Thornhill A, Tummon I, Dumesic D (2004) Transvaginal ultrasound detection of multifollicular ovaries in non-hirsute ovulatory women. Ultrasound Obstet Gynecol 23:183–187

Fruzzetti F, de Lorenzo D, Ricci C, Teti G (1995) Ovarian influence on adrenal androgen secretion in polycystic ovary syndrome. Fertil Steril 63:734–741

Ganie MA, Vasudevan V, Wani IA, Baba MS, Arif T, Rashid A (2019) Epidemiology, pathogenesis, genetics & management of polycystic ovary syndrome in India. Indian J Med Res 150:333

Gao Y, Liu H, Qiao L, Liang J, Yao H, Lin X, Gao Y (2023) Study of burden in polycystic ovary syndrome at global, regional, and national levels from 1990 to 2019. Healthcare. MDPI, p 562

Garad R, Teede HJ, Moran L (2011) An evidence-based guideline for polycystic ovary syndrome. Austral Nursing J 19:30–33

Garber AJ, Duncan TG, Goodman AM, Mills DJ, Rohlf JL (1997) Efficacy of metformin in type II diabetes: results of a double-blind, placebo-controlled, dose-response trial. Am J Med 103:491–497

Geldenhuys J, Rossouw TM, Lombaard HA, Ehlers MM, Kock MM (2018) Disruption in the regulation of immune responses in the placental subtype of preeclampsia. Front Immunol 9:1659

Gibson-Helm M, Teede H, Dunaif A, Dokras A (2017) Delayed diagnosis and a lack of information associated with dissatisfaction in women with polycystic ovary syndrome. J Clin Endocrinol Metabol 102:604–612

Gjönnaess H (1984) Polycystic ovarian syndrome treated by ovarian electrocautery through the laparoscope. Fertil Steril 41:20–25

Glendining KA, Campbell RE (2023) Recent advances in emerging PCOS therapies. Curr Opin Pharmacol 68:102345

Glintborg D, Mumm H, Hougaard D, Ravn P, Andersen M (2010) Ethnic differences in Rotterdam criteria and metabolic risk factors in a multiethnic group of women with PCOS studied in Denmark. Clin Endocrinol 73:732–738

Goodarzi MO, Jones MR, Li X, Chua AK, Garcia OA, Chen Y-DI, Krauss RM, Rotter JI, Ankener W, Legro RS (2012) Replication of association of DENND1A and THADA variants with polycystic ovary syndrome in European cohorts. J Med Genet 49:90–95

Greenblatt E, Casper RF (1987) Endocrine changes after laparoscopic ovarian cautery in polycystic ovarian syndrome. Am J Obstet Gynecol 156:279–285

Grundy SM (2002) Obesity, metabolic syndrome, and coronary atherosclerosis. Am Heart Assoc

Gu Y, Zhou G, Zhou F, Wu Q, Ma C, Zhang Y, Ding J, Hua K (2022) Life modifications and PCOS: old story but new tales. Front Endocrinol 13:808898

Guastella E, Longo RA, Carmina E (2010) Clinical and endocrine characteristics of the main polycystic ovary syndrome phenotypes. Fertil Steril 94:2197–2201

Gupta AK, Talukder M, Shemer A (2024) Efficacy and safety of low-dose oral minoxidil in the management of androgenetic alopecia. Expert opinion on pharmacotherapy

Hachey LM, Kroger-Jarvis M, Pavlik-Maus T, Leach R (2020) Clinical implications of polycystic ovary syndrome in adolescents. Nurs Womens Health 24:115–126

Hale GE, Robertson DM, Burger HG (2014) The perimenopausal woman: endocrinology and management. J Steroid Biochem Mol Biol 142:121–131

Handberg A, Levin K, Højlund K, Beck-Nielsen H (2006) Identification of the oxidized low-density lipoprotein scavenger receptor CD36 in plasma: a novel marker of insulin resistance. Circulation 114:1169–1176

Haqq L, Mcfarlane J, Dieberg G, Smart N (2014) Effect of lifestyle intervention on the reproductive endocrine profile in women with polycystic ovarian syndrome: a systematic review and meta-analysis. Endocr Connect 3:36–46

Harborne LR, Sattar N, Norman JE, Fleming R (2005) Metformin and weight loss in obese women with polycystic ovary syndrome: comparison of doses. J Clin Endocrinol Metabol 90:4593–4598

Hardy T, De Sousa S, Norman RJ (2013) Polycystic ovary syndrome: prognosis and risk of comorbidity. Future Medicine

Harris-Glocker M, Davidson K, Kochman L, Guzick D, Hoeger K (2010) Improvement in quality of life questionnaire measures (PCOSQ) in obese adolescent females with PCOS treated with lifestyle changes and oral contraceptives, with or without metformin. Fertil Steril 93:1016

Hiam D, Moreno-Asso A, Teede HJ, Laven JS, Stepto NK, Moran LJ, Gibson-Helm M (2019) The genetics of polycystic ovary syndrome: an overview of candidate gene systematic reviews and genome-wide association studies. J Clin Med 8:1606

Higgins SP, Freemark M, Prose NS (2008) Acanthosis nigricans: a practical approach to evaluation and management. Dermatol Online J 14

Hoeger KM, Dokras A, Piltonen T (2021) Update on PCOS: consequences, challenges, and guiding treatment. J Clin Endocrinol Metabol 106:e1071–e1083

Holoch KJ, Savaris RF, Forstein DA, Miller PB, Higdon HL III, Likes CE, Lessey BA (2014) Coexistence of polycystic ovary syndrome and endometriosis in women with infertility. J Endometriosis Pelvic Pain Disorders 6:79–83

Homburg R (2009) Androgen circle of polycystic ovary syndrome. Hum Reprod 24:1548–1555

Horn M, Geraci SA (2013) Polycystic ovary syndrome in adolescents:(women's health series). South Med J 106:570–576

Hřebíček JI, Janout VR, Malinčíková J, Horáková D, Čížek LK (2002) Detection of insulin resistance by simple quantitative insulin sensitivity check index QUICKI for epidemiological assessment and prevention. J Clin Endocrinol Metabol 87:144–144

Huang Z, Yong E-L (2016) Ethnic differences: is there an Asian phenotype for polycystic ovarian syndrome? Best Pract Res Clin Obstet Gynaecol 37:46–55

Huang JY, Chian R-C, Tan SL (2010) Ovarian hyperstimulation syndrome prevention strategies: in vitro maturation. Seminars in reproductive medicine. Thieme Medical Publishers, pp 519–531

Ibáñez L, de Zegher F (2006) Low-dose flutamide-metformin therapy for hyperinsulinemic hyperandrogenism in non-obese adolescents and women. Hum Reprod Update 12:243–252

Ibánez L, Lopez-Bermejo A, Díaz M, Marcos MV, de Zegher F (2008) Metformin treatment for four years to reduce total and visceral fat in low birth weight girls with precocious pubarche. J Clin Endocrinol Metabol 93:1841–1845

Ibáñez L, Oberfield SE, Witchel S, Auchus RJ, Chang RJ, Codner E, Dabadghao P, Darendeliler F, Elbarbary NS, Gambineri A (2017) An international consortium update: pathophysiology, diagnosis, and treatment of polycystic ovarian syndrome in adolescence. Horm Res Paediatr 88:371–395

Irani M, Minkoff H, Seifer DB, Merhi Z (2014) Vitamin D increases serum levels of the soluble receptor for advanced glycation end products in women with PCOS. J Clin Endocrinol Metabol 99:E886–E890

James WD, Elston D, Berger T (2011) Andrew's diseases of the skin E-book: clinical dermatology. Elsevier Health Sciences

Jehn M, Clark JM, Guallar E (2004) Serum ferritin and risk of the metabolic syndrome in US adults. Diabetes Care 27:2422–2428

Joham AE, Piltonen T, Lujan ME, Kiconco S, Tay CT (2022) Challenges in diagnosis and understanding of natural history of polycystic ovary syndrome. Clin Endocrinol 97:165–173

Johnson T, Kaplan L, Ouyang P, Rizza R (2012) National Institutes of Health evidence-based methodology workshop on polycystic ovary syndrome (PCOS). NIH EbMW report. Bethesda, National Institutes of Health 1:1–14

Kahsar-Miller MD, Nixon C, Boots LR, Go RC, Azziz R (2001) Prevalence of polycystic ovary syndrome (PCOS) in first-degree relatives of patients with PCOS. Fertil Steril 75:53–58

Kakoly NS, Earnest A, Teede HJ, Moran LJ, Joham AE (2019) The impact of obesity on the incidence of type 2 diabetes among women with polycystic ovary syndrome. Diabetes Care 42:560–567

Kalra B, Kalra S, Sharma J (2016) The inositols and polycystic ovary syndrome. Indian J Endocrinol Metab 20:720

Katulski K, Czyzyk A, Podfigurna-Stopa A, Genazzani AR, Meczekalski B (2015) Pregnancy complications in polycystic ovary syndrome patients. Gynecol Endocrinol 31:87–91

Kauffman RP, Baker VM, Dimarino P, Gimpel T, Castracane VD (2002) Polycystic ovarian syndrome and insulin resistance in white and Mexican American women: a comparison of two distinct populations. Am J Obstet Gynecol 187:1362–1369

Kaunitz AM (2019) Abnormal uterine bleeding in reproductive-age women. JAMA 321:2126–2127

Keen MA, Shah IH, Sheikh G (2017) Cutaneous manifestations of polycystic ovary syndrome: A cross-sectional clinical study. Indian Dermatol Online J 8:104

Kettel LM, Roseff SJ, Berga SL, Mortola JF, Yen SS (1993) Hypothalamic-pituitary-ovarian response to clomiphene citrate in women with polycystic ovary syndrome. Fertil Steril 59:532–538

Khan MJ, Ullah A, Basit S (2019) Genetic basis of polycystic ovary syndrome (PCOS): current perspectives. Appl Clin Genet 12:249

Kiddy DS, Hamilton-Fairley D, Bush A, Short F, Anyaoku V, Reed MJ, Franks S (1992) Improvement in endocrine and ovarian function during dietary treatment of obese women with polycystic ovary syndrome. Clin Endocrinol 36:105–111

Kim CH, Chon SJ, Lee SH (2020) Effects of lifestyle modification in polycystic ovary syndrome compared to metformin only or metformin addition: a systematic review and meta-analysis. Sci Rep 10:7802

Kitzinger C, Willmott J (2002) 'The thief of womanhood': women's experience of polycystic ovarian syndrome. Soc Sci Med 54:349–361

Kollmann M, Klaritsch P, Martins W, Guenther F, Schneider V, Herzog S, Craciunas L, Lang U, Obermayer-Pietsch B, Lerchbaum E (2015) Maternal and neonatal outcomes in pregnant women with PCOS: comparison of different diagnostic definitions. Hum Reprod 30:2396–2403

Köşüş N, Köşüş A, Kamalak Z, Hizli D, Turhan NÖ (2012) Impact of adrenal versus ovarian androgen ratio on signs and symptoms of polycystic ovarian syndrome. Gynecol Endocrinol 28:611–614

Landay M, Huang A, Azziz R (2009) Degree of hyperinsulinemia, independent of androgen levels, is an important determinant of the severity of hirsutism in PCOS. Fertil Steril 92:643–647

Langer RD, Hodis HN, Lobo RA, Allison MA (2021) Hormone replacement therapy—where are we now? Climacteric 24:3–10

Lanzo E, Monge M, Trent M (2015) Diagnosis and management of polycystic ovary syndrome in adolescent girls. Pediatr Ann 44:e223–e230

Laven JS, Imani B, Eijkemans MJ, Fauser BC (2002) New approach to polycystic ovary syndrome and other forms of anovulatory infertility. Obstet Gynecol Surv 57:755–767

Lebbi I, Ben Temime R, Fadhlaoui A, Feki A (2015) Ovarian drilling in PCOS: is it really useful? Front Surg 2:30

Legro RS, Finegood D, Dunaif A (1998) A fasting glucose to insulin ratio is a useful measure of insulin sensitivity in women with polycystic ovary syndrome. J Clin Endocrinol Metabol 83:2694–2698

Legro RS, Arslanian SA, Ehrmann DA, Hoeger KM, Murad MH, Pasquali R, Welt CK (2013) Diagnosis and treatment of polycystic ovary syndrome: an Endocrine Society clinical practice guideline. J Clin Endocrinol Metab 98:4565–4592

Lemieux S, Lewis GF, Ben-Chetrit A, Steiner G, Greenblatt EM (1999) Correction of hyperandrogenemia by laparoscopic ovarian cautery in women with polycystic ovarian syndrome is not accompanied by improved insulin sensitivity or lipid-lipoprotein levels. J Clin Endocrinol Metabol 84:4278–4282

Lempesis IG, Apple SJ, Duarte G, Palaiodimos L, Kalaitzopoulos DR, Dalamaga M, Kokkinidis DG (2023) Cardiometabolic effects of SGLT2 inhibitors on polycystic ovary syndrome. Diabetes Metab Res Rev 39:e3682

Li L, Feng Q, Ye M, He Y, Yao A, Shi K (2017) Metabolic effect of obesity on polycystic ovary syndrome in adolescents: a meta-analysis. J Obstet Gynaecol 37:1036–1047

Lim S, Kakoly N, Tan J, Fitzgerald G, Bahri Khomami M, Joham A, Cooray S, Misso M, Norman R, Harrison C (2019) Metabolic syndrome in polycystic ovary syndrome: a systematic review, meta-analysis and metaregression. Obes Rev 20:339–352

Liu M-M, Chen X-H, Lu X-M, Wang F-F, Wang C, Liu Y, Li P-L, Du B-T, Liang S, Gong P-D (2021a) Variations in the profiles of vascular-related factors among different sub-types of polycystic ovarian syndrome in northern China. Front Endocrinol 11:527592

Liu Y, Li J, Yan Z, Liu D, Ma J, Tong N (2021b) Improvement of insulin sensitivity increases pregnancy rate in infertile PCOS women: a systemic review. Front Endocrinol 12:657889

Livadas S, Diamanti-Kandarakis E (2013) Polycystic ovary syndrome: definitions, phenotypes and diagnostic approach. Polycystic Ovary Syndrome 40:1–21

Livadas S, Macut D, Bothou C, Kuliczkowska-Płaksej J, Vryonidou A, Bjekic-Macut J, Mouslech Z, Milewicz A, Panidis D (2020) Insulin resistance, androgens, and lipids are gradually improved in an age-dependent manner in lean women with polycystic ovary syndrome: insights from a large Caucasian cohort. Hormones (Athens) 19:531–539

Lizneva D, Gavrilova-Jordan L, Walker W, Azziz R (2016) Androgen excess: investigations and management. Best Pract Res Clin Obstet Gynaecol 37:98–118

Llurba E, Crispi F, Verlohren S (2015) Update on the pathophysiological implications and clinical role of angiogenic factors in pregnancy. Fetal Diagn Ther 37:81–92

Lo JC, Feigenbaum SL, Yang J, Pressman AR, Selby JV, Go AS (2006) Epidemiology and adverse cardiovascular risk profile of diagnosed polycystic ovary syndrome. J Clin Endocrinol Metabol 91:1357–1363

Ma T, Cui P, Tong X, Hu W, Shao LR, Zhang F, Li X, Feng Y (2018) Endogenous ovarian angiogenesis in polycystic ovary syndrome-like rats induced by low-frequency electro-acupuncture: the CLARITY three-dimensional approach. Int J Mol Sci 19:3500

Madnani N, Khan K, Chauhan P, Parmar G (2013) Polycystic ovarian syndrome. Indian J Dermatol Venereol Leprol 79:310

Magoffin DA (2007) Ovarian steroidogenic abnormalities in the polycystic ovary syndrome. Androgen excess disorders in women: polycystic ovary syndrome and other disorders, pp 203–211

Maharaj S, Amod A (2009) Polycystic ovary syndrome. J Endocrinol Metab Diabet South Afr 14:86–95

Malik SM, Traub ML (2012) Defining the role of bariatric surgery in polycystic ovarian syndrome patients. World J Diabetes 3:71

Mani H, Davies MJ, Bodicoat DH, Levy MJ, Gray LJ, Howlett TA, Khunti K (2015) Clinical characteristics of polycystic ovary syndrome: investigating differences in White and South Asian women. Clin Endocrinol 83:542–549

Mann T, Tomiyama AJ, Westling E, Lew A-M, Samuels B, Chatman J (2007) Medicare's search for effective obesity treatments: diets are not the answer. Am Psychol 62:220

March WA, Moore VM, Willson KJ, Phillips DI, Norman RJ, Davies MJ (2010) The prevalence of polycystic ovary syndrome in a community sample assessed under contrasting diagnostic criteria. Hum Reprod 25:544–551

Marchesan LB, Ramos RB, Spritzer PM (2021) Metabolic features of women with polycystic ovary syndrome in Latin America: a systematic review. Front Endocrinol 1364

Markopoulos MC, Rizos D, Valsamakis G, Deligeoroglou E, Grigoriou O, Chrousos GP, Creatsas G, Mastorakos G (2011) Hyperandrogenism in women with polycystic ovary syndrome persists after menopause. J Clin Endocrinol Metabol 96:623–631

Marshall JC, Dunaif A (2012) Should all women with PCOS be treated for insulin resistance? Fertil Steril 97:18–22

Mastorakos G, Lambrinoudaki I, Creatsas G (2006) Polycystic ovary syndrome in adolescents: current and future treatment options. Pediatr Drugs 8:311–318

Matthews DR, Hosker JP, Rudenski AS, Naylor BA, Treacher DF, Turner RC (1985) Homeostasis model assessment: insulin resistance and beta-cell function from fasting plasma glucose and insulin concentrations in man. Diabetologia 28:412–419

Mayer SB, Evans WS, Nestler JE (2015) Polycystic ovary syndrome and insulin: our understanding in the past, present and future. Womens Health 11:137–149

Meczekalski B, Niwczyk O, Kostrzak A, Maciejewska-Jeske M, Bala G, Szeliga A (2023) PCOS in adolescents—ongoing riddles in diagnosis and treatment. J Clin Med 12:1221

Mendoza N, Simoncini T, Genazzani AD (2014) Hormonal contraceptive choice for women with PCOS: a systematic review of randomized trials and observational studies. Gynecol Endocrinol 30:850–860

Meng Y-X, Ford ES, Li C, Quarshie A, Al-Mahmoud AM, Giles W, Gibbons GH, Strayhorn G (2007) Association of C-reactive protein with surrogate measures of insulin resistance among nondiabetic US adults: findings from national health and nutrition examination survey 1999–2002. Clin Chem 53:2152–2159

Millán-de-Meer M, Luque-Ramírez M, Nattero-Chávez L, Escobar-Morreale HF (2023) PCOS during the menopausal transition and after menopause: a systematic review and meta-analysis. Hum Reprod Update 29:741–772

Moran L, Norman RJ (2004) Understanding and managing disturbances in insulin metabolism and body weight in women with polycystic ovary syndrome. Best Pract Res Clin Obstet Gynaecol 18:719–736

Moran LJ, Pasquali R, Teede HJ, Hoeger KM, Norman RJ (2009) Treatment of obesity in polycystic ovary syndrome: a position statement of the androgen excess and polycystic ovary syndrome society. Fertil Steril 92:1966–1982

Moran C, Tena G, Moran S, Ruiz P, Reyna R, Duque X (2010a) Prevalence of polycystic ovary syndrome and related disorders in Mexican women. Gynecol Obstet Investig 69:274–280

Moran LJ, Misso ML, Wild RA, Norman RJ (2010b) Impaired glucose tolerance, type 2 diabetes and metabolic syndrome in polycystic ovary syndrome: a systematic review and meta-analysis. Hum Reprod Update 16:347–363

Moran L, Brown W, Mcnaughton S, Joham A, Teede H (2017) Weight management practices associated with PCOS and their relationships with diet and physical activity. Hum Reprod 32:669–678

Morley LC, Tang T, Yasmin E, Norman RJ, Balen AH (2017) Insulin-sensitising drugs (metformin, rosiglitazone, pioglitazone, D-chiro-inositol) for women with polycystic ovary syndrome, oligo amenorrhoea and subfertility. Cochrane Database Syst Rev 2018:CD003053

Motaghedi R, Gujral S, Sinha S, Sison C, Ten S, Maclaren NK (2007) Insulin-like growth factor binding protein-1 to screen for insulin resistance in children. Diabetes Technol Ther 9:43–51

Müderris İI, Bayram F, Şahin Y, Kelestimur F (1997) A comparison between two doses of flutamide (250 mg/d and 500 mg/d) in the treatment of hirsutism. Fertil Steril 68:644–647

Ndefo UA, Eaton A, Green MR (2013) Polycystic ovary syndrome: a review of treatment options with a focus on pharmacological approaches. Pharm Therapeut 38:336

Okoroh EM, Hooper WC, Atrash HK, Yusuf HR, Boulet SL (2012) Prevalence of polycystic ovary syndrome among the privately insured, United States, 2003-2008. Am J Obstet Gynecol 207:299. e1–299. e7

Ollila M-M, Arffman RK, Korhonen E, Morin-Papunen L, Franks S, Junttila J, Piltonen TT (2023) Women with PCOS have an increased risk for cardiovascular disease regardless of diagnostic criteria—a prospective population-based cohort study. Eur J Endocrinol 189:96–105

Ono YJ, Tanabe A, Nakamura Y, Yamamoto H, Hayashi A, Tanaka T, Sasaki H, Hayashi M, Terai Y, Ohmichi M (2014) A low-testosterone state associated with endometrioma leads to the apoptosis of granulosa cells. PLoS One 9:e115618

Oride A, Kanasaki H, Mijiddorj T, Sukhbaatar U, Ishihara T, Kyo S (2015) Regulation of kisspeptin and gonadotropin-releasing hormone expression in rat placenta: study using primary cultures of rat placental cells. Reprod Biol Endocrinol 13:1–9

Oriji VK, Onwuegbulam C (2019) Prevalence of polycystic ovary syndrome (PCOS) among infertile women attending fertility clinic at a university teaching hospital in Nigeria. J Gynecol Women's Health 15:555922

Orio JR, Palomba S, Spinelli L, Cascella T, Tauchmanovà L, Zullo F, Lombardi G, Colao A (2004) The cardiovascular risk of young women with polycystic ovary syndrome: an observational, analytical, prospective case-control study. J Clin Endocrinol Metabol 89:3696–3701

Osei K, Rhinesmith S, Gaillard T, Schuster D (2003) Is glycosylated hemoglobin A1c a surrogate for metabolic syndrome in nondiabetic, first-degree relatives of African-American patients with type 2 diabetes? J Clin Endocrinol Metabol 88:4596–4601

Osibogun O, Ogunmoroti O, Michos ED (2020) Polycystic ovary syndrome and cardiometabolic risk: opportunities for cardiovascular disease prevention. Trends Cardiovasc Med 30:399–404

Pangath M, Unnikrishnan L, Throwba PH, Vasudevan K, Jayaraman S, Li M, Iyaswamy A, Palaniyandi K, Gnanasampanthapandian D (2022) The epigenetic correlation among ovarian cancer, endometriosis and PCOS: a review. Crit Rev Oncol/Hematol 103852

Paradisi R, Fabbri R, Battaglia C, Venturoli S (2013) Ovulatory effects of flutamide in the polycystic ovary syndrome. Gynecol Endocrinol 29:391–395

Pasquali R (2018) Contemporary approaches to the management of polycystic ovary syndrome. Ther Adv Endocrinol Metab 9:123–134

Pasquali R, Gambineri A (2009) Targeting insulin sensitivity in the treatment of polycystic ovary syndrome. Expert Opin Ther Targets 13:1205–1226

Pasquali R, Gambineri A (2014) Therapy of endocrine disease: treatment of hirsutism in the polycystic ovary syndrome. Eur J Endocrinol 170:R75–R90

Phiske MM (2014) An approach to acanthosis nigricans. Indian Dermatol Online J 5:239

Pirotta S, Joham A, Grieger JA, Tay CT, Bahri-Khomami M, Lujan M, Lim SS, Moran LJ (2020) Obesity and the risk of infertility, gestational diabetes, and type 2 diabetes in polycystic ovary syndrome. Seminars in Reproductive Medicine. Thieme Medical Publishers, New York, pp 342–351

Pk HT, Unnikrishnan L, Pangath M, Vasudevan K, Jayaraman S, Li M, Iyaswamy A, Palaniyandi K, Gnanasampanthapandian D (2022) The epigenetic correlation among ovarian cancer, endometriosis and PCOS: a review. Crit Rev Oncol Hematol 180:103852

Practice Committee of the American Society for Reproductive Medicine (2013) Use of clomiphene citrate in infertile women: a committee opinion. Fertil Steril 100:341–348

Pramodh S (2020) Exploration of lifestyle choices, reproductive health knowledge, and polycystic ovary syndrome (PCOS) awareness among female Emirati university students. Int J Women's Health 12:927

Pringle D, Suliman S, Seedat S, van den Heuvel LL (2022) The impact of childhood maltreatment on women's reproductive health, with a focus on symptoms of polycystic ovary syndrome. Child Abuse Negl 133:105831

Pundir J, Sunkara SK, El-Toukhy T, Khalaf Y (2012) Meta-analysis of GnRH antagonist protocols: do they reduce the risk of OHSS in PCOS? Reprod Biomed Online 24:6–22

Purwar A, Nagpure S (2022) Insulin resistance in polycystic ovarian syndrome. Cureus 14

Puurunen J, Piltonen T, Morin-Papunen L, Perheentupa A, Järvelä I, Ruokonen A, Tapanainen JS (2011) Unfavorable hormonal, metabolic, and inflammatory alterations persist after menopause in women with PCOS. J Clin Endocrinol Metabol 96:1827–1834

Rababa'h AM, Matani BR, Yehya A (2022) An update of polycystic ovary syndrome: causes and therapeutics options. Heliyon 8:e11010

Raja-Khan N, Kunselman AR, Demers LM, Ewens KG, Spielman RS, Legro RS (2010) A variant in the fibrillin-3 gene is associated with TGF-β and inhibin B levels in women with polycystic ovary syndrome. Fertil Steril 94:2916–2919

Rana S, Burke SD, Karumanchi SA (2020) Imbalances in circulating angiogenic factors in the pathophysiology of preeclampsia and related disorders. Am J Obstet Gynecol

Rashid R, Mir SA, Kareem O, Ali T, Ara R, Malik A, Amin F, Bader G (2022) Polycystic ovarian syndrome-current pharmacotherapy and clinical implications. Taiwanese J Obstet Gynecol 61:40–50

Reichert RA (2012) Diagnostic gynecologic and obstetric pathology: an atlas and text. Lippincott Williams & Wilkins

Riaz Y, Parekh U (2022) Oligomenorrhea. StatPearls [Internet]. StatPearls Publishing

Richard S, Ricardo A (2003) Androgen excess disorders. Danforth's obstetrics and gynecology. Lippincott Williams & Wilkins, Philadelphia

Richardson CR, Borgeson JR, van Harrison R, Wyckoff JA, Yoo AS, Aikens JE, Griauzde DH, Tincopa MA, van Harrison R, Proudlock AL (2021) Management of type 2 diabetes mellitus

Roque M, Tostes AC, Valle M, Sampaio M, Geber S (2015) Letrozole versus clomiphene citrate in polycystic ovary syndrome: systematic review and meta-analysis. Gynecol Endocrinol 31:917–921

Rosenfield RL (2007) Identifying children at risk for polycystic ovary syndrome. J Clin Endocrinol Metabol 92:787–796

Rosenfield RL (2015) The polycystic ovary morphology-polycystic ovary syndrome spectrum. J Pediatr Adolesc Gynecol 28:412–419

Rosenfield RL, Ehrmann DA (2016) The pathogenesis of polycystic ovary syndrome (PCOS): the hypothesis of PCOS as functional ovarian hyperandrogenism revisited. Endocr Rev 37:467–520

Rotterdam & Eshre (2004) Revised 2003 consensus on diagnostic criteria and long-term health risks related to polycystic ovary syndrome. Fertil Steril 81:19–25

Rubin R, Strayer DS, Rubin E (2008) Rubin's pathology: clinicopathologic foundations of medicine. Lippincott Williams & Wilkins

Saboor Aftab SA, Halder L, Piya MK, Reddy N, Fraser I, Menon V, Bridgwater S, Kendrick D, Kumar S, Barber TM (2014) Predictors of weight loss at 1 year after laparoscopic adjustable gastric banding and the role of presurgical quality of life. Obes Surg 24:885–890

Safiri S, Noori M, Nejadghaderi SA, Karamzad N, Carson-Chahhoud K, Sullman MJ, Collins GS, Kolahi A-A, Avery J (2022) Prevalence, incidence and years lived with disability due to polycystic ovary syndrome in 204 countries and territories, 1990–2019. Hum Reprod 37:1919–1931

Saidunnisa B, Atiqulla S, Ayman G, Mohammad B, Housam R, Khaled N (2016) Prevalence of polycystic ovarian syndrome among students of rak medical and health sciences University United Arab Emirates. Int J Med Pharm Sci 6:109–118

Saini S, Gibson-Helm M, Cooney L, Teede H, Dokras A (2016) Gaps in knowledge in diagnosis and management of polycystic ovary syndrome. Fertil Steril 106:e100

Sangaraju SL, Yepez D, Grandes XA, Manjunatha RT, Habib S, Grandes X (2022) Cardio-metabolic disease and polycystic ovarian syndrome (PCOS): a narrative review. Cureus 14

Sawant S, Bhide P (2019) Fertility treatment options for women with polycystic ovary syndrome. Clin Med Insights 13:1179558119890867

Schneider D, Gonzalez JR, Yamamoto M, Yang J, Lo JC (2019) The association of polycystic ovary syndrome and gestational hypertensive disorders in a diverse community-based cohort. J Pregnancy 2019:2019

Schüring A, Welp A, Gromoll J, Zitzmann M, Sonntag B, Nieschlag E, Greb R, Kiesel L (2012) Role of the CAG repeat polymorphism of the androgen receptor gene in polycystic ovary syndrome (PCOS). Exp Clin Endocrinol Diabetes 120:73–79

Scotti L, Parborell F, Irusta G, de Zuñiga I, Bisioli C, Pettorossi H, Tesone M, Abramovich D (2014) Platelet-derived growth factor BB and DD and angiopoietin1 are altered in follicular fluid from polycystic ovary syndrome patients. Mol Reprod Dev 81:748–756

Sen A, Prizant H, Light A, Biswas A, Hayes E, Lee H-J, Barad D, Gleicher N, Hammes SR (2014) Androgens regulate ovarian follicular development by increasing follicle stimulating hormone receptor and microRNA-125b expression. Proc Natl Acad Sci 111:3008–3013

Sendur SN, Yildiz BO (2021) Influence of ethnicity on different aspects of polycystic ovary syndrome: a systematic review. Reprod Biomed Online 42:799–818

Shah D, Patil M (2018) Consensus statement on the use of oral contraceptive pills in polycystic ovarian syndrome women in India. J Hum Reprod Sci 11:96–118

Shahid R, Mahnoor, Awan KA, Iqbal MJ, Munir H, Saeed I (2022) Diet and lifestyle modifications for effective management of polycystic ovarian syndrome (PCOS). J Food Biochem 46:e14117

Shang K, Jia X, Qiao J, Kang J, Guan Y (2012) Endometrial abnormality in women with polycystic ovary syndrome. Reprod Sci 19:674–683

Shapiro J, Lui H (2001) Vaniqa--eflornithine 13.9% cream. Skin Therapy Lett 6:1–3

Sharif E, Rahman S, Zia Y, Rizk NM (2016) The frequency of polycystic ovary syndrome in young reproductive females in Qatar. Int J Women's Health:1–10

Shojaeian Z, Sadeghi R, Roudsari RL (2019) Calcium and vitamin D supplementation effects on metabolic factors, menstrual cycles and follicular responses in women with polycystic ovary syndrome: a systematic review and meta-analysis. Caspian J Intern Med 10:359

Siddiqui S, Mateen S, Ahmad R, Moin S (2022) A brief insight into the etiology, genetics, and immunology of polycystic ovarian syndrome (PCOS). J Assist Reprod Genet 39:2439–2473

Silfen ME, Manibo AM, Mcmahon DJ, Levine LS, Murphy AR, Oberfield SE (2001) Comparison of simple measures of insulin sensitivity in young girls with premature adrenarche: the fasting glucose to

insulin ratio may be a simple and useful measure. J Clin Endocrinol Metabol 86:2863–2868
Simbar M, Nazarpour S, Alavi Majd H, Dodel Andarvar K, Jafari Torkamani Z, Alsadat Rahnemaei F (2020) Is body image a predictor of women's depression and anxiety in postmenopausal women? BMC Psychiatry 20:202
Sin HY, Kim JY, Jung KH (2011) Total cholesterol, high density lipoprotein and triglyceride for cardiovascular disease in elderly patients treated with metformin. Arch Pharm Res 34:99–107
Singh G, Puckett Y (2023) Endometrial hyperplasia. StatPearls Publishing, Treasure Island (FL)
Singh B, Saxena A (2010) Surrogate markers of insulin resistance: a review. World J Diabetes 1:36–47
Singh S, Pal N, Shubham S, Sarma DK, Verma V, Marotta F, Kumar M (2023) Polycystic ovary syndrome: etiology, current management, and future therapeutics. J Clin Med 12:1454
Sinha B, Ghosal S (2022) A meta-analysis of the effect of sodium glucose cotransporter-2 inhibitors on metabolic parameters in patients with polycystic ovary syndrome. Front Endocrinol 13:830401
Sirmans SM, Pate KA (2014) Epidemiology, diagnosis, and management of polycystic ovary syndrome. Clin Epidemiol 6:1
Srikanthan P, Korenman S, Davis S (2006) Polycystic ovarian syndrome: the next cardiovascular dilemma in women? Endocrinol Metab Clin 35:611–631
Stein I, Leventhal M (1935) Amenorrhea associated with polycystic ovary syndrome. Am J Obstet Gynecol 29:181–191
Sultan C, Paris F (2006) Clinical expression of polycystic ovary syndrome in adolescent girls. Fertil Steril 86:S6
Szeliga A, Rudnicka E, Maciejewska-Jeske M, Kucharski M, Kostrzak A, Hajbos M, Niwczyk O, Smolarczyk R, Meczekalski B (2022) Neuroendocrine determinants of polycystic ovary syndrome. Int J Environ Res Public Health 19:3089
Tal R, Seifer DB, Grazi RV, Malter HE (2013a) Angiopoietin-1 and angiopoietin-2 are altered in polycystic ovarian syndrome (PCOS) during controlled ovarian stimulation. Vascular Cell 5:1–7
Tal R, Seifer DB, Shohat-Tal A, Grazi RV, Malter HE (2013b) Transforming growth factor-β1 and its receptor soluble endoglin are altered in polycystic ovary syndrome during controlled ovarian stimulation. Fertil Steril 100:538–543
Tal R, Seifer DB, Grazi RV, Malter HE (2014) Follicular fluid placental growth factor is increased in polycystic ovarian syndrome: correlation with ovarian stimulation. Reprod Biol Endocrinol 12:1–7
Tal R, Seifer DB, Arici A (2015) The emerging role of angiogenic factor dysregulation in the pathogenesis of polycystic ovarian syndrome. Seminars in reproductive medicine, 2015. Thieme Medical Publishers, pp 195–207
Talbott EO, Guzick DS, Sutton-Tyrrell K, Mchugh-Pemu KP, Zborowski JV, Remsberg KE, Kuller LH (2000) Evidence for association between polycystic ovary syndrome and premature carotid atherosclerosis in middle-aged women. Arterioscler Thromb Vasc Biol 20:2414–2421
Tanbo TG, Fedorcsak PZ (2021) Can time to menopause be predicted? Acta Obstet Gynecol Scand 100:1961–1968
Tang R, Ding X, Zhu J (2019) Kisspeptin and polycystic ovary syndrome. Front Endocrinol 10
Teede H, Deeks A, Moran L (2010) Polycystic ovary syndrome: a complex condition with psychological, reproductive and metabolic manifestations that impacts on health across the lifespan. BMC Med 8:1–10
Teede HJ, Misso ML, Costello MF, Dokras A, Laven J, Moran L, Piltonen T, Norman RJ (2018) Recommendations from the international evidence-based guideline for the assessment and management of polycystic ovary syndrome. Fertil Steril 110:364–379
Teede HJ, Tay CT, Laven JJ, Dokras A, Moran LJ, Piltonen TT, Costello MF, Boivin J, Redman LM, Boyle JA, Norman RJ (2023) Recommendations from the 2023 international evidence-based guideline for the assessment and management of polycystic ovary syndrome. Eur J Endocrinol 189(2):G43–G64
Thomas GN, Critchley JA, Tomlinson B, Anderson PJ, Lee ZS, Chan JC (2000) Obesity, independent of insulin resistance, is a major determinant of blood pressure in normoglycemic Hong Kong Chinese. Metab-Clin Exp 49:1523–1528
Tong C, Wu Y, Zhang L, Yu Y (2022) Insulin resistance, autophagy and apoptosis in patients with polycystic ovary syndrome: association with PI3K signaling pathway. Front Endocrinol 13:1091147
Ulloa-Aguirre A, Dias JA, Bousfield GR (2017) Gonadotropins. In: Simoni M, Huhtaniemi I (eds) Endocrinology of the testis and male reproduction. Springer International Publishing, Cham
Vanky E, Løvvik TS (2020) Polycystic ovary syndrome and pregnancy–from a clinical perspective. Curr Opin Endocrine Metabolic Res 12:8–13
Vrbikova J, Cibula D (2005) Combined oral contraceptives in the treatment of polycystic ovary syndrome. Hum Reprod Update 11:277–291
Vuguin P, Saenger P, Dimartino-Nardi J (2001) Fasting glucose insulin ratio: a useful measure of insulin resistance in girls with premature adrenarche. J Clin Endocrinol Metabol 86:4618–4621
Walters KA, Gilchrist RB, Ledger WL, Teede HJ, Handelsman DJ, Campbell RE (2018) New perspectives on the pathogenesis of PCOS: neuroendocrine origins. Trends Endocrinol Metab 29:841–852
Wan J, Gao Y, Zeng K, Yin Y, Zhao M, Wei J, Chen Q (2016) The levels of the sex hormones are not different between type 1 and type 2 endometrial cancer. Sci Rep 6:39744
Wang Y, Wang L, Chen C, Chu X (2018) New insights into the regulatory role of microRNA in tumor angiogenesis and clinical implications. Mol Cancer 17:1–10

Wekker V, van Dammen L, Koning A, Heida K, Painter R, Limpens J, Laven J, Roeters van Lennep J, Roseboom T, Hoek A (2020) Long-term cardiometabolic disease risk in women with PCOS: a systematic review and meta-analysis. Hum Reprod Update 26:942–960

Weyer C, Tataranni PA, Pratley RE (2000) Insulin action and insulinemia are closely related to the fasting complement C3, but not acylation stimulating protein concentration. Diabetes Care 23:779–785

Wickenheisser JK, Quinn PG, Nelson VL, Legro RS, Strauss JF III, Mcallister JM (2000) Differential activity of the cytochrome P450 17α-hydroxylase and steroidogenic acute regulatory protein gene promoters in normal and polycystic ovary syndrome theca cells. J Clin Endocrinol Metabol 85:2304–2311

Wijeyaratne CN, Balen AH, Barth JH, Belchetz PE (2002) Clinical manifestations and insulin resistance (IR) in polycystic ovary syndrome (PCOS) among south Asians and Caucasians: is there a difference? Clin Endocrinol 57:343–350

Wild RA, Carmina E, Diamanti-Kandarakis E, Dokras A, Escobar-Morreale HF, Futterweit W, Lobo R, Norman RJ, Talbott E, Dumesic DA (2010) Assessment of cardiovascular risk and prevention of cardiovascular disease in women with the polycystic ovary syndrome: a consensus statement by the androgen excess and polycystic ovary syndrome (AE-PCOS) society. J Clin Endocrinol Metabol 95:2038–2049

Wild RA, Rizzo M, Clifton S, Carmina E (2011) Lipid levels in polycystic ovary syndrome: systematic review and meta-analysis. Fertil Steril 95:1073–1079. e11

Williams T, Mortada R, Porter S (2016) Diagnosis and treatment of polycystic ovary syndrome. Am Fam Physician 94:106–113

Winters SJ, Talbott E, Guzick DS, Zborowski J, Mchugh KP (2000) Serum testosterone levels decrease in middle age in women with the polycystic ovary syndrome. Fertil Steril 73:724–729

Witchel SF, Tena-Sempere M (2013) The Kiss1 system and polycystic ovary syndrome: lessons from physiology and putative pathophysiologic implications. Fertil Steril 100:12–22

Witchel SF, Roumimper H, Oberfield S (2016) Polycystic ovary syndrome in adolescents. Endocrinol Metab Clin 45:329–344

Witchel SF, Oberfield SE, Peña AS (2019) Polycystic ovary syndrome: pathophysiology, presentation, and treatment with emphasis on adolescent girls. J Endocrine Soc 3:1545–1573

Witchel SF, Teede HJ, Peña AS (2020) Curtailing pcos. Pediatr Res 87:353–361

Wolf WM, Wattick RA, Kinkade ON, Olfert MD (2018) Geographical prevalence of polycystic ovary syndrome as determined by region and race/ethnicity. Int J Environ Res Public Health 15:2589

Xu Y, Wu Y, Huang Q (2017) Comparison of the effect between pioglitazone and metformin in treating patients with PCOS: a meta-analysis. Arch Gynecol Obstet 296:661–677

Yamamoto Y, Hirose H, Saito I, Nishikai K, Saruta T (2004) Adiponectin, an adipocyte-derived protein, predicts future insulin resistance: two-year follow-up study in Japanese population. J Clin Endocrinol Metabol 89:87–90

Yang R, Li Q, Zhou Z, Qian W, Zhang J, Wu Z, Jin L, Wu X, Zhang C, Zheng B (2022) Changes in the prevalence of polycystic ovary syndrome in China over the past decade. Lancet Regional Health–Western Pacific 25

Yasmin A, Roychoudhury S, Paul Choudhury A, Ahmed AF, Dutta S, Mottola F, Verma V, Kalita JC, Kumar D, Sengupta P (2022) Polycystic ovary syndrome: an updated overview foregrounding impacts of ethnicities and geographic variations. Life 12:1974

Yesiladali M, Yazici MGK, Attar E, Kelestimur F (2022) Differentiating polycystic ovary syndrome from adrenal disorders. Diagnostics (Basel) 12

Yildiz BO, Bolour S, Woods K, Moore A, Azziz R (2010) Visually scoring hirsutism. Hum Reprod Update 16:51–64

Youssef MA, van der Veen F, Al-Inany HG, Griesinger G, Mochtar MH, Aboulfoutouh I, Khattab SM, van Wely M (2011) Gonadotropin-releasing hormone agonist versus HCG for oocyte triggering in antagonist assisted reproductive technology cycles. Cochrane Database Syst Rev

Yu H-F, Chen H-S, Rao D-P, Gong J (2016) Association between polycystic ovary syndrome and the risk of pregnancy complications: a PRISMA-compliant systematic review and meta-analysis. Medicine 95:e4863

Zac-Varghese S, de Silva A, Bloom SR (2011) Translational studies on PYY as a novel target in obesity. Curr Opin Pharmacol 11:582–585

Zawadski J, Dunaif A (1992) Polycystic ovary syndrome: diagnostic criteria for polycystic ovarian syndrome: towards a rational approach. Blackwell, Boston

Zeng X, Xie Y-J, Liu Y-T, Long S-L, Mo Z-C (2020) Polycystic ovarian syndrome: correlation between hyperandrogenism, insulin resistance and obesity. Clin Chim Acta 502:214–221

Zhang J, Bao Y, Zhou X, Zheng L (2019) Polycystic ovary syndrome and mitochondrial dysfunction. Reprod Biol Endocrinol 17:1–15

Zhang J, Xu J-H, Qu Q-Q, Zhong G-Q (2020) Risk of cardiovascular and cerebrovascular events in polycystic ovarian syndrome women: a meta-analysis of cohort studies. Front Cardiovasc Med 184

Zhao Y, Qiao J (2013) Ethnic differences in the phenotypic expression of polycystic ovary syndrome. Steroids 78:755–760

Zhou S, Ji Y, Wang H (2021) The risk factors of gestational hypertension in patients with polycystic ovary syndrome: a retrospective analysis. BMC Pregnancy Childbirth 21:336

Zhu R-Y, Wong Y-C, Yong E-L (2016) Sonographic evaluation of polycystic ovaries. Best Pract Res Clin Obstet Gynaecol 37:25–37

Zhuang X-F, Zhao M-M, Weng C-L, Sun N-L (2009) Adipocytokines: a bridge connecting obesity and insulin resistance. Med Hypotheses 73:981–985

Zimmerman Y, Eijkemans M, Coelingh Bennink H, Blankenstein M, Fauser B (2014) The effect of combined oral contraception on testosterone levels in healthy women: a systematic review and meta-analysis. Hum Reprod Update 20:76–105

Zinman B, Hanley AJ, Harris SB, Kwan J, Fantus IG (1999) Circulating tumor necrosis factor-α concentrations in a native Canadian population with high rates of type 2 diabetes mellitus. J Clin Endocrinol Metabol 84:272–278

Zulian E, Sartorato P, Benedini S, Baro G, Armanini D, Mantero F, Scaroni C (2005) Spironolactone in the treatment of polycystic ovary syndrome: effects on clinical features, insulin sensitivity and lipid profile. J Endocrinol Investig 28:49–53

9 Visual Interpretation Using Multiplex Immunoassay of Serum Samples

Shoohana Singh and Thajasvarie Naicker

Abstract

Clinical proteomic research aims to identify biomarkers that may be useful in disease diagnosis, prognosis estimation, and treatment monitoring. An immunoassay that can test multiple analytes concurrently is known as a multiplex assay. The Enzyme-linked immunosorbent assay (ELISA) format is the foundation from which multiplex immunoassays were developed and generates several signal readings as opposed to a single signal measurement. This chapter aims to provide insight into serum sample analysis interpretation based on the data output from the multiplex immunoassay platform.

Although the research arena encompasses a variety of biological samples for testing, biological fluid is the matrix of choice for most multiplex immunoassays. Analysis on the immunoassay platform may allow multiplexing up to 100 different assays within a single sample. Fluorescent signals from each magnetic bead-based technology are simultaneously recorded by high-speed digital signal processors and software, which convert the signals into data. Tools for biomarker discovery-driven data analysis offer a variety of spectrum visualization options. Univariate analysis, receiver operating characteristic curves, and dot plots are used to examine differentially expressed proteins. Standard curves are automatically generated and easily optimized by the user upon visual inspection. Graphical representation of the analyte concentration levels present in the serum samples provides an easy avenue for the researcher to convey their findings. These results can be further supported with analysis using statistical software packages. Multiplex immunoassays provide several major benefits, including screening biomarkers for further assessment and confirming data for drug development. Conservation of restricted sample volume and reduction in cost per data point. Furthermore, this platform ensures increased accuracy due to reduced sample handling.

This chapter further discusses the significance of selecting the best-suited commercially available kits based on optimal methods and the design of researcher-specific needs. Including taking into account present and future advancements for the multiplex immunoassay platform.

Keywords

Bio-Plex · Multiplex · Immunoassay · ELISA

S. Singh (✉)
Discipline of Human Physiology, College of Health Sciences, University of KwaZulu-Natal, Durban, South Africa
e-mail: singhs5@ukzn.ac.za; naickera@ukzn.ac.za

T. Naicker
Optics and Imaging Centre, Doris Duke Medical Research Institute, College of Health Sciences, University of KwaZulu-Natal, Durban, South Africa
e-mail: singhs5@ukzn.ac.za; naickera@ukzn.ac.za

L. Shapiro (ed.), *How to use 3D Printing Innovations and Digital Storage to Democratize Anatomy Education*, Biomedical Visualization 6, https://doi.org/10.1007/978-3-031-68501-9_9

9.1 Annals of Immunoassays

The foundation of this bioanalytical immunoassay technique relies on the interaction between an antibody and antigen (or analyte), which allows for the quantification of the desired analyte.

Dating back to the 1940s, detection of the antibody–antigen interaction was essentially dependent on visualized banding precipitate on agar. The Ouchterlony twofold immunodiffusion test is attributed to the Swedish bacteriologist and immunologist Örjan Thomas Ouchterlony (1948). This technique quantified analytes in the minute range of milligrams to micrograms. Limitations of Ouchterlony's method were its labor-intensive protocol, low sensitivity, and subjectiveness flaws, addressed in 1959 by medical physicist Rosalyn Sussman Yalow and physician Solomon Aaron Berson (Yalow n.d.). These researchers developed radioimmunoassay, a method of radioisotope tracing that makes it possible to measure minuscule amounts for both low- and high-molecular weight analytes in human blood. It usually provides detection limits from 10^{-12} to 10^{-15} mol/L. This method was first used to examine insulin levels in people with diabetes mellitus. Yet, in the process, they could distinguish between two types of diabetes, Type I and Type II, which have significantly different mechanisms (do Nascimento Carvalho et al. 2023). Subsequently, radioimmunoassays have been used for several different substances, including enzymes, vitamins, and hormones. Rosalyn Yalow went on to be the winner of the 1977 Nobel Prize in Physiology or Medicine for this significant advance in clinical biochemistry (Meites 1977). However, the limitations of Yalow's method include concerns over the safe handling and disposal of radioactive reagents and waste. At this stage, scientists were working on an alternate method of analysis giving rise to the ELISA. The singleplex ELISA test is a technique independently developed by the Swedish scientists Peter Perlman and Eva Engvall that completely revolutionized medicine (Sten et al. 2006). The technique looks for the presence of hormones or viruses using antibodies specific to the protein to be evaluated; the test employs a solid-phase form of enzyme immunoassay to find the presence of a ligand, usually a protein, in a liquid sample (Engvall and Perlmann 1971). ELISA has been applied as a quality control measure in numerous sectors and a diagnostic tool in medicine, biotechnology, and plant pathology. Antigens from the sample to be analyzed are adhered to the surface of a microplate in the most basic type of Direct ELISA (Clark et al. 1986; Lin 2015). The surface is then covered with a corresponding antibody to bind the antigen. After attaching this antibody to an enzyme, any unbound antibodies are eliminated. The last stage involves adding a reagent containing the enzyme's substrate. The ensuing reaction yields an observable indication, most typically a color change, if binding occurs (Crowther 2008). A direct ELISA uses only one antibody, making it less specific. The detection capability of an ELISA test can range from 2 to 3000 pg/ml. In the following decade, improvements to the ELISA method gave rise to the Indirect, Competitive, and Sandwich techniques (Alhajj et al. 2023). An additional amplification detection phase is included in an indirect ELISA, comparable to a direct ELISA in that it involves immobilizing an antigen on a plate. An unconjugated primary detection antibody is first applied and attaches itself to the particular antigen. A conjugated secondary antibody that targets the primary antibody's host species is added. Subsequently, the substrate produces a signal corresponding to the amount of antigen bound in the well (Boguszewska et al. 2019). Sandwich techniques are the most widely used type of ELISA. The antigen is sandwiched between two distinct antibodies, often known as matched antibody pairs. A microplate is coated with a capture antibody, the sample is added, and the target protein binds and becomes immobile on the plate. Next, an extra epitope on the target protein is bound by a conjugated-detection antibody. When the substrate is added, the signal it generates is proportionate to the analyte concentration in the sample. Because two antibodies are needed to bind to the target protein, sandwich ELISAs tend to be highly specific (Huang et al. 2004). When the target protein is extremely small to sandwich with two antibodies effectively, competitive ELISAs are frequently utilized for small compounds. A microplate is coated with a capture antibody, similar to a sandwich ELISA. A conjugated antigen is used to complete binding with the antigen present in the sample, as opposed to a conjugated-detection antibody. Less conjugated antigens will attach to the capture antibody in samples with higher antigen concentrations. When substrate is introduced, the signal generated is inversely correlated with the protein content of the sample (Shah and Maghsoudlou 2016). A few major drawbacks of an ELISA include inconsistent performance, insufficient data analysis, and non-standard experimental methods. Furthermore, only one antigen can be detected per microplate.

Due to increasing demand to improve disease diagnosis, prognosis, and therapy, there is a growing need for technology to extract a significant quantity of bioinformation from a small sample volume. This has resulted in the creation of high throughput, adaptable, and automated technologies for the simultaneous detection of various proteins. Multiplex immunoassays were developed to test many analytes simultaneously in a single biological sample while requiring minimal sample volume, assay time, and expense (Ahsan and Ahmad 2020).

9.2 Multiplex Immunoassay

Multiplex technology is well-established in developing clinical biomarkers, proteomic circuits and pathways, and genetic discovery validation. Target analytes in the sample bind to the immobilized ligands in multiplex immunoassays, either on microspheres in suspension or planar format, which is a flat surface protein chip. There are two types available for producing planar arrays: slide and microtiter. The format for slides offers multiple layout options, enabling the printing of individual or repeatable experiments using distinct antibody sets automatically onto the surface of the activated slide (Ahsan and Ahmad 2020). Discrete assays are regarded as distinct microtiter wells and segregated spatially once the sample matrix is applied. Using particular secondary antibodies conjugated to chemiluminescent/fluorescent reporter systems allows detection. Standard protein-binding plate wells are dedicated to regularly printed antibody sets in microtiter-based immunoassay. Thus, the plate can be handled much like in a traditional ELISA, which involves blocking, incubating, and washing it before using a pair of reporter-conjugated detection antibodies to detect it (Kingsmore 2006).

The suspension assay incorporates thousands of micrometer-sized plastic microbeads that have been functionally activated and injected with one or more chemiluminescent or fluorescent dyes before being linked with a particular capture antibody (Lin et al. 2015). Several sets of these beads are created, each with a distinct fluorescence signal that allows for identification and distinct capture antibodies based on the analyte. The detection antibodies are added after the incubation and washing phases are completed. Each antibody is uniquely tagged with a single chemiluminescent/fluorescent reporter (different from those found within the beads). As a result, each bead can hold an assembly comprising the reporter-conjugated detection antibody and the collected target analyte. Lasers are used to activate the chemiluminescent/fluorescent reporters in the bead analyte reporter structures, and the produced light is then collected by a series of detectors for quantitative analysis in a flow chamber that facilitates bead separation (Vafajoo et al. 2018). A wide range of biological samples from humans and certain animals are suitable for analysis using the suspension microbead assay. These include tissue culture supernatant, blister fluid, bronchoalveolar lavage fluid, cerebrospinal fluid, fat (adipose) interstitial fluid, nasal lavage fluid, peritoneal fluid, urine, serum, plasma, synovial fluid, and tissue (colon, kidney, lung, nervous system and spleen). However, the most commonly used are plasma and serum samples. The bio-fluid that remains after blood clots is called serum. It is blood without clotting factors and cellular components; however, it contains various substances like electrolytes, antibodies, hormones, and other proteins. The ease of collection and storage of serum is one of its main benefits. Centrifugation and blood coagulation are all that are needed to extract the serum from the cellular components (Flores et al. 2020). Serum is also frequently utilized in diagnostic procedures, including biomarker identification and serological testing, which makes it a dependable and standardized option for many labs. In serological assays, serum is frequently utilized to gauge the amount of antibodies produced in response to immunizations or diseases (Haddad et al. 2021).

In this chapter, we reflect on our research team's experience using the magnetic microbeads-based technology of the Bio-Plex® MAGPIX system, licensed from Luminex Corporation. This company has patented xMAP (Multi-Analyte Profiling) technology, a multiplex assay format that generates up to 100 distinct bead identities by infusing a panel of microspheres with two fluorescent dyes, enabling the use across a variety of applications (Platchek et al. 2020).

9.3 Bio-Plex® MAGPIX System: Requirements and Work Flow

Up to 50 distinct tests can be multiplexed within a single sample using the Bio-Plex® MAGPIX technology. This method uses two fluorescent dyes (red and infrared) at different ratios to make 50 different colored bead sets. A distinct target analyte is coupled to every bead region. Analyzing time for 50 analyte kits on a 96-well plate will take approximately 60 min. The Bio-Plex® MAGPIX system illuminates and images immobilized magnetic beads using inexpensive light-emitting diodes and a charge-coupled device imager.

The instrumentation provided with the Bio-Plex® MAGPIX system includes the analyzer and a desktop computer installed with Bio-Plex Manager software 6.1 and Bio-Plex Manager MP (Fig. 9.1). Additional equipment required for workflow are a vortex, ultrasonic bath, refrigerated centrifuge, refrigerated shaker, automated magnetic plate washer, and a probe height adjustment plate (Fig. 9.2a–f).

On the maintenance menu of the Bio-Plex Manager 6.1 software (Fig. 9.3), the first panel indicates "Daily Startup" including Calibration and Verification that should be performed daily. Calibration and verification ensure that the MAGPIX correctly detects and sorts the beads and targets, thereby certifying the instrument's performance. If the instrument fails either calibration or verification, troubleshooting options are suggested to address before repeating this quality control step. The screen's middle panel will indicate the required reagents and positions allocated in the reagent block and well strips. Depending on the option chosen, the reagent block would require one or all of the following: distilled water, 70% isopropanol, and 0.1 M sodium hydroxide (NaOH). The third panel indicates the MAGPIX status with regard to calibration and verification, as well as the Calibration and Verification kit lot numbers. These kits are commercially available, and the kit lot numbers are imported using the CD provided with each kit.

Additionally, the third panel allows for the probe height adjustment to be performed using the adjustment plate (Fig. 9.2f). A pop-up screen will indicate successful probe height adjustment or calibration and verification of the system once complete. The status bar at the bottom of the screen will indicate the duration of all these maintenance steps.

Creating or running a protocol is the tab on the far left-hand panel, below the maintenance tab. On this panel, important information about the 96-well plate layout is input. The location, lot numbers, and expiry date of the standards, as well as the kit-specific controls, blanks, unknown samples, and analyte bead regions, are entered. At this point, it is important to enter the standard dilution factor and indicate which standard is the most concentrated (Fig. 9.4). Once the information has been entered, the plate is set up with the labeled wells, the bead map with the regions of the beads of interest highlighted, and the assay panel will list the wells and the sample type in each well. The prepared 96-well plate (according to the kit protocol insert) can be loaded into the analyzer and the plate label prompt will allow the data file to be saved for the run.

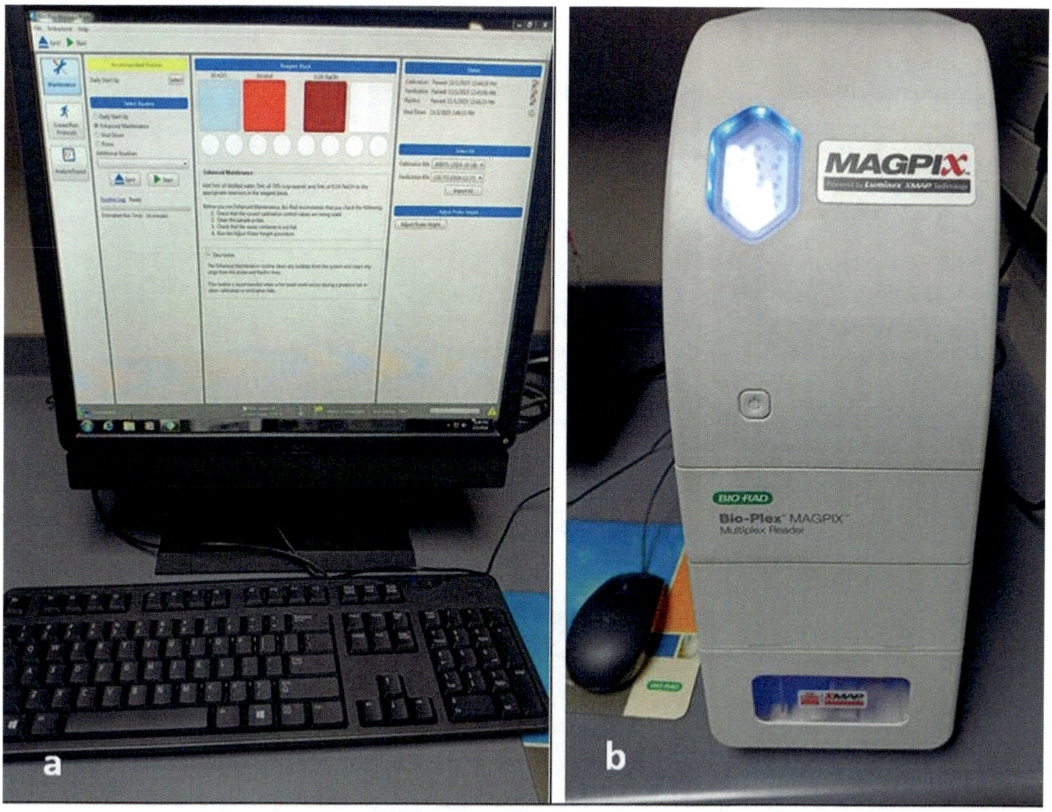

Fig. 9.1 (**a**) Desktop computer displaying the maintenance page of the Bio-Plex Manager software 6.1; (**b**) Bio-Plex® MAGPIX analyzer with the illuminated blue light indicating that the instrument is powered on

Visual confirmation of the data being collected is the presence of blue to green ticked wells on the plate layout and the bead events on the bead map. In the assay panel, there will be several data parameters from each well that are automatically updated as each well is analyzed. It is acceptable to step away from the analyzer once the protocol has started, but advisable to check on the instrument during the run time in the event of any errors or warnings that may occur (Fig. 9.5). When the run is complete, a pop-up alert will appear indicating that the protocol is complete. The data can be exported as a default "rbx" file into Bio-Plex Manager MP software for analysis. Additional information can be added to the data file in Bio-Plex Manager 6.1, such as unknown sample dilution (if the protocol required sample dilution), unknown sample renaming, and control values.

The Bio-Plex Manager MP main screen will indicate on the left-hand panel the "Raw data'' tab in which the raw fluorescent intensity (FI) values for each of the unknown samples, standards, and controls are shown with the bead map on the top right-hand corner. The "Standard curve" tab is also on the main screen panel. When selected, it will populate the standard curves for all the analytes of the kit. The advantage of multiplexing is that the standard of multiple analytes is plexed together into one standard. Qualifying a standard 5 Parameter Logistic (PL) regression curve is calculated using the following equation: (Observed concentration/Expected concentration) ×100. For curve-fitting analysis in immunoassays, the 5PL regression model is frequently utilized. Some refer to the standard dose-response curve as the five-parameter logistic equation. It is distinguished by its traditional sigmoidal shape, which fits the slope factor (Hill's slope), the

Fig. 9.2 (**a**) An orbital vortex used to vortex the serum samples, kit standards, and controls; (**b**) an ultrasonic bath used to sonicate the magnetic beads prior to reconstitution; (**c**) refrigerated centrifuge for serum extraction; (**d**) refrigerated shaker to resuspension and incubation steps; (**e**) automated magnetic plate washer to perform washing steps without loss of magnetic microbeads; and (**f**) probe height adjustment plate to assist with the analyzer probe alignment

EC50 (represents the concentration of a compound where 50% of its maximal effect is observed), and the bottom and top plateaus of the curve. The curve exhibits symmetry at the point of inflection. The 5PL equation includes an extra parameter that quantifies the asymmetry to expand the model to accommodate curves that are not symmetrical (Davis et al. 2005). Ideally, if the observed concentration is equal to the expected concentration to provide a ratio of 1, multiplied by 100 would equate to 100. This would make the best-fit standard curve if all the data points are closer to the value of 100. However, it is possible to allow these values to range between 70 and 130. By selecting the "Optimize" tab and setting the range, Bio-Plex Manager does this automatically, allowing for best-fit standard curves for each analyte such that the standard curve fit between the upper limit of quantitation (ULOQ) and lower limit of quantitation (LLOQ) (Fig. 9.6).

Upon completion of optimizing the standard curves, one can confidently view the "Report Table" tab. The observed concentration (Obs. Conc.) values are automatically extrapolated for the unknown samples from the associated standard curves and displayed in the report table. Of note, the values that have an asterisk (*) next to them indicate that the point is within the detection range of the instrument but below the last point on the standard curve. Some data points may display a negative out-of-range (OOR<)

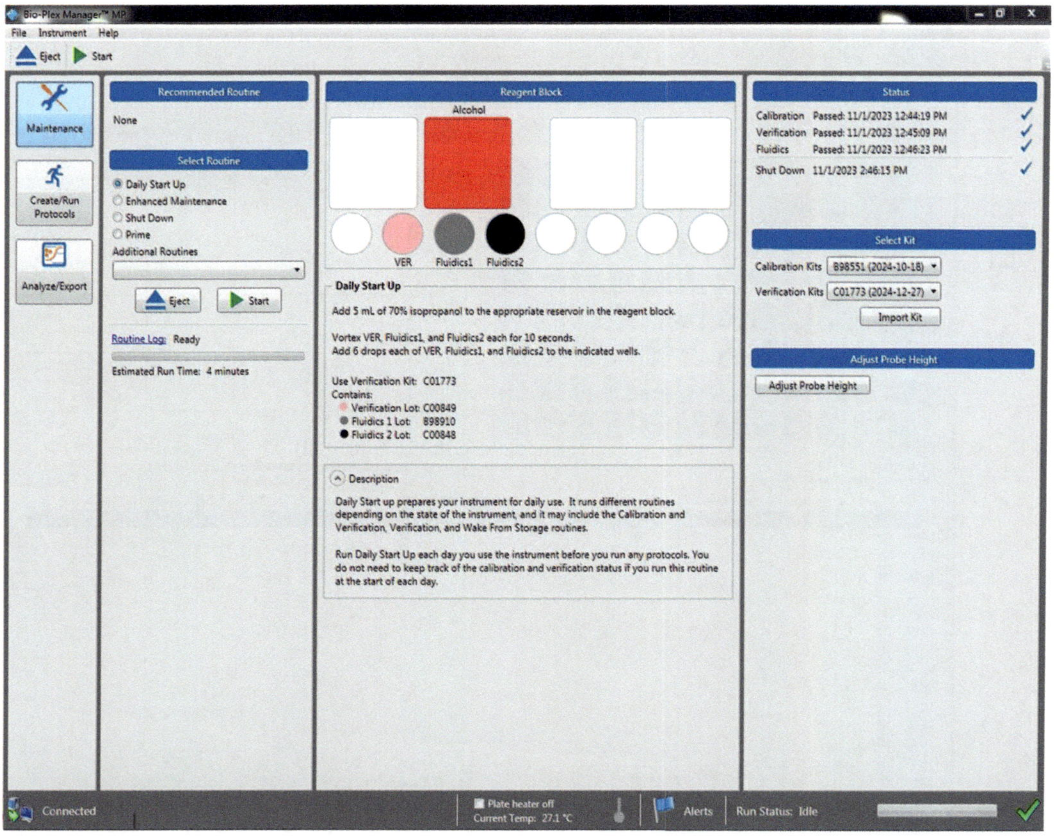

Fig. 9.3 Bio-Plex Manager 6.1 software is on the main screen showing that Maintenance has been selected. The user-friendly software provides brief instructions on the user's requirements (middle panel). The blue ticks on the right-hand side panel indicate that the Calibration and Verification, including the fluidic check, have passed

indicating that this point was undetected by the instrument, hence below the instrument's detection limit. If the data point displays a positive out-of-range (OOR>), it would indicate that this point was beyond the detection limit. A low coefficient of variability (CV or %CV) between sample replicates and standard replicates is essential for proving that an experiment was carried out correctly and the precise and accurate data it produced. On observation, the %CV should ideally be lower than 20%. Bio-Plex Manager allows for the Report table and the standard curves to be exported to Microsoft Excel format for easier data handling for statistical analysis (Fig. 9.7).

The "Graph" tab allows complete data sets for all unknown samples to be exported to Excel format. There is an option to select subsets of data to compare across all analytes, and once exported to Excel, the data points can be highlighted, and a graph of choice can be populated (Fig. 9.8).

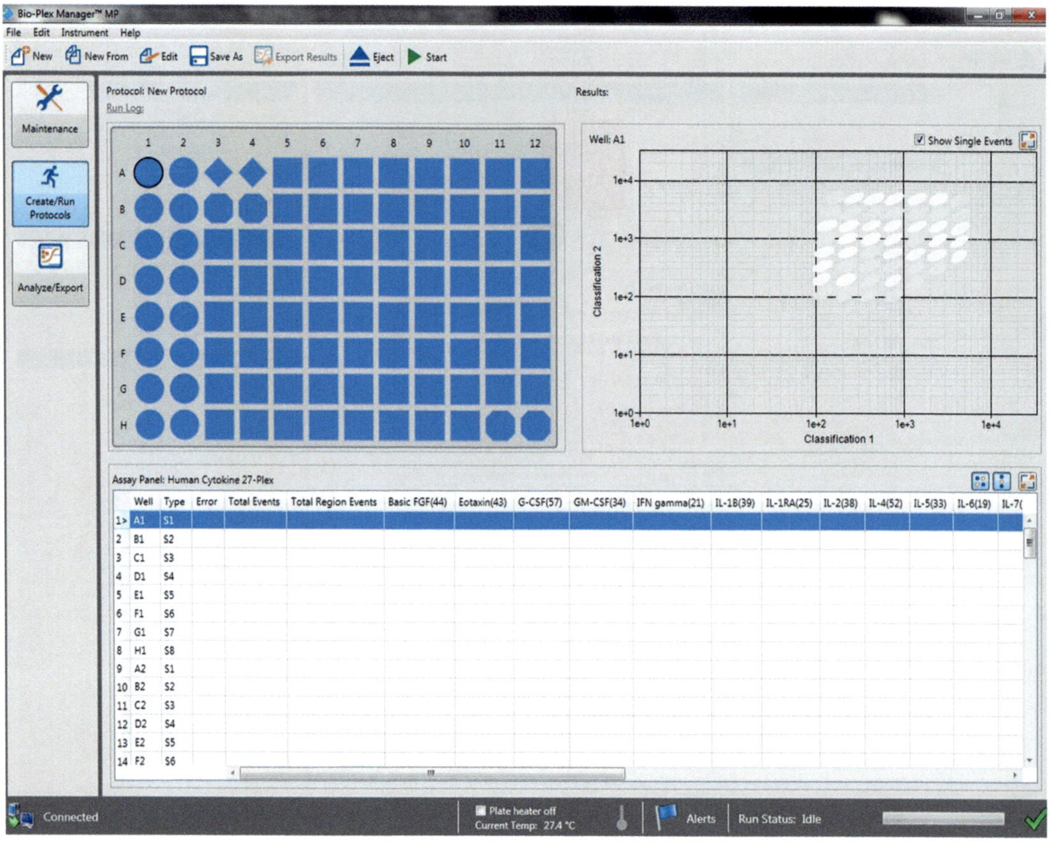

Fig. 9.4 Plate layout created for a Human Cytokine 27-Plex assay using serum samples

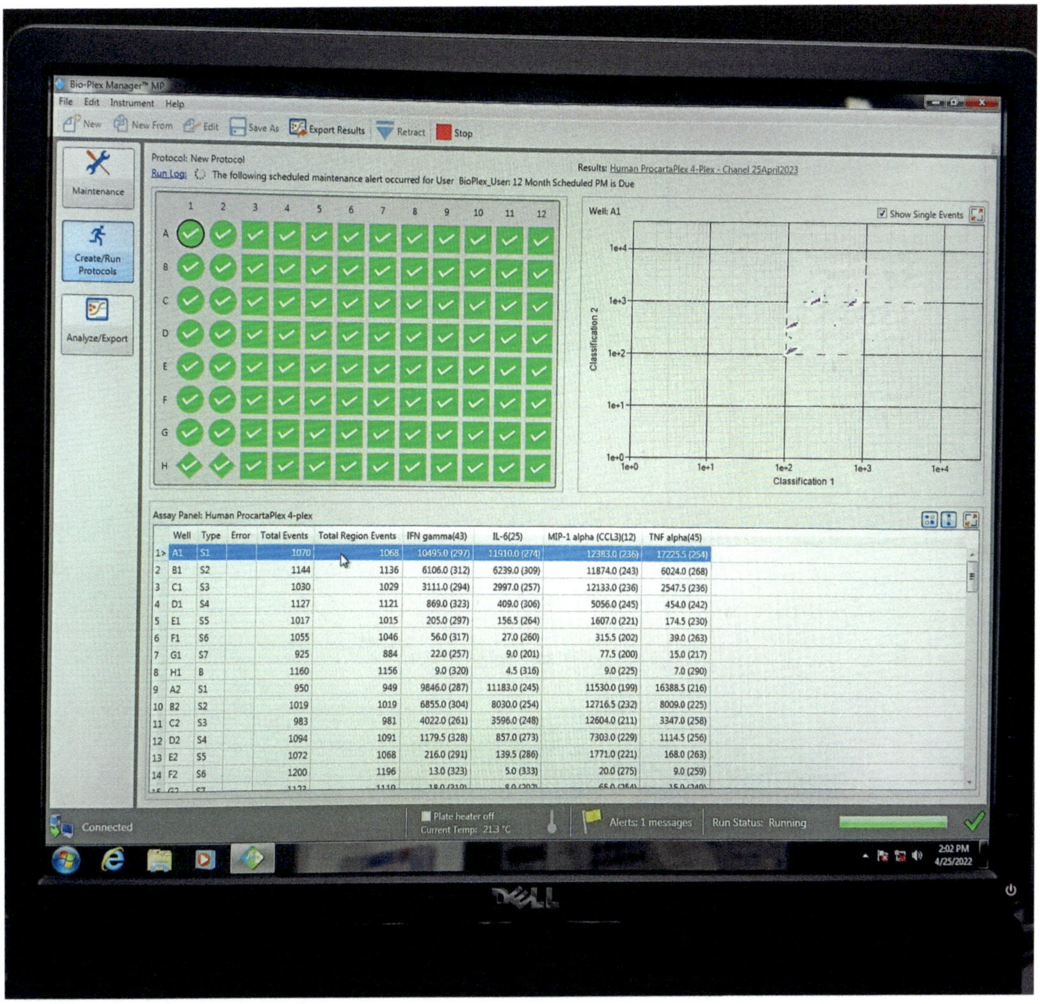

Fig. 9.5 Visual confirmation of the protocol complete with Total events, Bead Events count, and the target areas

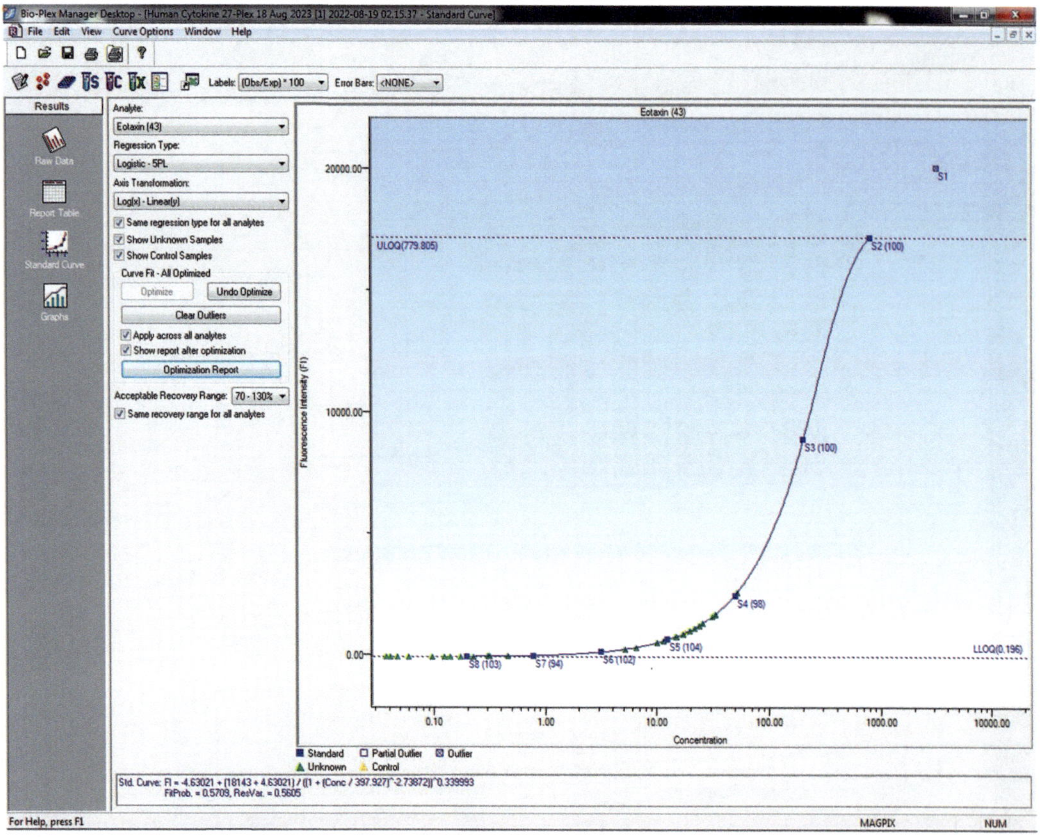

Fig. 9.6 Displays one of the 27 standard curves for the analyte Eotaxin. Standard 1 was eliminated as an outlier while Standards 2–8 constitute the best-fit curve in an acceptable recovery range of 70–130% and lie between the ULOQ and LLOQ. The green triangles on the Standard curve indicate the extrapolated serum samples being analyzed

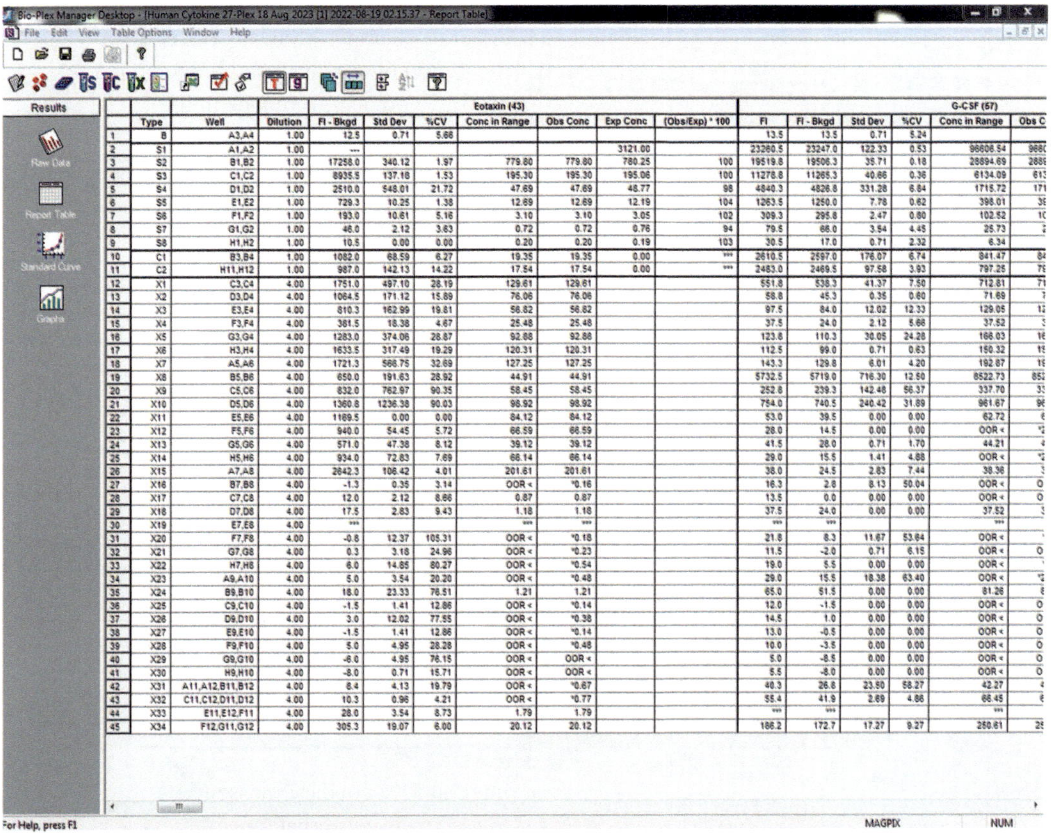

				Eotaxin (43)							G-CSF (67)				
	Type	Well	Dilution	FI - Bkgd	Std Dev	%CV	Conc in Range	Obs Conc	Exp Conc	(Obs/Exp) * 100	FI	FI - Bkgd	Std Dev	%CV	Conc in Range
1	B	A3,A4	1.00	12.5	0.71	5.66					13.5	13.5	0.71	5.24	
2	S1	A1,A2	1.00	---					3121.00		23260.5	23247.0	122.33	0.53	96606.54
3	S2	B1,B2	1.00	17258.0	340.12	1.97	779.80	779.80	780.25	100	19519.8	19506.3	35.71	0.18	28894.69
4	S3	C1,C2	1.00	8935.5	137.18	1.53	195.30	195.30	195.06	100	11278.8	11265.3	40.66	0.36	6134.09
5	S4	D1,D2	1.00	2510.0	548.01	21.72	47.69	47.69	48.77	98	4840.3	4826.8	331.28	6.84	1715.72
6	S5	E1,E2	1.00	729.3	10.25	1.38	12.69	12.69	12.19	104	1263.5	1250.0	7.78	0.62	398.01
7	S6	F1,F2	1.00	193.0	10.61	5.16	3.10	3.10	3.05	102	309.3	295.8	2.47	0.80	102.52
8	S7	G1,G2	1.00	46.0	2.12	3.63	0.72	0.72	0.76	94	79.5	66.0	3.54	4.45	25.73
9	S8	H1,H2	1.00	10.5	0.00	0.00	0.20	0.20	0.19	103	30.5	17.0	0.71	2.32	6.34
10	C1	B3,B4	1.00	1082.0	68.59	6.27	19.35	19.35	0.00	***	2610.5	2597.0	176.07	6.74	841.47
11	C2	H11,H12	1.00	987.0	142.13	14.22	17.54	17.54	0.00	***	2483.0	2469.5	97.58	3.93	797.25
12	X1	C3,C4	4.00	1751.0	497.10	28.19	129.61	129.61			551.8	538.3	41.37	7.50	712.81
13	X2	D3,D4	4.00	1064.5	171.12	15.89	76.06	76.06			58.8	45.3	0.35	0.60	71.69
14	X3	E3,E4	4.00	810.3	162.99	19.81	56.82	56.82			97.5	84.0	12.02	12.33	129.05
15	X4	F3,F4	4.00	381.5	18.38	4.67	25.48	25.48			37.5	24.0	2.12	5.66	37.52
16	X5	G3,G4	4.00	1283.0	374.06	28.87	92.88	92.88			123.8	110.3	30.05	24.28	166.03
17	X6	H3,H4	4.00	1633.5	317.49	19.29	120.31	120.31			112.5	99.0	0.71	0.63	150.32
18	X7	A5,A6	4.00	1721.3	566.75	32.69	127.25	127.25			143.3	129.8	6.01	4.20	192.87
19	X8	B5,B6	4.00	650.0	191.63	28.92	44.91	44.91			5732.5	5719.0	716.30	12.50	8522.73
20	X9	C5,C6	4.00	832.0	762.97	90.35	58.45	58.45			252.8	239.3	142.48	56.37	337.70
21	X10	D5,D6	4.00	1360.8	1236.38	90.03	98.92	98.92			754.0	740.5	240.42	31.89	961.67
22	X11	E5,E6	4.00	1169.5	0.00	0.00	84.12	84.12			53.0	39.5	0.00	0.00	62.72
23	X12	F5,F6	4.00	940.0	54.45	5.72	66.59	66.59			28.0	14.5	0.00	0.00	OOR <
24	X13	G5,G6	4.00	571.0	47.38	8.12	39.12	39.12			41.5	28.0	0.71	1.70	44.21
25	X14	H5,H6	4.00	934.0	72.83	7.69	66.14	66.14			29.0	15.5	1.41	4.88	OOR <
26	X15	A7,A8	4.00	2642.3	106.42	4.01	201.61	201.61			38.0	24.5	2.83	7.44	38.36
27	X16	B7,B8	4.00	-1.3	0.35	3.14	OOR <	*0.16			16.3	2.8	8.13	50.04	OOR <
28	X17	C7,C8	4.00	12.0	2.12	8.66	0.87	0.87			13.5	0.0	0.00	0.00	OOR <
29	X18	D7,D8	4.00	17.5	2.83	9.43	1.18	1.18			37.5	24.0	0.00	0.00	37.52
30	X19	E7,E8	4.00	***			***	***			***	***			***
31	X20	F7,F8	4.00	-0.8	12.37	105.31	OOR <	*0.18			21.8	8.3	11.67	53.64	OOR <
32	X21	G7,G8	4.00	0.3	3.18	24.96	OOR <	*0.23			11.5	-2.0	0.71	6.15	OOR <
33	X22	H7,H8	4.00	6.0	14.85	80.27	OOR <	*0.54			19.0	5.5	0.00	0.00	OOR <
34	X23	A9,A10	4.00	5.0	3.54	20.20	OOR <	*0.48			29.0	15.5	18.38	63.40	OOR <
35	X24	B9,B10	4.00	18.0	23.33	76.51	1.21	1.21			65.0	51.5	0.00	0.00	81.26
36	X25	C9,C10	4.00	-1.5	1.41	12.86	OOR <	*0.14			12.0	-1.5	0.00	0.00	OOR <
37	X26	D9,D10	4.00	3.0	12.02	77.55	OOR <	*0.38			14.5	1.0	0.00	0.00	OOR <
38	X27	E9,E10	4.00	-1.5	1.41	12.86	OOR <	*0.14			13.0	-0.5	0.00	0.00	OOR <
39	X28	F9,F10	4.00	5.0	4.95	28.28	OOR <	*0.48			10.0	-3.5	0.00	0.00	OOR <
40	X29	G9,G10	4.00	-6.0	4.95	76.15	OOR <	OOR <			5.0	-8.5	0.00	0.00	OOR <
41	X30	H9,H10	4.00	-8.0	0.71	15.71	OOR <	OOR <			5.5	-8.0	0.00	0.00	OOR <
42	X31	A11,A12,B11,B12	4.00	8.4	4.13	19.79	OOR <	*0.67			40.3	26.8	23.50	58.27	42.27
43	X32	C11,C12,D11,D12	4.00	10.3	0.96	4.21	OOR <	*0.77			55.4	41.9	2.69	4.86	66.45
44	X33	E11,E12,F11	4.00	28.0	3.54	8.73	1.79	1.79			***	***			***
45	X34	F12,G11,G12	4.00	305.3	19.07	6.00	20.12	20.12			186.2	172.7	17.27	9.27	250.61

Fig. 9.7 The Report table provides the concentration in range and the Observed concentration for the unknown serum samples X1–X34. These values are easily exported to Excel format for further statistical analysis

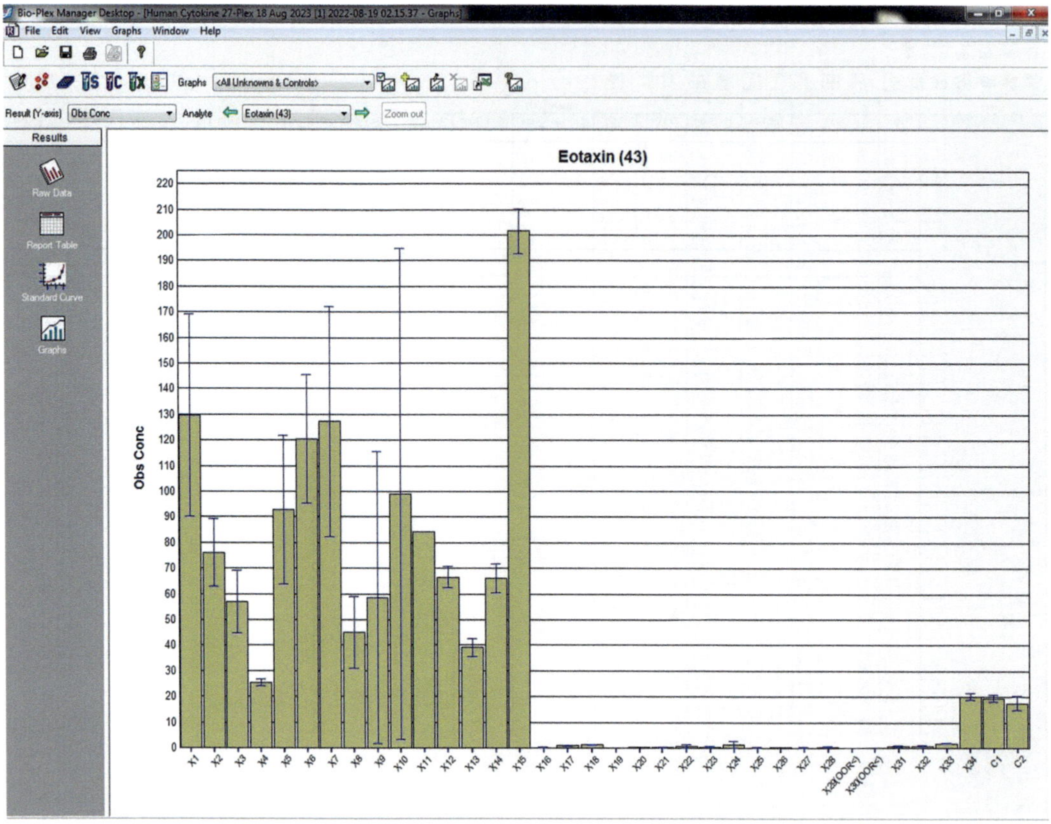

Fig. 9.8 The graph illustrates the observed concentration of Eotaxin between the analyzed samples. Comparative graphs can be created with subsets of data to visualize the difference across all analytes processed for the run

9.4 Sourcing a Multiplex Immunoassay Kit

Fortunately, the Bio-Plex® MAGPIX instrument is not a closed system, which allows users to analyze samples with kits from various suppliers. It is possible to source custom-made or premade immunoassay panels from various suppliers, including ThermoFisher, Merck, Biocom, and Bio-Rad. The most common panels available are Acute phase proteins, Apolipoproteins, Apoptosis pathway, Chemokines, Cell signaling Assays, Cytokines and Growth factors, Diabetes markers, Hormone markers, Inflammation markers, Immunotherapy panel, Isotyping, Kidney toxicity markers, severe acute respiratory syndrome coronavirus 2 (SARS CoV2) Serology and T helper 17 (Th17) Cytokines. Catalogues with detailed descriptions of the various analytes offered for the available panels are easily accessible on the supplier's website with the option of quick online purchase. Although the kit cost increases with more analytes requested from a panel, having multiple suppliers introduces competitive pricing.

9.5 Future of Multiplex Immunoassays

Improved versions of Bio-Plex® MAGPIX are the Bio-Plex® 200 and Bio-Plex® 3D, with upgraded capabilities and throughput. The Bio-Plex® 200 can process 100 different analytes in a single sample with a read time of 45 min for a 96-well plate. In comparison, Bio-Plex® 3D can perform a 20-min read on a 96-well plate for a 500-plex and 75 min for a 384-well plate. Both platforms are compatible with magnetic and non-magnetic beads. These upgraded instruments can easily quantify biomarkers with multiplex immunoassays, receptor-ligand assays, nucleic acid hybridization assays, and enzyme assays. Trending with more than 35,000 peer-reviewed publications, 15,500 deployed instruments, and over 70 Luminex Partners offering over 1300 research-based kits and customized assay solutions, xMAP Technology is currently the most frequently used bead-based multiplexing technology. The term biomarker has gained popularity in recent years and is now used in an increasing number of clinical contexts. A biomarker is a biological molecule that can be measured and utilized to indicate any biological event. This may occur during diagnosing, treating, or predicting an illness or in the pharmacodynamics, safety, or effectiveness of a medication (Califf 2018). Thus, almost all tests in a clinical laboratory are included in this general category. Its application in ongoing studies typically relates to the specific analytes utilized to track the condition. As an alternative, biomarker testing can grow into a more comprehensive panel that examines biological issues for inflammation or cancer. This provides access to a far greater range of analytes often assessed as panels of various distinct analytes but less frequently measured in clinical laboratories. The groups of biological molecules known as chemokines and cytokines are often the source of the relevant analytes (Liu et al. 2021).

The xMAP platform's open architecture has allowed it to be used in numerous applications, including personalized medicine, neurodegeneration, cancer research, pathogen detection, drug discovery, vaccine development, transplant medicine, and biomarker identification and validation (Graham et al. 2019). In a patient with some degree of intact liver function, the liver transaminases aspartate transaminase (AST or SGOT) and alanine transaminase (ALT or SGPT) are helpful biomarkers of liver injury (Poynard and Imbert-Bismut 2012). In this case, the analytes being examined are common clinical laboratory tests that can serve as biomarkers depending on their use. Accurately diagnosing neurodegenerative disorders like Multiple Sclerosis and Alzheimer's can be challenging since clinicians rely on differential diagnosis for the former and limited approved testing for the latter (Sand 2015; Mason et al. 2010). Robust research into cytokine and chemokine biomarkers has revealed an association with these diseases, and early detection will improve diagnosis and patient care (Puthenparampil et al. 2024; de la Monte et al. 2023). A recent study discovered a set of eight analytes that can be employed as biomarkers of inflammatory breast cancer, allowing for early identification, prevention of metastasis, and lymph node invasion (Bessaad et al. 2023). It has been documented that overexpression of immune-related and inflammatory markers in colorectal cancer (CRC) tissues is associated with unfavorable clinicopathological characteristics and a poor prognosis in CRC patients. These indicators may act as diagnostic or prognostic biomarkers for CRC (An et al. 2023). Furthermore, several clusters of co-expressing cytokines have been identified, indicating shared patterns of immune activation. The cytokine clusters' common expression profiles and early connections with specific diseases or disease categories were observed. These findings will improve our understanding of the immunological microenvironment that underpins the clinical presentations of patients with inflammatory, autoimmune, and neoplastic disorders, potentially improving diagnosis and identifying evidence-based therapeutic options (Østergaard et al. 2024; Polley et al. 2023). Therefore, using a holistic approach, clinicians can tailor personalized medicine for treatment by identifying clear biomarkers to assist in stratifying related patients. The xMAP technology was employed in novel ways to evaluate neutralizing antibodies during the COVID-19 pandemic, giving a rapid, high-throughput, culture-free technique of detection in sera (Heaps et al. 2023; Olvera-Collantes et al. 2023). Another study found that peptide stimulation can induce antigen-specific T cells when exposed to SARS-CoV-2 peptide pools. Assessing particular cytokines can be used to examine a variety of disorders, including chronic inflammatory diseases, and provide an alternate way to determine vaccine reactions. This flexible platform can adapt to changing vaccine compositions, making it applicable to future vaccination programs (Taylor et al. 2023).

Scientific and technological advancements have made it possible to examine the molecules that make up biological systems and pathways at the molecular, tissue, and cellular levels by measuring the molecules that make up biological systems and pathways. All of these high-throughput molecular measurements are referred to as omics (Vailati-Riboni et al. 2017). Importantly, the application in the field of omics has resulted in identifying disease markers and unraveling complex disorders. Metabolomics refers to the high-throughput qualitative and qualitative characterization of biomolecules or metabolites. Metabolites are usually the building blocks of macromolecules and include amino acids, alcohols, vitamins, polyols, organic acids, lipids, carbohydrates, and other intermediate compounds. While other omics methods (genomics, transcriptomics, and proteomics) offer insights into genetic and functional possibilities (Karczewski and Snyder 2018), metabolomics offers a more comprehensive view by revealing phenotypes resulting from the intricate and complicated interplay among metabolic processes (Misra et al. 2019). There are two techniques for researching the metabolome: targeted and untargeted. Targeted metabolomics studies aim to measure a specific number of metabolites in the pathway of interest. This method enables researchers to formulate a precise biochemical question based on a hypothesis. It is beneficial to gain deeper insights into a certain hypothesis; nevertheless, the study's success depends on the strength of existing data and expertise. Untargeted metabolomics is an unbiased screening strategy to find thousands of metabolites in a single experiment, allowing for exploratory studies on unknown metabolites (Karnovsky and Li 2020). The metabolic profile is then evaluated under two situations or across a population to identify a new metabolite. The untargeted technique enables a comprehensive investigation of the metabolome. Analyzing the metabolite profiles of common bio-fluids, like urine, serum, cerebrospinal fluid, saliva, and tissue extracts, provides diagnostic and predictive information regarding diseases, dietary habits, and drug toxicity (Bingol 2018; Singh et al. 2023). Metabolomics has significantly transformed the area of cancer research. It is one of the most effective omics techniques, primarily used in cancer research, to detect metabolites with levels affected by cancer progression in a biological sample. Applications range from biomarker identification to drug discovery or development, clinical toxicology, nutritional studies, and quantitative phenotyping (Califf 2018; Bodaghi et al. 2023). Metabolomics, like all other omics approaches, depends extensively on the availability and accuracy of public databases. In this respect, the most significant milestone in the field occurred in 2007 with the completion of the first draft of the human metabolome, the chemical counterpart of the human genome (Wishart 2007). The Human Metabolome Database (HMDB) is a freely accessible electronic database designed to link data to other databases (GenBank, PubChem, Kyoto Encyclopedia of Genes and Genomes—KEGG, MetaCyc, and UniProt), combining chemical data, clinical information, and molecular biology and biochemistry models. The database offers approximately 220,000 metabolite entries, which include both water- and lipid-soluble metabolites.

The latest efforts to identify novel biomarkers that may be used to either determine the likelihood of acquiring a chronic illness or allow for the very early detection of changing disease processes are clinically more significant. The goals are to ensure that early therapeutic intervention slows the progression of the disease in high-risk persons and minimizes clinical intervention while reassuring low-risk individuals. Furthermore, in the modern day, it is expected that novel clinical biomarkers would support the economical evaluation of novel treatments.

References

Ahsan H, Ahmad R (2020) Multiplex technology for biomarker immunoassays. Innate Immunity in Health and Disease, pp 1–10

Alhajj M, Zubair M, Farhana A (2023) Enzyme linked immunosorbent assay. StatPearls

An S, Kim S-K, Kwon HY, Kim CS, Bang H-J, Do H et al (2023) Expression of immune-related and inflammatory markers and their prognostic impact in colorectal cancer patients. Int J Mol Sci 24(14):11579

Bessaad M, Habel A, Ahmed MH, Xu W, Stayoussef M, Bouaziz H et al (2023) Assessing serum cytokine profiles in inflammatory breast cancer patients using Luminex® technology. Cytokine 172:156409

Bingol K (2018) Recent advances in targeted and untargeted metabolomics by NMR and MS/NMR methods. High-throughput 7(2):9

Bodaghi A, Fattahi N, Ramazani A (2023) Biomarkers: promising and valuable tools towards diagnosis, prognosis and treatment of Covid-19 and other diseases. Heliyon 9(2):e13323

Boguszewska K, Szewczuk M, Urbaniak S, Karwowski BT (2019) Immunoassays in DNA damage and instability detection. Cell Mol Life Sci 76(23):4689–4704

Califf RM (2018) Biomarker definitions and their applications. Exp Biol Med 243(3):213–221

Clark MF, Lister RM, Bar-Joseph M (1986) ELISA techniques. Methods in enzymology. Elsevier, pp 742–766

Crowther JR (2008) ELISA: theory and practice. Springer Science & Business Media

Davis D, Zhang A, Etienne C, Huang I. Bio-Plex® suspension array system tech note 2861. 2005

de la Monte SM, Tong M, Hapel AJ (2023) Concordant and discordant cerebrospinal fluid and plasma cytokine and chemokine responses in mild cognitive impairment and early-stage Alzheimer's disease. Biomedicines 11(9):2394

do Nascimento Carvalho MS, de Luna Costa AS, Santana AVA, de Luna Batista RS, de Mesquita Neto FP, de Lima JC et al (2023) The Nobel prize in physiology 1977: a literature review of the radioimmunoassay technique. Res Soc Dev 12(6):e14712642090-e

Engvall E, Perlmann P (1971) Enzyme-linked immunosorbent assay (ELISA) quantitative assay of immunoglobulin G. Immunochemistry 8(9):871–874

Flores CFY, ÁdlMH P, Bonilla VMC, Sáenz-Flor K (2020) Sample management: stability of plasma and serum on different storage conditions. Ejifcc 31(1):46

Graham H, Chandler DJ, Dunbar SA (2019) The genesis and evolution of bead-based multiplexing. Methods 158:2–11

Haddad NS, Nguyen DC, Kuruvilla ME, Morrison-Porter A, Anam F, Cashman KS et al (2021) One-stop serum assay identifies COVID-19 disease severity and vaccination responses. Immunohorizons 5(5):322–335

Heaps AL, Sobolewski MD, Jacobs JL, Gordon KC, Haidar G, Mellors JW et al (2023) Rapid determination of SARS-CoV-2 antibody neutralization titer using bio-rad bio-Plex correlates strongly with pseudovirus-determined neutralization titer. J Virol Methods 316:114726

Huang R-P, Lin Y, Chen L-P, Yang W, Huang R (2004) ELISA-based protein arrays: multiplexed sandwich immunoassays. Curr Proteomics 1(3):199–210

Karczewski KJ, Snyder MP (2018) Integrative omics for health and disease. Nat Rev Genet 19(5):299–310

Karnovsky A, Li S (2020) Pathway analysis for targeted and untargeted metabolomics. Computational methods and data analysis for metabolomics, pp 387–400

Kingsmore SF (2006) Multiplexed protein measurement: technologies and applications of protein and antibody arrays. Nat Rev Drug Discov 5(4):310–321

Lin A (2015) Direct ELISA. Methods Molecular Biol (Clifton, NJ) 1318:61–67

Lin A, Salvador A, Carter JM (2015) Multiplexed microsphere suspension array-based immunoassays. ELISA: methods and protocols, pp 107–118

Liu C, Chu D, Kalantar-Zadeh K, George J, Young HA, Liu G (2021) Cytokines: from clinical significance to quantification. Adv Sci 8(15):2004433

Mason S, McShane R, Ritchie C (2010) Diagnostic tests for Alzheimer′ s disease: rationale, methodology, and challenges. Int J Alzheimers Dis 2010(1):972685

Meites J (1977) The 1977 nobel prize in physiology or medicine. Science 198(4317):594–596

Misra BB, Langefeld C, Olivier M, Cox LA (2019) Integrated omics: tools, advances and future approaches. J Mol Endocrinol 62(1):R21–R45

Olvera-Collantes L, Moares N, Fernandez-Cisnal R, Muñoz-Miranda JP, Gonzalez-Garcia P, Gabucio A et al (2023) Development and validation of a highly sensitive multiplex immunoassay for SARS-CoV-2 humoral response monitorization: a study of the antibody response in COVID-19 patients with different clinical profiles during the first and second waves in Cadiz, Spain. Microorganisms 11(12):2997

Østergaard AA, Feddersen S, Barnkob MB, Lynggaard RB, Karstoft ACA, Borup M et al (2024) Whole-blood culture-derived cytokine combinations for the diagnosis of tuberculosis. Front Immunol 15:1397941

Ouchterlony O (1948) In vitro method for testing the toxin-producing capacity of diphtheria bacteria. Acta Patholog Microbiolog Scandinavica 25(1–2):186–191

Platchek M, Lu Q, Tran H, Xie W (2020) Comparative analysis of multiple immunoassays for cytokine profiling in drug discovery. SLAS Discov 25(10):1197–1213

Polley DJ, Latham P, Choi MY, Buhler KA, Fritzler MJ, Fritzler ML (2023) Identification of novel clusters of co-expressing cytokines in a diagnostic cytokine multiplex test. Front Immunol 14:1223817

Poynard T, Imbert-Bismut F (2012) Laboratory testing for liver disease. Zakim and Boyer's hepatology. Elsevier Saunders, pp 201–215

Puthenparampil M, Marin A, Zanotelli G, Mauceri V, De Napoli F, Gaggiola M et al (2024) Blood-brain barrier damage associates with glia-related cytokines in the cerebrospinal fluid of patients with multiple sclerosis. Mult Scler Relat Disord 82:105403

Sand IK (2015) Classification, diagnosis, and differential diagnosis of multiple sclerosis. Curr Opin Neurol 28(3):193–205

Shah K, Maghsoudlou P (2016) Enzyme-linked immunosorbent assay (ELISA): the basics. Br J Hosp Med 77(7):C98–C101

Singh S, Sarma D, Verma V, Nagpal R, Kumar M (2023) Unveiling the future of metabolic medicine: omics technologies driving personalized solutions for precision treatment of metabolic disorders. Biochem Biophys Res Commun 682:1–20

Sten H, Klavs B, Peter B, Eva E, Marie-Louise H, Göran H et al (2006) Peter Perlmann 1919–2005. Wiley Online Library

Taylor JV, Callery EL, Rowbottom A (2023) Optimisation of SARS-CoV-2 peptide stimulation and measurement of cytokine output by intracellular flow cytometry and bio-plex analysis. J Immunol Methods 522:113556

Vafajoo A, Rostami A, Foroutan Parsa S, Salarian R, Rabiee N, Rabiee G et al (2018) Multiplexed microarrays based on optically encoded microbeads. Biomed Microdevices 20:1–14

Vailati-Riboni M, Palombo V, Loor JJ (2017) What are omics sciences? Periparturient diseases of dairy cows: a systems biology approach 1–7

Wishart DS (2007) Proteomics and the human metabolome project. Expert Rev Proteomics 4(3):333–335

Yalow R (n.d.) Rosalyn Sussman Yalow. Journeys of women in science and engineering: no universal constants, pp 441–444

GPSR Compliance

The European Union's (EU) General Product Safety Regulation (GPSR) is a set of rules that requires consumer products to be safe and our obligations to ensure this.

If you have any concerns about our products, you can contact us on ProductSafety@springernature.com

In case Publisher is established outside the EU, the EU authorized representative is:

Springer Nature Customer Service Center GmbH
Europaplatz 3
69115 Heidelberg, Germany

Batch number: 10740022

Printed by Printforce, the Netherlands